T0276821

Advances in Wireless Mesh Networks

Edited by **Timothy Kolaya**

New Jersey

Published by Clanrye International,
55 Van Reypen Street,
Jersey City, NJ 07306, USA
www.clanryeinternational.com

Advances in Wireless Mesh Networks
Edited by Timothy Kolaya

International Standard Book Number: 978-1-63240-056-7 (Hardback)

Printed in the United States of America.

Contents

Preface

This book aims to highlight the current researches and provides a platform to further the scope of innovations in this area. This book is a product of the combined efforts of many researchers and scientists, after going through thorough studies and analysis from different parts of the world. The objective of this book is to provide the readers with the latest information of the field.

The latest advances in the field of wireless mesh networks have been described in this profound book. It presents an insight into current developments related to new design techniques and algorithms to enhance performance and functionality of Wireless Mesh Networks (WMNs). Information in this book has been contributed by reputed veterans in wireless mesh networking. The book highlights link scheduling schemes for selection of a subset of links for simultaneous transitions under interference restraints in an effective and appropriate manner to guarantee a specific level of network connectivity. It also elucidates channel assignment techniques for enhancement of network throughput in multi-channel multi-radio wireless mesh networks through effective channel utilization and minimization of the interference. The book addresses a few significant network planning issues regarding effective routing protocols in dynamic large-scale mesh environment, the accuracy of the mesh security architecture, attainable capacity limit of a single wireless link between two multi-interface mesh nodes and fault-tolerant mesh network topology planning.

I would like to express my sincere thanks to the authors for their dedicated efforts in the completion of this book. I acknowledge the efforts of the publisher for providing constant support. Lastly, I would like to thank my family for their support in all academic endeavors.

Editor

Section 1

Efficent Link Scheduling and Channel Assignment Strategies in WMNs

Chapter 1

Stability-Based Topology Control in Wireless Mesh Networks

Gustavo Vejarano

Additional information is available at the end of the chapter

1. Introduction

Topology control[1] in wireless mesh networks is an important problem due to the effects it has on the different layers of the protocol stack [1]. For example, the network connectivity, energy consumption, total physical-link throughput, spatial reuse, and total end-to-end throughput as a function of the network topology have been investigated in [2-6] respectively. In this chapter, we look at the problem of topology control for adapting the stability region of the link-scheduling policy of the network. Therefore, we start by defining the problem of link scheduling and the stability region.

The goal when designing link-scheduling policies is to achieve maximum throughput while making the policies amenable for implementation [7, 8]. Link scheduling refers to the selection of a subset of links for simultaneous transmission that have the following characteristic: When the links are activated simultaneously, the interference between them is low enough to allow successful reception for every activated link. A link-scheduling policy specifies the mechanism that determines, for every time slot, a subset of links that fits this characteristic. For example, consider the network and the link (i,j) shown in Figure 1. Let this network operate under the frame structure shown in Figure 2. Therefore, in the network, time is divided into frames; each frame is divided into a control subframe and a data subframe, and each subframe is further divided into a series of time slots. Whenever link (i,j) is activated by the link-scheduling policy during a data-time slot, the link transmits a data packet. In order for the packet to be received successfully, none of the links that interfere with (i,j) can be active while (i,j) is active. Otherwise, the packet transmitted by node i is not received successfully by node j. This is known as a packet collision at

[1]In this chapter, topology control refers to the problem of controlling the creation and elimination of wireless links and the interference between them by controlling the transmission power of the nodes.

node j, i.e., the packet transmitted over (i,j) collides with the packet transmitted over the interfering link. The set of links that interfere with (i,j) is denoted by $\mathcal{I}^{(i,j)}$ in Figure 1. Therefore, when (i,j) is active, none of the links in $\mathcal{I}^{(i,j)}$ can be active. Given that every link has a set of interfering links, only subsets of the set of all links in the network can be active at a given time. The task of the link scheduling policy is to select one of these subsets for every data-time slot. This selection is done by exchanging control information during the control-time slots.

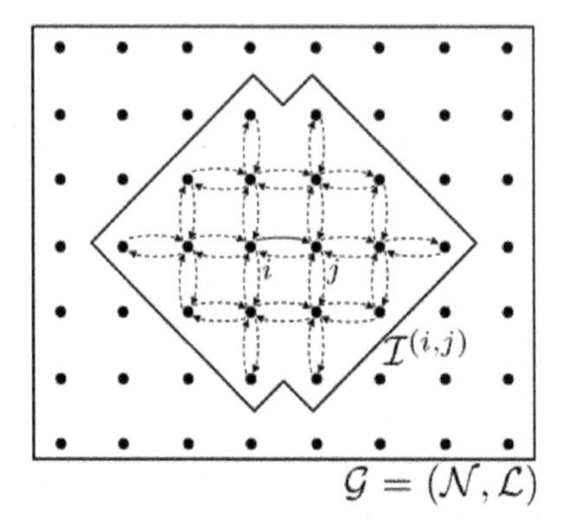

Figure 1. Interfering Links

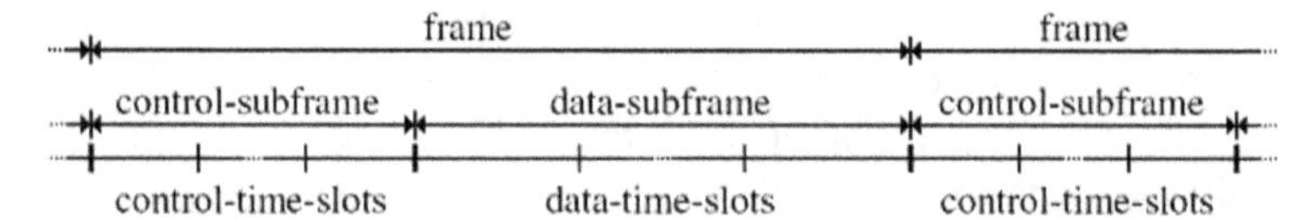

Figure 2. Frame Structure

Besides considering the interfering link sets of every link, the link-scheduling policy needs to consider the queue length of every link. In a wireless mesh network, when data packets are being transported over the flow's path, the links that form the path need to store the packets temporarily from the moment the node receives the packet until the moment the node forwards the packet to the next node in the path. Therefore, each link maintains queues of data packets for every flow that it belongs to. This is shown in Figure 3, which includes the queues of both link (i,j) and link (j,i). Each link has two queues. These are the input and output queues, which are denoted by $Q_i^{(i,j)}$ and $Q_o^{(i,j)}$ respectively for link (i,j). When node i receives a data packet that needs to be forwarded to node j, it stores the packet in $Q_i^{(i,j)}$ first. Then, it exchanges control packets with neighboring nodes in order to determine the data subframe and data-time slot when the data packet can be transmitted to node j without collisions. This is done according to the link-scheduling policy of the network. Once the transmission schedule of the data packet has been determined, the packet is moved to $Q_o^{(i,j)}$ where it waits for the data-time slot scheduled for its transmission. Finally, node i forwards the packet to node j at the scheduled data-time slot. At this point, the packet leaves $Q_o^{(i,j)}$. When node j receives the data packet, it checks whether it is the packet's destination. If it is, the packet is no longer stored in any queue and leaves the network[2]. If it is not, it starts the link-scheduling process again in order to forward the packet to the next node in the data flow's path.

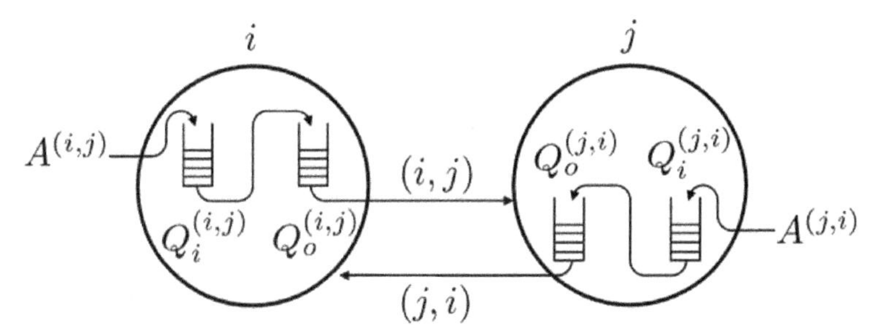

Figure 3. Data-packet transmissions over links (i,j) and (j,i)

When designing a link-scheduling policy, the goal is to support the largest set of data-packet rates for all the flows established in the network, and this should be done while guaranteeing the following conditions:

- There are no packet collisions
- The queues do not grow indefinitely
- A given level of fairness is guaranteed for all the flows

Packet collisions need to be avoided in order to guarantee the completeness of the information being delivered to the user. Given the limited amount of memory that nodes have, the queue lengths need to be guaranteed not to grow indefinitely. Otherwise, the nodes will drop data packets when they have run out of memory to store the packets while the transmission schedules are being determined. The fairness among data flows guarantees that each flow is assigned some part of the total capacity of the network to transport information[3].

The mathematical formulation of this problem is based on Markovian systems [9]. In order to do this formulation, the following definitions for each node's queues need to be considered first. In Equations 1 and 2, S_1^j is the set of 1-hop neighbors of node j. These are the nodes that have links with node j. Therefore, Q_i^j is the total number of packets stored in node j's 1-hop neighbors that need to be forwarded to node j and that are waiting to be scheduled, and Q_o^j is the maximum number of scheduled packets waiting to be forwarded to node j among all of j's 1-hop neighbors[4]. The time indexes n and m_n represent the nth

[2]Actually, node j sends the data packet to its application layer so that the packet's content can be finally delivered to the user.

[3]It should be noted that there is not an absolute definition for fairness. For example, the network operator may be interested in assigning the same maximum data-packet rates to all flows or different maximum rates to different flows depending on the demands of the users.

[4]The actual length of Q_o^j has a more involved definition. However, for the sake of clarity, we do not consider the exact definition until Section 4.2.1.

time that at least one control packet is transmitted in the network and the control-time slot m_n when this takes place.

$$Q_i^j(n) \triangleq \sum_{i \in \mathcal{S}_1^j} Q_i^{(i,j)}(m_n) \tag{1}$$

$$Q_o^j(n) \triangleq \max_{i \in \mathcal{S}_1^j} \{Q_o^{(i,j)}(m_n)\} \tag{2}$$

Consider the following measure of the queue lengths of all links in the network, where $\mathcal{N}$ is the set of all nodes in the network, and the indexes i and o indicate whether input or output queues are being considered.

$$V_s^{i,o}(n) \triangleq \sum_{j \in \mathcal{N}} (Q_{i,o}^j(n))^2 \tag{3}$$

Intuitively, $V_s^{i,o}(n)$ can be interpreted as a total volume occupied by all of the input or output queues[5], depending on whether the index is i or o, and that is updated at every control-time slot in which there is at least one control-packet transmission. $V_s^{i,o}(n)$ increases and decreases randomly in time. It increases due to the data packets that the flows input into the network, and it decreases when data packets reach their destination and leave the network. This is shown graphically in Figure 4, which includes a network of 7 nodes. The volume of the network, shown in circles, increases and decreases according to the queue lengths in the network.

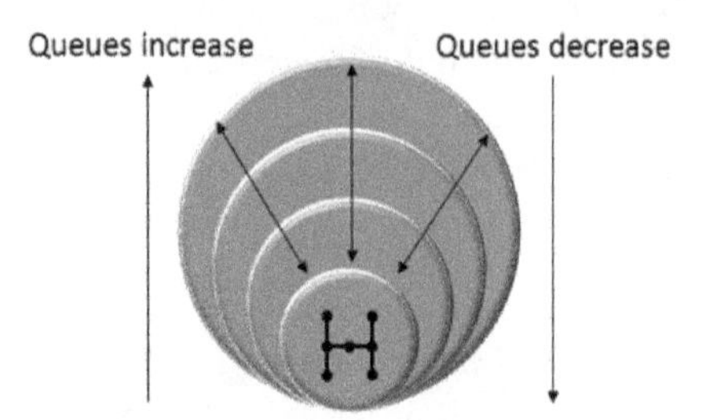

Figure 4. Network stability

Based on the concept of $V_s^{i,o}(n)$, the stability of a network can be defined as follows. A network is stable if $V_s^{i,o}(n)$ decreases to zero with some probability greater than zero at some finite future time $n+m$, i.e., there is a probability that the volume of the network decreases to zero within some finite time independently of the current volume. It can be shown that this condition is met if the expectation that $V_s^{i,o}(n)$ decreases is greater than zero [9]. Therefore, a network is stable if Equation 4 holds[6].

[5]In the theory of Markovian processes [9], $V_s^{i,o}(n)$ is known as a Lyapunov function.

[6] $E[X \mid Y]$ denotes the expected value of X given Y.

$$\mathrm{E}[V_s^{i,o}(n+1) - V_s^{i,o}(n) \mid V_s^{i,o}(n)] < 0 \quad (4)$$

A network becomes unstable when the rate at which data flows input packets into the network increases to a point in which the link-scheduling policy is not able decrease queue lengths fast enough to guarantee the condition given by Equation 4. Therefore, the task of the link-scheduling policy is to maintain the network stable under the constraints that there should not be data-packet collisions and that data-flows are fairly serviced.

The performance of the link-scheduling policy in performing this task is measured in terms of the set of data-packet rates for which it guarantees that the network is stable. The largest set of data-packet rates supported by the link-scheduling policy is known as the stability region. In order to compare different link-scheduling policies, these are usually compared against the optimal stability region, which is the largest region that any policy can achieve[7]. This comparison is done using the concept of efficiency ratio, which is defined as the fraction of the optimal stability region in which a suboptimal link-scheduling policy guarantees the stability of the network. Therefore, an optimal link-scheduling policy has an efficiency ratio of unity. When the link-scheduling policy has an optimal efficiency ratio, the network is able to support the largest set of data-packet rates, and so it achieves maximum throughput.

The stability region of most link-scheduling policies depends on the interference sets of the links in the network. This can be observed, for example, in the following case that considers two links of a network. If the two links interfere with each other, only one of them can be active at a time. However, if they do not interfere with each other, they can be active simultaneously. Therefore, when they do not interfere, the links are able to support higher data-packet rates for the flows that they belong to, and this increases the size of the stability region. Given that the interference sets are determined from the network topology, i.e., from the relative distance between nodes and their transmission powers, the stability region can be modified by controlling the network topology. Therefore, for a given network with a given link-scheduling policy and a given set of end-to-end data flows, the stability region can be adapted by means of topology control in order to increase the data-packet rates supported by the links for the flows that they belong to. An example of this adaptation is shown in Figure 5. This example considers two flows. There are an initial stability region and a final stability region. The coordinates of the operating point indicate the data-packet rates of the two flows. Therefore, as the flows increase their data-packet rates, the operating point moves further away from the origin. Given that the operating point has not crossed the boundary of the initial stability region, the network is stable. After controlling the network topology, the stability region is modified such that the distance from the boundary of the region to the operating point is increased. Therefore, the final stability region allows the operating point to be moved further away from the origin without crossing the boundary. In this way, the flows are able to operate at higher data-packet rates without destabilizing the network.

---

[7]The optimal stability region and the link-scheduling policy that achieves it were characterized in [10].

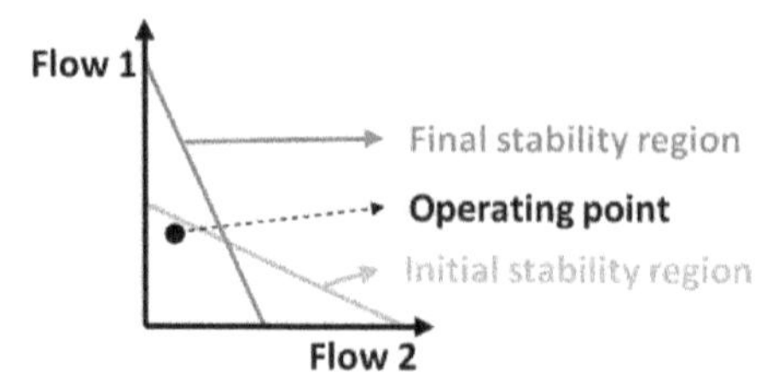

Figure 5. An example of stability-region adaptation

In the following, the operation and performance of the different link-scheduling policies is discussed. Special attention is given to reservation-based scheduling (RBDS) policies [8,11]. Then, based on the stability region of RBDS policies, the topology-control mechanisms are discussed.

2. Link-scheduling policies[8]

The challenge in link scheduling is that the policies are highly complex. The scheduling problem in general is nondeterministic polynomial time (NP) hard [12]. Therefore, the research literature has focused on policies of lower complexity that are more amenable to implementation [7].

Most distributed scheduling policies that achieve provable efficiency ratios calculate, at the onset of every frame, a subset of links that is allowed to transmit data in the immediately following frame only. In this chapter, we refer to these policies as non-RBDS policies, i.e., they do not reserve any future frame but only the following one. On the other hand, RBDS policies [8, 11] select links to transmit data in any future frames by means of frame reservations. Since this framework considers reservations of any future frames, non-RBDS policies correspond to a special case within the RBDS framework, i.e., the case that links are allowed to reserve the next frame only.

It should be noticed that non-RBDS policies require the input queue only (i.e., $Q_i^{(i,j)}$). They do not need the output queue (i.e., $Q_o^{(i,j)}$) because data packets do not need to wait for future data subframes. In non-RBDS policies, once a data packet is scheduled at the onset of the data subframe, the packet is transmitted immediately.

2.1. Non-RBDS policies

The concept of optimal stability region and a centralized scheduling policy with efficiency ratio of unity were introduced in [10]. The centralized scheduling policy attempts to solve a complex global optimization problem so that the entire network is stable for the largest possible set of input data-packet rates. Under the 1-hop interference model[9], the problem is

---

[8]The material presented in this section is based on the material presented in [8, 11].

[9]In the 1-hop interference model, only the links that the 1-hop neighbors of a node belong to interfere with the links that the node belongs to.

shown to correspond to a maximum weighted matching (MWM), where the weights of the links are determined from the length of their queues. The solution to MWM has complexity $O(N^3)$ [13, 14], where N is the number of nodes. Under the k-hop interference model, the problem has been proven to be NP-Hard [12]. Therefore, the optimal scheduling policy is not convenient for implementation due to its high complexity. As a consequence, less complex scheduling policies that achieve only a fraction of the optimal stability region for general network topologies have been developed [12, 15-28].

The different suboptimal scheduling policies proposed in the literature can be classified according to the techniques they use to calculate the next schedule. These techniques usually depend on the interference model assumed for the network and the links' weights at the onset of every frame. Also, the suboptimal scheduling policies can be further classified according to their centralized or distributed mechanism (Unless otherwise specified, the scheduling policies reviewed in this section consider 1-hop traffic only, i.e., the data flows' paths have one link only.).

2.1.1. Centralized policies

In [15], a centralized scheduling approach known as pick-and-compare [17] that achieves the optimal efficiency ratio is defined. The pick-and-compare scheduling policy selects the optimal schedule at every frame with some probability greater than zero. First, the scheduling algorithm randomly picks a new schedule such that the links can satisfy the interference model constraints. Then, the newly picked schedule is compared with the current schedule. If the picked schedule reduces the total weight of the network (i.e., queue lengths) more than the current schedule, then the picked schedule is selected as the next schedule; otherwise the current schedule is used again. The pick-and-compare policy requires the calculation and comparison of the updated total weight for every frame. Therefore, the complexity of this technique grows linearly with N, which makes it difficult to implement in networks with a high number of nodes or in networks where nodes have low processing capabilities.

Greedy maximal scheduling (GMS) is a suboptimal, centralized scheduling policy. In GMS, the links of the network are ordered according to their weights, where the link with maximum weight is placed at the top of this globally ordered list. A valid schedule is found by selecting links from the list from top to bottom that do not interfere with each other. The complexity of GMS is $O(L\log(N))$, where L is the number of links [29]. GMS has efficiency ratio of 1/2 under the 1-hop interference model [7], and under the k-hop interference model, GMS has efficiency ratio of 1, 1/6, and 1/49 for tree, geometric, and general network graphs respectively [12, 27].

2.1.2. Distributed policies

A distributed version of the pick-and-compare scheduling policy was proposed in [19]. In this policy, a node is selected with some probability less than one to initiate the calculation of a schedule for the links in its neighborhood. The new schedule is selected for the next frame if the new schedule reduces the neighborhood's weight by more than the current schedule. The algorithm has constant complexity, so it does not depend on the number of nodes of the network. It does depend, however, on the diameter of the neighborhood. The efficiency ratio increases as the diameter of the neighborhood increases. The algorithm assumes the 1-hop interference model, so it can only be directly used on networks with physical layers such as frequency-hopping code-division-multiple-access (FH-CDMA) that allow that assumption to be made.

Greedy scheduling (GS) policies [29] have been developed that achieve the same efficiency ratio of GMS [17, 25, 28]. In the GS policies, nodes calculate locally the next schedule based on the links that have the maximum local weights.

In [17, 20-23], a maximal scheduling (MS) approach is described. In this approach, maximum weight is not required to schedule a link. A link is eligible for the next schedule as long as it has enough packets in the queue to transmit during the entire duration of a frame. The efficiency ratio of MS scheduling policies is $1/\kappa$, where κ is the maximum number of non-interfering links in the interference set of any link in the network. MS policies have also been adapted to multi-hop flow scenarios[10] in which a set of flows with their respective rates and routes are given [16, 20-23].

Lastly, distributed scheduling policies of complexity $O(1)$ have been developed in [18, 24, 30]. These are known as constant time (CT) scheduling policies [17]. The CT approach differs from the MS approach in that when a link does not interfere with the links in a schedule, it is selected with probability less than one. Therefore, in CT scheduling policies, frames can be wasted with some probability greater than zero. In [30], CT policies are proposed for the 1-hop and 2-hop interference models[11]. The efficiency ratios of these policies were improved in [18, 24]. In [25], the improved efficiency ratios are $\frac{1}{2}-\frac{1}{\sqrt{m}}$ and $\frac{2}{\hat{n}}\left(\frac{1}{2}-\frac{1}{\sqrt{m}}\right)$ for the 1-hop and 2-hop interference models respectively, where $\hat{n}$ is the maximum number of 1-hop neighboring links for any link of the network.

2.2. Reservation-based distributed scheduling

In an RBDS wireless network, the nodes negotiate with their neighbors the reservation of future data-time slots for their links. This negotiation is based on a three-way handshake that consists of a request, a grant, and a grant confirmation. Requests, grants, and grant confirmations are transmitted in scheduling packets. The nodes access the control-time slots for transmitting scheduling packets using an election algorithm. Therefore, in an RBDS wireless network, the nodes access the wireless channel using two different algorithms: the

[10]A multi-hop flow has a path that is at least 2 links long.

[11]In the 2-hop interference model, only the links that the 1-hop or 2-hop neighbors of a node belong to interfere with the links that the node belongs to. The 2-hop neighbors of a node are the nodes that have a shortest path to the node of length 2 links.

election algorithm and the RBDS algorithm, whose roles are to avoid collisions and wasted time slots in the control and data subframes respectively.

In this chapter we assume that the election algorithm is given, and we focus on the RBDS algorithm only. For example, in the IEEE 802.16 standard [31], the election algorithm is completely specified while the link-scheduling algorithm is not. The standard only specifies the control messages that can be used for the implementation of RBDS policies. We adopt the election algorithm of IEEE 802.16 wireless mesh networks with coordinated distributed scheduling. Also, it is assumed that the RBDS wireless network follows the 2-hop interference model, which is the model considered in the IEEE 802.16 standard [31].

In the IEEE-802.16 election algorithm, the nodes in every 2-hop neighborhood take turns by competing between them to access the control-time slots and transmit scheduling packets. Let the 2-hop neighborhood of node i, i.e., node i, its 1-hop neighbors, and its 2-hop neighbors, be denoted by $\mathcal{S}^i_{\leq 2}$. We model the operation of this election algorithm as follows[12].

- In order to avoid scheduling-packet collisions, no more than one node is selected in every $\mathcal{S}^i_{\leq 2}$ at any control-time slot.
- The nodes in $\mathcal{S}^i_{\leq 2}$, where i can be any node in $\mathcal{N}$, are selected in cycles. We refer to these cycles as scheduling cycles.
- Within a scheduling cycle, the nodes in $\mathcal{S}^i_{\leq 2}$ are selected once and only once each. The order in which they are selected is uniformly distributed among all the possible orders of selection.
- The order that nodes in $\mathcal{S}^i_{\leq 2}$ are selected is independent across scheduling cycles.

When nodes i and j exchange scheduling messages to perform the three-way handshake, they schedule data packets on link (i,j) and multicast the negotiated schedule to all links in $\mathcal{I}^{(i,j)}$. The handshake consists of the following steps[13].

1. Node i sends a request to node j for a certain number of data-time slots along with a set of data-time slot numbers that are available for reservation at node i.
2. Node j sends a grant to node i for the requested number of data-time slots according to its set of data-time slots available for reservation and those of node i.
3. Node i confirms the successful reception of the grant by echoing the grant in its next scheduling-packet transmission.

The reservation of the data-time slots takes place at steps 2 and 3. When node j transmits its scheduling packet, j's 1-hop neighbors receive the grant and mark the granted data-time slots as unavailable. When node i confirms the grant, i's 1-hop neighbors receive the grant and mark the granted data-time slots as unavailable too. Therefore, at the end of step 3, all

[12]The operation of the election algorithm for IEEE 802.16 mesh networks with coordinated distributed scheduling is described in detail in [32, 33].

[13]It is assumed that in this handshake node j grants node i's request and that the data-packet-slot reservation is successful at both i and j.

links in $\mathcal{I}^{(i,j)}$ have made the granted data-time slots unavailable (i.e., the grant has been multicast to all links in $\mathcal{I}^{(i,j)}$).

The requests and grants transmitted by the nodes are defined as follows.

Definition 1. Request $r_m^{(i,j)} \triangleq (f_s, f_x, z)$, where $(f_s, f_x, z) \in N^3$, is the request transmitted by node i at control-time slot m that requests for link (i,j) the data-time slots of z consecutive data-subframes starting at frame f_s or any other frame after f_s. Request $r_m^{(i,j)}$ expires at the onset of frame f_x.

Definition 2. Grant $g_m^{(i,j)} \triangleq (f_s, f_e)$, where $(f_s, f_e) \in \mathbf{N}^2$, is the grant transmitted by node j at control-time slot m that assigns to link (i,j) the data-time slots of the series of frames that starts and ends with frames f_s and f_e respectively. Grant $g_m^{(i,j)}$ expires at the end of frame f_e.

Definition 3. The length of grant $g_m^{(i,j)}$, denoted by $|g_m^{(i,j)}|$, is the number of data-subframes assigned in the grant. Therefore,

$$|g_m^{(i,j)}| \triangleq f_e - f_s + 1.$$

In order to implement RBDS policies, each node maintains two tables per link that the node belongs to. These are the unavailable-data-time-slots table and the requested-data-time-slots table. The tables are updated with the grants and requests exchanged with the node's 1-hop neighbors. An unavailable-data-time-slots table contains the set of unexpired grants that interfere with the link that the table belongs to. This set is denoted by $\mathcal{T}_u^{(i,j)}(m)$ for link (i,j) and is given by Equation 5, where $(g_m^{(x,y)})_{f_e}$ is the f_e component of $g_m^{(x,y)}$, and f_m is the current frame number (i.e., the frame that control-time slot m belongs to). The requested-data-time-slots table contains the set of unexpired requests made for the link the table belongs to. This set is denoted by $\mathcal{T}_r^{(i,j)}(m)$ for link (i,j) and is given by Equation 6, where $(r_m^{(i,j)})_{f_x}$ is the f_x component of $r_m^{(i,j)}$. $\mathcal{T}_u^{(i,j)}(m)$ and $\mathcal{T}_r^{(i,j)}(m)$ are functions of m given that the tables are updated with the grants and requests transmitted at every control-time slot.

$$\mathcal{T}_u^{(i,j)}(m) \triangleq \{g_l^{(x,y)} : (g_l^{(x,y)})_{f_e} \geq f_m, (x,y) \in \mathcal{I}^{(i,j)}, l \leq m\} \tag{5}$$

$$\mathcal{T}_r^{(i,j)}(m) \triangleq \{r_l^{(i,j)} : (r_l^{(i,j)})_{f_x} > f_m, l \leq m\} \tag{6}$$

In RBDS policies, two grants overlap with each other if the frame ranges given by their respective f_s and f_e frame numbers have one or more frame numbers in common.

2.2.1. RBDS Markovian system model

In order for RBDS policies to be mathematically characterized under the framework of network stability proposed in [10], it is necessary to show how networks that use RBDS policies can be modeled as Markovian systems [9].

In an RBDS network, each link has an input-queue and an output-queue as described in Section 3. The length of an input-queue (i.e., $Q_i^{(i,j)}(m)$) is defined as the number of data packets in the queue. The length of an output-queue (i.e., $Q_o^{(i,j)}(m)$) corresponds to the number of data-subframes in the following frame range: from the current frame to the last frame scheduled for the packets in the output-queue. Therefore, the length of output-queues does not depend on the number of scheduled packets waiting to be transmitted but on the schedules of those packets. The length of output-queues is given by Equation 7[14], where $\mathcal{T}_g^{(i,j)}$ is the set of unexpired grants of link (i,j) (i.e., $\mathcal{T}_g^{(i,j)}(m) \triangleq \{g_l^{(i,j)} : (g_l^{(i,j)})_{f_e} \geq f_m, l \leq m\}$).

$$Q_o^{(i,j)}(m) \triangleq [\max\left(\left\{(g)_{f_e} : g \in \mathcal{T}_g^{(i,j)}(m)\right\}\right) - f_m + 1]^+ \quad (7)$$

A node transmits scheduling packets by accessing control-time slots according to the election algorithm. The next control-time slot that node i is going to access is determined by this algorithm. This control-time slot is denoted by $M^i(m)$, i.e., at control-time slot m, the future control-time slot that node i transmits a scheduling packet is control-time slot $M^i(m)$.

Based on the previous definitions, RBDS wireless network $\mathcal{G} = (\mathcal{N}, \mathcal{L})$, where $\mathcal{N}$ and $\mathcal{L}$ are the sets of nodes and links respectively, can be represented as a Markovian system whose state $\mathcal{S}^{\mathcal{G}}$ is given by the lengths of the input and output queues of all the links and the scheduling control-time slots of all the nodes. That is,

$$\mathcal{S}^{\mathcal{G}} \triangleq \left\{Q_i^{(i,j)}(m), Q_o^{(i,j)}(m), M^i(m) : (i,j) \in \mathcal{L}, i \in \mathcal{N}\right\}. \quad (8)$$

Within this framework for RBDS networks, the stability analysis of different RBDS policies can be performed. In the following, a greedy-maximal RBDS policy is considered.

2.2.2. The greedy-maximal RBDS policy and its stability region

The GM-RBDS policy is as follows. When any node i in $\mathcal{N}$ transmits a scheduling packet,

- It grants the longest request among all the unexpired requests made by its incoming links, and sets the grant's f_s component at the frame following the interfering grant that expires the latest.
- For every of its outgoing links, it requests as many consecutive data-subframes as unscheduled data packets cover entirely, sets every request's f_s component at the frame following the interfering grant that expires the latest, and sets every request's f_x component at the frame scheduled for its next scheduling-packet transmission.

When any node i in $\mathcal{N}$ receives a scheduling packet, it checks whether there is a grant in the packet and whether the grant is directed to one of its outgoing links. If that is the case, it confirms the grant only if the grant does not overlap with any of the grants in the link's unavailable-data-time-slots table.

---

[14] $[\cdot]^+$ is the positive-part operator.

The GM-RBDS policy is greedy maximal in the sense that the requests that are granted are the longest requests and each request corresponds to the maximum integer number of data-subframes that are covered by a link's unscheduled data packets (i.e., each request corresponds to $\left\lfloor \frac{Q_i^{(i,j)}}{m_{\mathrm{ds}}} \right\rfloor$, where $Q_i^{(i,j)}$ is the number of unscheduled data packets to be transmitted on link (i,j), and m_{ds} is the number of data-time slots per data-subframe).

The size of the stability region of the GM-RBDS policy depends on the ability of the links to perform the three-way handshakes successfully. If the probability that a link finishes successfully a three-way handshake is low, the link's queue will decrease at a lower rate. Therefore, the link's ability to forward data packets within some time range is going to be lower (i.e., the highest packet rate supported by the link is lowered), and this reduces the size of the stability region. In [8], it was shown that the probability that a three-way handshake of link (i,j) is successful depends on the following aspects of the 2-hop neighborhoods of nodes i and j.

- The set of active nodes that i can listen to but j cannot, where an active node is a node that either forwards data-packets or is the destination node for at least one flow. This set is given by $\mathcal{S}_{\mathrm{a}}^i \setminus \mathcal{S}_{\mathrm{a}}^j$, where $\mathcal{S}_{\mathrm{a}}^i$ is the set of active 1-hop neighbors of node i, and $\setminus$ refers to the relative complement, i.e., $\mathcal{S}_{\mathrm{a}}^i \setminus \mathcal{S}_{\mathrm{a}}^j \triangleq \{k \in \mathcal{S}_{\mathrm{a}}^i : k \notin \mathcal{S}_{\mathrm{a}}^j\}$.
- The degree $d^{(i,j)}$ of link (i,j), which is defined as the number flows that traverse link (i,j).
- The direct 1-hop neighborhood of a node which is defined as the set of 1-hop neighbors that send data packets to the node. Therefore, the direct 1-hop neighbors of a node always precede the node in at least one flow's path. Node j's direct 1-hop neighborhood is denoted by $\mathcal{S}_{\mathrm{d}}^j$.

Based on the probability of successful three-way handshakes, sufficient conditions that guarantee queue stability under the GM-RBDS policy are given as follows[15].

Theorem 1. Let $\mathcal{G} = (\mathcal{N}, \mathcal{L})$ be a wireless mesh network that operates under GM-RBDS, shortest-path routing, and the 2-hop interference model, where $\mathcal{N}$ and $\mathcal{L}$ are the sets of nodes and links of the network respectively. Let λ_{f}^j be the maximum packet rate that node j can support for each of the flows for which it is an intermediate or destination node. $\mathcal{G}$ is stable if the packet rate λ_{f}^j supported by every node j in $\mathcal{N}$ satisfies Equation 9.

$$\lambda_{\mathrm{f}}^j < \frac{1}{5 \sum_{i \in \mathcal{S}_{\mathrm{d}}^j} d^{(i,j)} \, | \mathcal{S}_{\mathrm{a}}^i \setminus \mathcal{S}_{\mathrm{a}}^j |} \quad \forall \; j \in \mathcal{N} \tag{9}$$

[15]The proof of Theorem 1 is given in [34].

Therefore, in order to guarantee stability under shortest-path routing and GM-RBDS, the data-packet rate of a flow must be less than the following rate: the minimum packet rate among all the packet rates that nodes along the flow's path can assign to the flow. This is shown in Equation 10, where $\mathcal{F}$ is the set of data flows in the network, f_n is the nth flow in $\mathcal{F}$, p_n is the path followed by f_n, λ_n is the data-packet rate of flow f_n, and λ_{max}^j is the upper-bound for node j's rate λ_{f}^j according to Equation 9 (i.e., $\lambda_{\text{max}}^j \triangleq (5\sum_{i\in \mathcal{S}_{\text{d}}^j} d^{(i,j)} \mid \mathcal{S}_{\text{a}}^i \setminus \mathcal{S}_{\text{a}}^j \mid)^{-1}$).

$$\lambda_n < \min\{\lambda_{\text{max}}^j : j \in p_n\} \quad \forall\ f_n \in \mathcal{F} \tag{10}$$

Remark. Notice that the sufficient condition for stability given by Equation 9 is of the same form of the condition for the non-RBDS greedy policies analyzed in [23] (Equation 4 in [23]). That is, the total packet-arrival rate of a set of interfering links needs to be lower than some constant in order to guarantee stability, and the constant depends on some characteristic of the network topology (i.e., $\mathcal{S}_{\text{a}}^i \setminus \mathcal{S}_{\text{a}}^j$ for the GM-RBDS policy, and κ for the greedy policies in [23]). Other policies have the same behavior as well. For example, the stability properties of GMS [27] and the bipartite simulation (BP-SIM) [24] policies depend on the local-pooling factor[16] and the maximum node degree[17] of the network respectively, and these are determined by the network topology.

3. Stability-based topology control[18]

In this section, we look at the problem of topology control for adapting the stability region of the backbone of the wireless mesh network to a given set of flows such that the total throughput is improved. This topology-control framework was originally studied in [34-36]. Specifically, we ask the question of what are the nodes' transmission powers (TP) that adapt the stability region to the flows in the network when a set of source-destination pairs, the routing algorithm, and the link-scheduling policy are given. Notice that by adapting the TPs of the nodes (i.e., wireless mesh routers and gateways), the topology of the network is being controlled due to the creation and elimination of links. Also, notice that the flows correspond to the traffic established across the wireless mesh routers and gateways of the network.

By adapting the stability region of the network, the queue lengths across the network are decreased in average for a given set of flows' data-packet rates. In this way, the flows among the source-destination pairs are able to maintain higher levels of end-to-end throughput and lower levels of end-to-end delay while guaranteeing queue stability. Therefore, the problem considered in this chapter is of particular interest for applications that establish non-bursty sessions between source-destination pairs such as audio/video calls.

---

[16] The local-pooling factor is a topological property of the network that indicates how different the effectiveness of the different maximal link schedules is from each other [27]. When the different maximal link schedules are similarly effective, GMS policies are able to support packet rates that are closer to the boundaries of the optimal stability region.

[17] The node degree is defined as the number of links that the node belongs to.

[18] The material presented in this section is based on the material presented in [34, 36].

In order to adapt the stability region, we propose an algorithm that is executed by the flows established between the source-destination pairs. The idea behind the algorithm is to adapt a lower-bound region of the stability region (i.e., a region covered by the stability region) by modifying the TPs. The lower-bound region is a widely accepted theoretical performance metric used for comparing different link-scheduling policies [23][19]. In the algorithm, once the flows' paths are determined by the routing algorithm, the flows calculate the maximum data-packet rate they can support within the lower-bound region; then, each flow tries to stretch the lower-bound region by modifying the TP of nodes surrounding it. The effect that the stretch of the lower-bound region has on the stability region is another stretch on this region. Therefore, the result is a stability region adapted to the flows that allows them to support higher data-packet rates while guaranteeing the stability of the network. A graphical example of this adaptation was shown in Figure 5.

We consider IEEE 802.16 wireless mesh networks that operate under shortest-path routing and the GM-RBDS policy. However, our results can be readily extended to other networks, routing algorithms, and link-scheduling policies.

3.1. Stability-region expansion algorithms

The main idea presented in this chapter (i.e., adapting the stability region of a given link-scheduling policy by means of TP control) is based on the results obtained in [37, 38, 39]. In [37], the network is partitioned based on the notion of local pooling, and each partition is assigned to a channel of the network. In this way, the GMS policy is guaranteed to achieve the optimal stability region in each channel. In [38, 39], network topologies are identified for which distributed link-scheduling policies achieve the optimal stability region. However, these network topologies are not suitable for real scenarios [27] because of their sufficient conditions that guarantee the optimal stability region. These conditions include [38] 1-hop interference, 1-hop traffic, and a topology that is a graph that belongs to one of the following perfect-graph classes: chordal graphs, chordal bipartite graphs, cographs, and a subgroup of co-comparability graphs. In real scenarios, these conditions limit the suitability of wireless mesh networks. For example, only a few physical-layer technologies such as code-division-multiple-access (CDMA) can be approximated with the 1-hop interference model, and the traffic in wireless mesh networks is multihop by definition. Also, making the topology fall within the previous graph families imposes constraints on the locations and TPs of the nodes and the available routes. The multihop traffic case was considered in [38], and it was shown that only a subset of the previous graph families guarantee the optimal stability region in the multihop-traffic scenario. These were identified as forest of stars, where every connected component of the network graph is a star graph. Also, the results in [37, 38] are valid only for GMS policies under 1-hop traffic or backpressure routing-scheduling policies under multihop traffic[20]. In

[19]The reason for this is that the exact formulation of the stability region is not actually available. The stability region is usually characterized with the lower-bound region because its exact characterization is not feasible due to its complexity. See [8, 10, 11, 16, 18, 19, 23-25, 29, 30] for the literature on the problem of characterizing the stability region of link-scheduling policies.

[20]It should be noted that the objective in [37, 38] was mainly to identify the topologies that enable the optimality of the GMS policy, and not to design topology-control algorithms.

[40], a random-power-selection algorithm for random-access scheduling policies was proposed. It is shown that it achieves maximal throughput in the following sense: the throughput achieved by any fixed power selection is at most equal to the throughput achieved by the random-power-selection algorithm.

Our approach is built upon the idea of [37, 38] that under certain topologies a link scheduling policy performs better. We modify realistically the network topology using TP control to adapt the policy's stability region to the flows. The algorithm receives any set of end-to-end paths, node locations, and scheduling policy, and adapts the policy's stability region to the paths. Such an approach is beneficial because it improves the end-to-end throughput and delay without the restrictions previously discussed. In this chapter, we consider the case of shortest-path routing, GM-RBDS scheduling, and randomly chosen source-destination pairs of nodes in IEEE 802.16 mesh networks.

Other heuristic algorithms have been proposed in the literature that improve the performance of the link-scheduling policy in terms of throughput by means of TP control. These algorithms include the ones reported in [41-43] whose basic idea is to increase the total throughput in the network by means of spatial reuse. The spatial reuse is increased by reducing the TP of the nodes. The algorithms differ between them in the way they are adapted to request-to-send (RTS)/clear-to-send (CTS) based protocols. In [44, 45], it is shown that better throughput improvements can be achieved not only by decreasing the TP to increase the spatial reuse but also by considering the hidden and exposed nodes. The algorithms proposed in [44] perform TP control with the objective of avoiding hidden nodes. In this way, the links in the network are able to sustain higher data-packet rates. In [46], a TP control algorithm for RTS/CTS-based protocols is proposed that decreases the area occupied by links during their transmissions, which is defined as the area in which other nodes must remain silent during the time the link is active. Then, it is shown that with this scheme, routing algorithms that favor short hops achieve higher levels of throughput. The goal of our algorithm is similar to the goal of the previous algorithms [41-46], i.e., to increase the data-packet rates that a given link-scheduling policy can support by means of TP control. However, our approach differs in that it is directly based on a quantitative metric which is the stability region. It is not based on qualitative observations of the operation of the link-scheduling policy such as the hidden and exposed nodes in RTS/CTS-based policies. Therefore, it can be readily adapted to any link-scheduling policy whose stability region has been characterized such as the ones discussed in Section 4.

A different type of TP control algorithms, which are based on optimization techniques, are discussed in [47, 48]. In [47], the problem of integrated link scheduling and TP control for throughput optimization is shown to be nondeterministic polynomial time (NP) complete. Therefore, a heuristic algorithm is developed. The goal of the algorithm is to minimize the schedule length necessary to satisfy all the link loads determined by a given routing algorithm. By minimizing the schedule length, the total throughput of the network is increased because more scheduling cycles can be performed per time unit. In [48], the problem of jointly optimizing the flow routes, link schedules, TP, modulation and coding schemes is addressed. This is a more general problem than the one considered in [47] given that it does not only include the calculation of TPs and link schedules but also includes the routing and physical layers (i.e., flow routes, modulation, and coding schemes). In our algorithm, we are only concerned in the TP control problem when the flows and link-scheduling policy are given. That is, for a given set of flows, we determine TPs that improve the performance of the link-scheduling policy in terms of throughput and end-to-end delay.

3.2. The HSRA-topology-control algorithm

The goal of our TP control algorithm is to expand the lower-bound region given by Equation 10. By expanding this region, the flow rates λ_n can take higher values while guaranteeing stability, and therefore, the maximum total throughput the network can support for the given flows is increased. Let the maximum total throughput be denoted by λ_{T} and defined in terms of the lower-bound region for the flows' data-packet rates given by Equation 10 as follows.

$$\lambda_{\mathrm{T}} \triangleq \sum_{p_n \in \mathcal{F}} \min\{\lambda_{\max}^{j} : j \in p_n\} \tag{11}$$

According to Equations 11, 10, and 9, λ_{T} depends on the direct 1-hop neighborhoods (i.e., $\{\mathcal{S}_{\mathrm{d}}^{j} : j \in \mathcal{N}\}$), the link degrees (i.e., $\{d^{(i,j)} : (i,j) \in \mathcal{L}\}$), and the active 1-hop neighborhoods (i.e., $\{\mathcal{S}_{\mathrm{a}}^{j} : j \in \mathcal{N}\}$) as follows.

$$\lambda_{\mathrm{T}} = \sum_{p_n \in \mathcal{F}} \left(\min_{j \in p_n} \frac{1}{5 \sum_{i \in \mathcal{S}_{\mathrm{d}}^{j}} d^{(i,j)} \mid \mathcal{S}_{\mathrm{a}}^{i} \setminus \mathcal{S}_{\mathrm{a}}^{j} \mid} \right) \tag{12}$$

Given that the flows are determined by the shortest-path routing algorithm, the following parameters in Equation 12 are fixed: $\{p_n \in \mathcal{F}\}$, $\{\mathcal{S}_{\mathrm{d}}^{j} : j \in \mathcal{N}\}$, and $\{d^{(i,j)} : (i,j) \in \mathcal{L}\}$. Therefore, in order to increase λ_{T}, the only parameters that can be modified are the active 1-hop neighborhoods (i.e., $\{\mathcal{S}_{\mathrm{a}}^{j} : j \in \mathcal{N}\}$). They can be modified by means of TP control such that λ_{T} is maximized. This optimization problem, which we call stability region adaptation for throughput maximization (SRA-TM), is given as follows.

Definition 4. Given a set of flows $\mathcal{F}$ calculated by the shortest-path routing algorithm, the SRA-TM problem consists of the maximization of λ_{T} by means of TP control such that none of the nodes exceed the maximum TP and none of the paths are broken. That is,

$$\begin{aligned} \text{maximize} \quad & \sum_{p_n \in \mathcal{F}} \left(\min_{j \in p_n} \frac{1}{5 \sum_{i \in \mathcal{S}_{\mathrm{d}}^{j}} d^{(i,j)} \mid \mathcal{S}_{\mathrm{a}}^{i} \setminus \mathcal{S}_{\mathrm{a}}^{j} \mid} \right) \\ \text{subject to} \quad & 0 \le r^{i} \le r^{\max} \quad \forall\ i \in \mathcal{N} \\ & r^{i}, r^{j} \ge \| i,j \| \quad \forall\ i \in \mathcal{S}_{\mathrm{d}}^{j}, j \in \mathcal{N}, \end{aligned} \tag{13}$$

where r^i is the transmission radius of node i, $r^{\max}$ is the maximum transmission radius, and $||i,j||$ is the Euclidean distance between nodes i and j.

Remark. In the SRA-TM problem, the flow paths are given and left unmodified. Higher values for λ_T could be achieved if the flow paths were modified by including them as decision variables. For example, a routing scheme can uniformly distribute the traffic loads across the links of the network so that links with high levels of congestion are avoided. This problem corresponds to a joint optimization of the topology and flow paths based on the stability region. This problem can be further studied due to its potential benefits on λ_T. However, this chapter deals only with the stability-region-based topology control as a first step towards the problem of stability-region-based joint topology and routing control.

Remark. If the data traffic in the network changes dynamically, the flow paths may change as well. In this scenario, the SRA-TM problem needs to be solved for every flow-path change. Therefore, the speed of convergence of algorithms that solve the SRA-TM problem is an important metric for such a scenario. The algorithms should be able to keep up with the rate of change of the flow paths. On the other hand, if the data-traffic levels of a set of flows change but the flow paths do not change, the SRA-TM problem does not need to be solved again. The reason is that the solution of the SRA-TM problem is the topology that allows those flows to support the maximum level of data traffic while guaranteeing stability. This means that the data-traffic levels in the flows may vary as long as they do not exceed such maximum levels (i.e., $\min\{\lambda_{\max}^j : j \in p_n\} \quad \forall\ f_n \in \mathcal{F}$), and this can be guaranteed by means of call-admission-control algorithms.

In order to solve the SRA-TM problem, the following TP algorithm is proposed. It is called heuristic stability region adaptation (HSRA)[21].

The following definitions are necessary for the operation of the HSRA algorithm.

Definition 5. The bottleneck node of flow f_n is the node with the lowest maximum rate among all the intermediate and destination nodes of the flow, i.e., let j be the bottleneck node of f_n, then $j = argmin_{i \in \{p_n(m): 2 \le m \le |p_n|\}} \lambda_f^i$, where p_n is the set of nodes in the path of flow f_n, $p_n(m)$ is the mth node in p_n, and $|p_n|$ is the number of nodes in p_n.

Definition 6. Node h is hidden from node j if and only if $h \in \mathcal{S}_a^i \setminus \mathcal{S}_a^j$ for some $i \in \mathcal{S}_d^j$.

Definition 7. The MinPower setup is the set of minimum TPs whose transmission ranges guarantee that none of the links of the flows in $\mathcal{F}$ is broken.

The operation of the algorithm is as follows. First, the nodes' TPs (i.e., $\{r^i : i \in \mathcal{N}\}$) are set according to the MinPower setup (line 2 of the HSRA Algorithm). By reducing the TPs (Definition 7), the spatial reuse in the network is increased, and as a consequence, the total throughput is increased as well[22]. Then, the maximum throughput that intermediate and

[21]The SRA-TM problem is formulated as a mixed integer program with non-linear constraints in [34]. This formulation is used in Section 6 for calculating the optimal solution of the simulated instances of the SRA-TM problem.
[22]This spatial-reuse-based TP control is the basis of the algorithms proposed [39-43, 46].

HSRA Algorithm

```
1: procedure HSRA(𝒩, ℱ, M)
2:     {r^i} ← MinPowerSetup(𝒩)
3:     {λ^i_max} ← NodeMaxRates(𝒩)
4:     T ← TotalThroughput(ℱ, {λ^i_max})
5:     for m ← 0, m < M, m ← m + 1 do
6:         f_n ← PickFlowRandomly(ℱ)
7:         j ← BottleneckNode(f_n, {λ^i_max})
8:         S^j_d ← Direct1HopNeigh(j)
9:         for every i in S^j_d do
10:            S^i_a \ S^j_a ← HiddenActiveNodes((i, j))
11:        end for
12:        i ← NodeHiddenTheMost({S^i_a \ S^j_a})
13:        if ||i, j|| < r^max then
14:            r_aux ← r^i
15:            r^i ← ||i, j||
16:            {λ^i_aux} ← NodeMaxRates(𝒩)
17:            T_aux ← TotalThroughput(ℱ, {λ^i_aux})
18:            if T_aux > T then
19:                {λ^i_max} ← {λ^i_aux}
20:                T ← T_aux
21:            else
22:                r^i ← r_aux
23:            end if
24:        end if
25:    end for
26: end procedure
```

destination nodes can support for the flows they belong to is calculated (line 3 of the HSRA Algorithm). This is done using Equation 9, which defines the nodes' maximum throughput. Based on these maximums, the total throughput the network can support is calculated (line 4 of the HSRA Algorithm).

Once the total throughput under the MinPower setup is known, flows are selected randomly one-by-one for a number of M times (line 5 of the HSRA Algorithm). Every time a flow is selected, the maximum throughput the flow can support is increased if this causes that the total throughput be increased as well. Otherwise, the flow is left unmodified. The throughput of the selected flow is increased as follows.

Let the selected flow be denoted by f_n (line 6 of the HSRA Algorithm). The bottleneck node of f_n is found first by tracking the node of the flow with the lowest maximum throughput (Equation 10). Let this node be denoted by j (line 7 of the HSRA Algorithm). The maximum rate of j (i.e., λ_f^j) is increased by increasing the TP of one of j's 2-hop neighbors (lines 8 to 15 of the HSRA Algorithm). However, this TP increase is confirmed only if the total throughput (i.e., λ_T) is increased as well (lines 16 to 20 of the HSRA Algorithm). Otherwise, the TP of j's 2-hop neighbor is left unmodified (line 22 of the HSRA Algorithm). The total throughput may be decreased given that the TP increase of j's 2-hop neighbor may decrease the maximum rate of other bottleneck nodes in the network, and this maximum-rate decrease may be higher than the increase on j's maximum rate.

The 2-hop neighbor of node j whose TP is increased is selected so that the factor $|\mathcal{S}_{\mathrm{a}}^{i} \setminus \mathcal{S}_{\mathrm{a}}^{j}|$ on the denominator of the upper-bound for λ_{f}^{j} is decreased (Equation 9). Qualitatively, this TP increase can be explained as follows. Node j (i.e., the bottleneck node) has a set of 1-hop neighbors that are sending data packets to it (i.e., $\mathcal{S}_{\mathrm{d}}^{j}$). Let i be one of these nodes, and consider the link (i,j) and the input and output queues $Q_{i}^{(i,j)}$ and $Q_{o}^{(i,j)}$ of node i as shown in Figure 3. In order for i to transmit packets to j, a reservation of future data-time slots is required. When nodes i and j finish this reservation successfully, data packets in node i's input-queue (i.e., $Q_{i}^{(i,j)}$) are moved to node i's output-queue (i.e., $Q_{o}^{(i,j)}$), and these packets are later pulled from $Q_{o}^{(i,j)}$ for their transmission. Therefore, for the queues $Q_{i}^{(i,j)}$ and $Q_{o}^{(i,j)}$ to have their lengths decreased, the reservation performed by nodes i and j needs to be successful, i.e., the three-way handshake for scheduling data-packet transmissions on link (i,j) needs to be successful. The probability that the handshake is successful and that the queues decrease their length depends on the grants received by node i and not received by node j. In the following, we refer to these grants as hidden-grants. If i requests future data-time slots to j and a hidden grant is received by i before j transmits its grant to i, j's grant may not be confirmed by i. This is because the hidden grant may interfere with j's grant. On the other hand, if j is able to listen to the hidden grant, j is able to generate its grant such that it does not interfere with the hidden grant, and i will be able to confirm j's grant[23]. Therefore, in order to increase the probability of handshake success and queue decrease, the TP of the node that transmits the hidden grant (i.e., the node hidden from j) can be increased such that node j is able to listen to the hidden node's transmissions.

Node j may have more than 1 hidden nodes in every incoming link from the nodes in its direct 1-hop neighborhood. The HSRA algorithm chooses only one of those hidden nodes for increasing its TP. The node that is chosen is the node that is hidden from the highest number of nodes (i.e., node j and all the other intermediate or destination nodes unable to listen to the hidden node). This is performed in lines 8 to 12 of the HSRA Algorithm. In this way, the maximum rate is increased for all those nodes so that, if one or more of those nodes are bottleneck nodes, higher improvements on the total throughput can be achieved.

The role that the objective function of the SRA-TM problem (Equation 13) plays in the HSRA Algorithm is the quantification of the throughput improvement by the TP increase on hidden nodes. By increasing the TP of a node hidden from a bottleneck node, the factor $\mathcal{S}_{\mathrm{a}}^{i} \setminus \mathcal{S}_{\mathrm{a}}^{j}$ in the denominator of Equation 13 is decreased for the bottleneck node, and as a consequence the bottleneck node's maximum rate is increased. However, the TP increase on the hidden node may also cause an increase on the $\mathcal{S}_{\mathrm{a}}^{i} \setminus \mathcal{S}_{\mathrm{a}}^{j}$ factor of other bottleneck nodes. Therefore, the objective function allows the algorithm to trade off between decreasing the number of hidden nodes by increasing TP and maintaining spatial reuse by not increasing

[23]The problem of node j not being able to listen to hidden grants is the hidden-node problem version for reservation-based distributed scheduling policies. This problem is studied in detail in [8, 11].

TP. In the algorithm, this tradeoff is achieved by testing the improvement on the total throughput (lines 14 to 23 of the HSRA Algorithm).

4. Simulation results

The performance evaluation of the HSRA algorithm was performed by means of simulation using the simulator proposed in [49]. The simulated network is an IEEE 802.16 mesh network with distributed scheduling under the configuration shown in Table 1. The number of nodes was specified as a simulation parameter. The nodes were uniformly distributed in a square area such that the node density was always kept at 15 nodes per unit area. The maximum transmission range of the nodes was set at 0.3 (i.e., $r^{\max} = 0.3$). The connectivity of the network under bidirectional links and with the nodes' transmission ranges set at $r^{\max}$ was confirmed before executing the shortest-path routing algorithm. The number of flows was specified as a simulation parameter. The source and destination nodes of every flow were uniformly distributed among all the nodes in the network. The shortest-path routing algorithm calculated the flow paths under the MaxPower setup which is the power assignment when all the nodes' transmission ranges are set at the maximum (i.e., $r^{\max}$). Once the paths were calculated, the transmission ranges of the nodes were found using the HSRA algorithm. Also, the optimal transmission ranges (i.e., the solution to the SRA-TM problem (Equation 13)), which we call OptPower, were found in [34] using the formulation of the SRA-TM problem as a mixed integer program with non-linear constraints (MIP-NLC). The MIP-NLC was solved using the Branch And Reduce Optimization Navigator (BARON) Solver [50], which is a system for solving non-convex optimization problems to global optimality. Finally, the network was simulated under the MaxPower, MinPower, OptPower, and HSRA setups.

Parameter	Value
Frame length	10 ms
Control-time-slot length	63 μs
Number of control-time slots per frame	4
Number of data-time slots per frame	256
NextXmtMx †	7
XmtHoldoffExponent †	6
Link scheduling	GM-RBDS
Routing	shortest-path

Table 1. IEEE 808.16 mesh network configuration

[†] This is a parameter of the election algorithm used t specify the frequency that nodes transmit scheduling packets.

Figure 6 shows the average output-queue length for three networks with 20 nodes each and increasing number of flows (i.e., 10 flows in Figure 6a, 15 flows in Figure 6b, and 20 flows in

Figure 6c). The input-queues have been omitted because they are guaranteed to always be stable [8, 11]. The flow rates in each network were all set at the same value. These are 8, 6, and 5 packets per frame for the networks in Figures 6a, 6b, and 6c respectively. These values were set so that their corresponding HSRA network became unstable if they were increased by at least one point. In this way, the network operates at a point inside the stability region and close to its boundary. Therefore, when any of the rates is increased by at least one point, the network operates outside the stability region, and therefore, it is unstable.

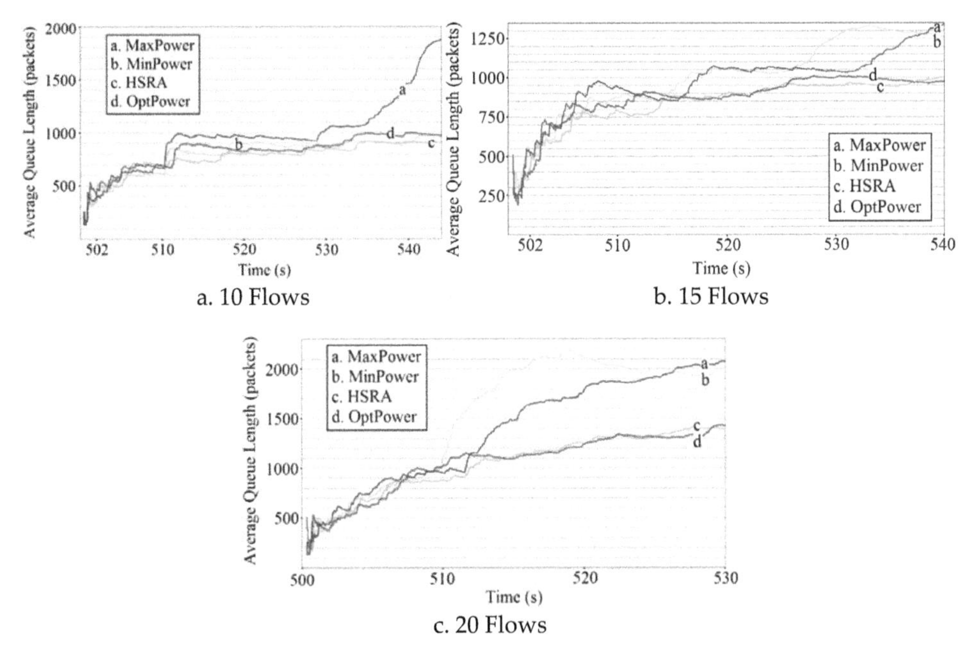

Figure 6. Average output-queue comparison for the HSRA, MinPower, MaxPower, and OptPower configurations

Close to the end of the simulation time, when the transient behavior of the queues is over, the average queue lengths of the different power setups (i.e., MaxPower, MinPower, OptPower, and HSRA) can be compared. In Figure 6a, the MaxPower setup has the worst performance (i.e., the largest average queue length), and the MinPower, OptPower, and HSRA have similar performance. Therefore, when the number of flows is low (i.e., 10 flows), the MinPower and the HSRA algorithms are able to achieve queue lengths that are close to the lengths achieved by the optimal solution (i.e., OptPower). On the other hand, when the number of flows increases, the MinPower algorithm does not achieve a performance close to the optimal one while the HSRA algorithm does. This is shown in Figures 6b and 6c. The MaxPower and MinPower algorithms have similar performance which is worse when compared with the HSRA algorithm. The HSRA algorithm achieves average queue lengths that are close to the lengths achieved by the optimal solution. Therefore, the HSRA algorithm enables the flows to carry more traffic while guaranteeing stability than the MaxPower and

MinPower algorithms do. Also, it is confirmed that the technique of only maximizing the spatial reuse by reducing the transmission ranges (i.e., MinPower) does not perform well when the flow density increases (i.e., when the number of flows increases and the number of nodes is kept constant.). On the other hand, the technique of adapting the stability region to the given set of flows by means of TP control (i.e., HSRA) does perform well when the flow density increases.

5. Conclusion

A new framework for the stability analysis of scheduling policies for wireless networks that allow the reservation of future data-subframes has been proposed. The concepts of input-queue and output-queue were introduced into the framework in order to account for the packets waiting to be scheduled and the schedules assigned to these packets. Based on these concepts, sufficient conditions for the stability of RBDS wireless networks were found.

Within the proposed framework, an RBDS policy which uses the concept of greedy-maximal scheduling was analyzed. The nodes implement this policy by exchanging scheduling packets using the IEEE 802.16 election algorithm. A region in which the proposed reservation-based scheduling policy is stable was found using the framework. It was shown that the size of this region depends on a characteristic of the network topology (i.e., Q_o^j).

The HSRA algorithm has been proposed for transmission power control. This algorithm increases the data-packet rates that flows can support and decreases the end-to-end delays. It is based on the adaptation of the stability region of a given link-scheduling policy when only the links that belong to a given set of flows are considered. The algorithm can be readily adapted to any link-scheduling policy whose stability region has been characterized, so it is not limited to any specific scheduling approach such as RTS/CTS-based policies. The improvement on throughput achieved by our algorithm was evaluated by means of. It was shown that it outperforms the classical solution of reducing transmission powers to increase spatial reuse.

Future lines of research include the development of a new framework for distributed topology-control algorithms. For example, this framework could based on a game-theoretical approach in which a given set of flows act as players that collaborate to maximize the packet rates they can support while guaranteeing stability. Also, based on the new framework, new distributed topology-control algorithms should be developed for IEEE 802.16 WMNs. Finally, the algorithms should be implemented and tested on WMN testbeds in order to evaluate the improvement in throughput they achieve in a real scenario.

Author details

Gustavo Vejarano

Department of Electrical Engineering and Computer Science,
Loyola Marymount University, Los Angeles, CA, USA

6. References

[1] Kawadia, V. and Kumar, P. R. (2005) Principles and protocols for power control in wireless ad hoc networks. IEEE Journal on Selected Areas in Communications. 23: 76–88.
[2] Santi, P. and Blough, D. M. (2003) The critical transmitting range for connectivity in sparse wireless ad hoc networks. IEEE Transactions on Mobile Computing. 2: 25–39.
[3] Sridhar, A. and Ephremides, A. (2008) Energy optimization in wireless broadcasting through power control. Ad Hoc Networks. 6: 155–167.
[4] Chiang, M., Tan, C.W., Palomar, D. P., O'Neill, D., and Julian, D. (2007) Power control by geometric programming. IEEE Transactions on Wireless Communications. 6: 2640–2651.
[5] Kim, T.-S., Lim, H., and Hou, J. C. (2008) Understanding and improving the spatial reuse in multihop wireless networks. IEEE Transactions on Mobile Computing. 7: 1200–1212.
[6] Chiang, M. (2005) Balancing transport and physical layers in wireless multihop networks: Jointly optimal congestion control and power control. IEEE Journal on Selected Areas in Communications. 23: 104–116.
[7] Lin, X. and Shroff, N. (2006) The impact of imperfect scheduling on cross-layer congestion control in wireless networks. IEEE/ACM Transactions on Networking. 14: 302–315.
[8] Vejarano, G. and McNair, J. (2012) Stability analysis of reservation-based scheduling policies in wireless networks. IEEE Transactions on Parallel and Distributed Systems. 23: 760–767.
[9] Asmussen, S. (2003) Applied Probability and Queues. New York: Springer. 438 p.
[10] Tassiulas, L. and Ephremides, A. (1992) Stability properties of constrained queuing systems and scheduling policies for maximum throughput in multihop radio networks. IEEE Transactions Automatic Control. 37: 1936–1948.
[11] Vejarano, G. and McNair, J. (2010) Reservation-based distributed scheduling in wireless networks. Proc. 11th IEEE International Symposium on a World of Wireless, Mobile and Multimedia Networks (WoWMoM'10). Montreal, QC, Canada. pp. 1–9.
[12] Sharma, G., Mazumdar, R., and Shroff, N. (2006) On the complexity of scheduling in wireless networks. Proc. Annual International Conference on Mobile Computing and Networking (MobiCom'06). Los Angeles, CA. pp. 227–238.
[13] Hajek, B. and Sasaki, G. (1988) Link scheduling in polynomial time. IEEE Transactions on Information Theory. 34: 910–917.
[14] Papadimitriou, C. and Steiglitz, K. (1982) Combinatorial Optimization: Algorithms and Complexity. Prentice-Hall.
[15] Tassiulas, L. (1998) Linear complexity algorithms for maximum throughput in radio networks and input queued switches. Proc. Annual Joint Conference of the IEEE Computer and Communications Societies (INFOCOM'98). San Francisco, CA. pp. 533–539.
[16] Wu, X. and Srikant, R. (2005) Regulated maximal matching: A distributed scheduling algorithm for multi-hop wireless networks with node-exclusive spectrum sharing. Proc. IEEE Conference on Decision and Control, and the European Control Conference (CDC-ECC'05). Seville, Spain. pp. 5342–5347.

[17] Sharma, G., Joo, C., and Shroff, N. (2006) Distributed scheduling schemes for throughput guarantees in wireless networks. Proc. Annual Allerton Conference on Communication, Control, and Computing (Allerton'06). Monticello, IL. pp. 1–10.

[18] Joo, C. and Shroff, N. (1997) Performance of random access scheduling schemes in multihop wireless networks. Proc. Annual Joint Conference of the IEEE Computer and Communications Societies (INFOCOM'97). Kobe, Japan. pp. 19–27.

[19] Sanghavi, S., Bui, L., and Srikant, R. (2007) Distributed link scheduling with constant overhead. Proc. International Conference on Measurement and Modeling of Computer Systems (SIGMETRICS'07). San Diego, CA. pp. 313–324.

[20] Chaporkar, P., Kar, K., and Sarkar, S. (2005) Throughput guarantees through maximal scheduling in multi-hop wireless networks. Proc. 43rd Annual Allerton Conference on Communications, Control, and Computing (Allerton'05). Urbana-Champaign, IL. pp. 1–11.

[21] Sarkar, S., Chaporkar, P., and Kar, K. (2006) Fairness and throughput guarantees with maximal scheduling in wireless networks. Proc. 4th International Symposium on Modeling and Optimization in Mobile, Ad Hoc, and Wireless Networks (WiOpt'06). Boston, MA. pp. 1–13.

[22] Chaporkar, P., Kar, K., and Sarkar, S. (2006) Achieving queue-length stability through maximal scheduling in wireless networks. Proc. Information Theory and Applications: Inaugural Workshop (ITA'06). San Diego, CA. pp. 1–4.

[23] Wu, X., Srikant, R., and Perkins, J. (2007) Scheduling efficiency of distributed greedy scheduling algorithms in wireless networks. IEEE Transactions on Mobile Computing. 6: 595–605.

[24] Gupta, A., Lin, X., and Srikant, R. (1997) Low-complexity distributed scheduling algorithms for wireless networks. Proc. Annual Joint Conference of the IEEE Computer and Communications Societies (INFOCOM'97). Kobe, Japan. pp. 1631–1639.

[25] Joo, C. (2008) A local greedy scheduling scheme with provable performance guarantee. Proc. ACM International Symposium on Mobile Ad Hoc Networking and Computing (MOBIHOC' 08). Hong Kong SAR, China. pp. 111–120.

[26] Bui, L., Eryilmaz, A., and Srikant, R. (2008) Asynchronous congestion control in multi-hop wireless networks with maximal matching-based scheduling. IEEE/ACM Transactions on Networking. 16: 826–839.

[27] Joo, C., Lin, X., and Shroff, N. B. (2009) Understanding the capacity region of the greedy maximal scheduling algorithm in multihop wireless networks. IEEE/ACM Transactions on Networking. 17: 1132–1145.

[28] Chen, L., Low, S. H., Chiang, M., and Doyle, J. C. (2006) Cross-layer congestion control, routing and scheduling in ad hoc wireless networks. Proc. Annual Joint Conference of the IEEE Computer and Communications Societies (INFOCOM'06). Barcelona, Spain. pp. 1–13.

[29] Penttinen, A., Koutsopoulos, I., and Tassiulas, L. (2006) Low-complexity distributed fair scheduling for wireless multi-hop networks. Proc.Workshop on Resource Allocation in Wireless Networks (RAWNET'06). Boston, MA. pp. 1–6.

[30] Lin, X. and Rasool, S. B. (2006) Constant-time distributed scheduling policies for ad hoc networks. Proc. IEEE Conference on Decision and Control (CDC'06). San Diego, CA. pp. 1258–1263.

[31] IEEE (2004), IEEE standard for local and metropolitan area networks - part 16: Air interface for fixed broadband wireless access systems. IEEE Std. 802.16. 857 p.

[32] Cao, M., Zhang, Q., and Wang, X. (2007) Analysis of ieee 802.16 mesh mode scheduler performance. IEEE Transactions on Wireless Communication. 6: 1455–1464.

[33] Wang, S.-Y., Lin, C.-C., Chu, H.-W., Hsu, T.-W., and Fang, K.-H. (2008) Improving the performance of distributed coordinated scheduling in ieee 802.16 mesh networks. IEEE Transactions on Vehicular Technology. 4: 2531–2547.

[34] Vejarano, G., Wang, D., and McNair, J. (2011) Stability region adaptation using transmission power control for transport capacity optimization in ieee 802.16 wireless mesh networks. Computer Networks (Elsevier). 55: 3694–3704.

[35] Vejarano, G. and McNair, J. (2007) An intelligent wireless mesh network backbone. Proc. 3rd International Conference on Wireless Internet (WICON'7). Austin, TX. pp. 1–5.

[36] Vejarano, G. and McNair, J. (2010) Queue-stability-based transmission power control in wireless multihop networks. Proc. Global Communications Conference, IEEE (GLOBECOM' 10). Miami, FL. pp. 1–5.

[37] Brzezinski, A., Zussman, G., and Modiano, E. (2006) Enabling distributed throughput maximization in wireless mesh networks: A partitioning approach. Proc. 12th Annual International Conference on Mobile Computing and Networking (MobiCom'06). Los Angeles, CA. pp. 26–37.

[38] Zussman, G., Brzezinski, A., and Modiano, E. (2008) Multihop local pooling for distributed throughput maximization in wireless networks. Proc. 27th Conference on Computer Communications IEEE (INFOCOM'08). Phoenix, AZ. pp. 1139–1147.

[39] Birand, B., Chudnovsky, M., Ries, B., Seymour, P., Zussman, G., and Zwols, Y. (2010) Analyzing the performance of greedy maximal scheduling via local pooling and graph theory. Proc. 29th Conference on Computer Communications IEEE (INFOCOM'10). San Diego, CA. pp. 26-37.

[40] Gao, Y., Zeng, Z., and Kumar, P. R. (2010) Joint Random Access and Power Selection for Maximal Throughput in Wireless Networks. Proc. 29th Conference on Computer Communications IEEE (INFOCOM'10). San Diego, CA. pp. 1-5.

[41] Monks, J. P., Bharghavan, V., and Hwu, W. (2001) A power controlled multiple access protocol for wireless packet networks. Proc. 20th Conference on Computer Communications. IEEE (INFOCOM'01). Anchorage, AK. pp. 219–228.

[42] Muqattash, A. and Krunz, M. (2003) Power controlled dual channel (pcdc) medium access protocol for wireless ad hoc networks. Proc. 22nd Conference on Computer Communications. IEEE (INFOCOM'03). San Francisco, CA. pp. 470–480.

[43] Muqattash, A. and Krunz, M. (2004) A single-channel solution for transmission power control in wireless ad hoc networks. Proc. 5th ACM International Symposium on Mobile Ad Hoc Networking and Computing (MobiHoc'04). Roppongi, Japan. pp. 210–221.

[44] Ho, I.W.-H. and Liew, S. C. (2007) Impact of power control on performance of IEEE 802.11 wireless networks. IEEE Transactions on Mobile Computing. 6: 1245–1258.

[45] Jiang, L. B. and Liew, S. C. (2008) Improving throughput and fairness by reducing exposed and hidden nodes in 802.11 networks. IEEE Transactions on Mobile Computing. 7: 34–49.

[46] Wang, W., Srinivasan, V., and Chua, K.-C. (2008) Power control for distributed mac protocols in wireless ad hoc networks. IEEE Transactions on Mobile Computing. 7: 1169–1183.

[47] Behzad, A. and Rubin, I. (2007) Optimum integrated link scheduling and power control for multihop wireless networks. IEEE Transactions on Vehicular Technology. 56: 194–205.

[48] Karnik, A., Iyer, A., and Rosenberg, C. (2008) Throughput-optimal configuration of fixed wireless networks. IEEE/ACM Transactions on Networking. 16: 1161–1174.

[49] Vejarano, G. and McNair, J. (2010) WiMax-RBDS-Sim: An OPNET simulation framework for IEEE 802.16 mesh networks. Proc. 3rd International ICST Conference on Simulation Tools and Techniques (SIMUTools'10). Torremolinos, Malaga, Spain. pp. 1–10.

[50] Sahinidis Optimization Group (2012) Baron global optimization software. Available: http://archimedes.cheme.cmu.edu/baron/baron.html. Accessed 2012 March 23.

Chapter 2

Application of Genetic Algorithms in Scheduling of TDMA-WMNs

Vahid Sattari Naeini and Naser Movahhedinia

Additional information is available at the end of the chapter

1. Introduction

WMN (Wireless Mesh Network) is usually built on fixed stations and interconnected via wireless links to form a multi-hop network. Typical and inexpensive deployment of WMNs use some MSSs (Mesh Subscriber Station) and one MBS (Mesh Base Station), where their multi-hop feature can be utilized to increase their range of accessibility in rural areas effectively. Moreover, since they are dynamically self-organized and self-configuring, these networks turn to be more reliable.

TDMA-WMN (Time Division Multiple Access-WMN) is a special WMN which has some special features: TDM (Time Division Multiplexing) is adopted between MSSs and the MBS to access the air interface; frames are defined and divided into some equal duration subframes to provide better timing and synchronization to MSSs. As these subframes (called transmission opportunities) are taken by MSSs to transmit packets on unidirectional links, it's more preferable to schedule each link rather than each node connection.

Four sources of interference are defined in TDMA-WMNs:

- Transmitter-Transmitter: each node can't receive data flows from more than one source.
- Receiver-Receiver: each node can't send data flows to more than one destination.
- Transmitter-Receiver: each node can't receive and transmit simultaneously.
- Transmitter-Receiver-Transmitter: two sources can't transmit at the same time while the transmitter and the receiver share a neighbor which can hear both transmissions. For example, in Figure 1 where the two conflicting links (e_1 and e_7) are shown with bold lines, nodes cannot transmit simultaneously as they share the same neighbor (Node 2) which can here both transmissions.

The first three sources of interference are known as first hop conflict (primary conflicts) and the fourth one is knows as second hop conflict (secondary conflicts). Second hop conflict is disregarded in most of the presented works [1], [2].

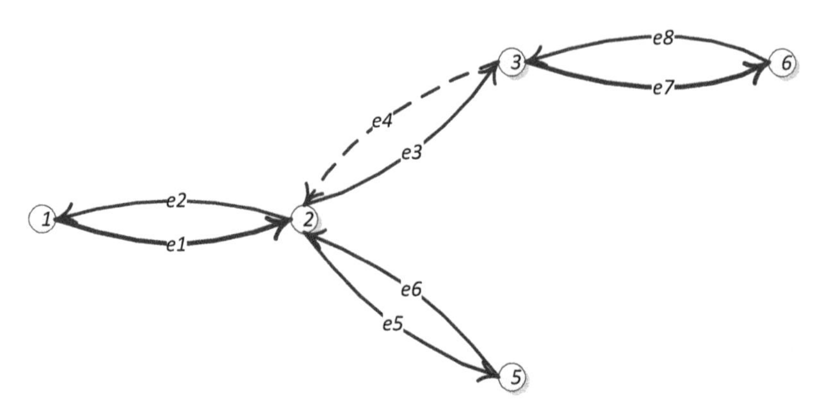

Figure 1. A TDMA-WMN with its conflicting links

A major challenge in WMNs is to provide QoS support and fair rate allocation among data flows. Almost all of the routing and scheduling algorithms presented in the literature have one common weak point: when the MBS collects requests larger than the frame length from all the MSSs, these algorithms shrink link durations to fit in the frames. Scaling down the link durations may cause some drawbacks in guaranteeing the QoS requirements of voice and video traffic. Two schemes can be exploited to overcome this problem: 1) A call admission control mechanism can be deployed to avoid link duration shrinkage; 2) A new scheduling method may be proposed to schedule the packets received from the underlying network. In this chapter we focus on the second solution with respect to the first solution.

Scheduling in WMNs is divided in two categories: centralized and distributed scheduling. In centralized scheduling, there exists one MBS and the other stations (MSSs) relay packets of other stations to/from end points (in this chapter we call these end points as MTs, while MSSs assume to be fixed). The main purpose of this chapter is related to centralized scheduling and admission control.

On the other hand, rapid growth of wireless networks has commenced challenging issues in co-deployment of various technologies including WiFi, WiMAX. While WiFi networks are very popular for providing data services to Internet users in LAN environments, WiMAX technology has been adopted for MAN networks to provide urban accessibility to hot spots or end users. These two technologies seem to be competitors; however, they can interwork to gain metro-networks performance, cost effectiveness and coverage area. This configuration can be used in TDMA-WMNs, however when the same frequency band is employed with different network elements (e.g., the U-NII frequency at 5GHz may be shared among IEEE 802.16d and IEEE 802.11a or IEEE 802.11n), more complex strategies are required for scheduling and packet translation from one technology to another.

In this chapter, with respect to the interoperability of WiFi and TDMA-WMNs networks, we develop a scheduling and admission control mechanism among data flows such that the QoS requirements of delay sensitive traffic types can be provisioned and elastic traffic types get a fair duration of bandwidth.

To provide QoS support for delay sensitive traffic over WiFi, IEEE 802.11e introduces two types of channel access methods: EDCA (Enhanced Distributed Channel Access) and HCCA (Hybrid Coordination Channel Access). Since the HCCA function deployed in the MTs is essentially designed to meet the negotiated QoS requirements of admitted flows, we apply this function to the WiFi network [3], [4].

The chapter is organized as follows: in sections 2, some research activities in scheduling mechanisms in WMNs and IEEE 802.11 are summarized. In section 3, we introduce an overview of IEEE 802.16 and IEEE 802.11(e) standards. QoS comparison between IEEE 802.11 and IEEE 802.16 mesh modes are described in this section. Fourth section is devoted to describe the system model. In this section we introduce the basic assumptions of the system and formally describe the system. In fifth section, the genetic algorithm is briefly described and its application to our problem is discussed. The proposed method is evaluated using simulation results in section 6. Finally, conclusions are drawn in seventh section.

2. Related works

Centralized scheduling mechanism in WMNs has been investigated in [1], [2], [5]-[10], [32-33]. Most of the research activities in this area are not suitable for TDMA mesh networks (e.g., IEEE 802.16d). They consider only primary conflicts in which the connections share a neighbor, while TDMA-WMN is faced with secondary conflicts where the transmitter and the receiver share a neighbor, which can hear both transmissions.

The main algorithm in IEEE 802.16d finds a link ranking during a breath-first traversal of the routing tree. This algorithm has no idea for spatial reuse in the network. Spatial reuse in these networks has been investigated in [5], [7]-[10]. Ref. [9] uses Transmission-Tree Scheduling (TTS) algorithm that is based on graph coloring. This algorithm don't consider the protocol overhead of TDMA scheduling. While [10] uses the load-balancing algorithm to increase spatial reuse, [8] considers Bellman-Ford method for both spatial reuse and minimum TDMA delay. These schemes don't take into account the underlying network behavior which can affect scheduling of traffic flows of other MSSs. On the other hand, these algorithms shrink the link duration when the frame is short for scheduling the links.

Application of intelligent scheduling methods in wireless mesh networks has been inspired by the fact that finding a schedule in TDMA scheduling is *NP*-complete [11]. Ref. [12] uses fuzzy hopfield neural network technique to solve the TDMA broadcast scheduling problem in wireless sensor networks. Artificial neural network with reinforcement learning has been introduced in [13] to schedule downlink traffic of wireless networks. A genetic algorithm approach is used in [2] to find the schedule related to each link in a WMN. Here again, their scheduling method merely considers the traffic flown on the links; however, how these links empty their queues has not been elaborated.

None of the above research activities, consider neither the underling network behavior nor the types of traffic streams flown on the links. Our system model is different from the previous works in two aspects. First, we take into account the underlying network traffic related to each MSS. Second, the algorithm proposed in this chapter is such that shrinking the link duration doesn't affect the minimum QoS requirements of real-time traffic types.

3. IEEE 802.16 and IEEE 802.11e: An overview

3.1. Overview of IEEE 802.11e

The multiple access mechanism in 802.11e is arisen in super-frames which start with beacon frames having the same duration as beacon intervals. The super-frame comprises an optional CFP (Contention Free Period) followed by a CP (Contention Period) divided into equal duration SIs as shown in Figure 2. At each SI (Service Interval), each QSTA (QoS Station) should transmit its own traffic streams with respect to its QoS constraints. This mechanism is called HCCA function which defines a centrally-controlled polling-based medium access scheme for IEEE 802.11e WLANs. Each SI is divided into a CAP (Controlled Access Phase) period and an optional EDCA period in which the traffic streams having less stringent QoS constraints contend for access to the medium. Usually best effort traffic streams such as HTTP use this period which offers no QoS guarantee. The CAP period is further divided into a number of TXOPs (Transmission Opportunity). Each TXOP is granted by QAP (QoS Access Point) to each QSTA and each QSTA is responsible for sharing this period among its traffic streams.

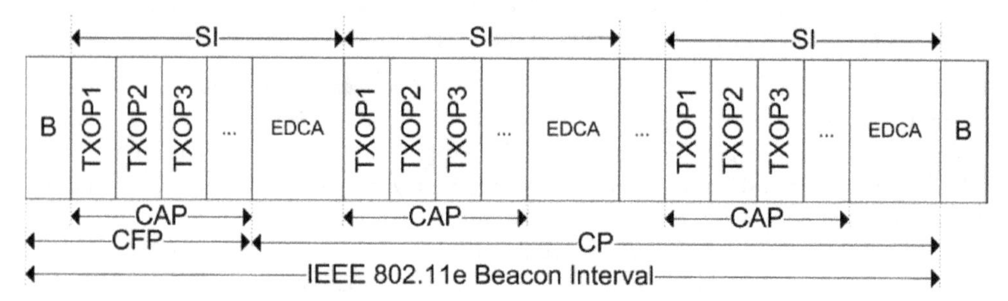

Figure 2. HCF super-frame structure

3.2. Overview of IEEE 802.16 mesh mode

IEEE 802.16 MAC PDUs (Protocol Data Unit) (Figure 3) begin with a fixed-length generic MAC header (6 bytes). The MAC header field contains a 2 bytes CID (Connection Identifier) field which carries 8 bits *Link ID* used for addressing nodes in the local neighborhood. The header is followed by the Mesh sub-header (2 bytes) which includes *Xmt Node Id*. Mesh BS grants *Node Ids* to candidate nodes when authorized to the network. After the variable length payload there exists a 4 bytes CRC. The medium in IEEE 802.16 mesh mode is divided into equal duration frames (Figure 4), consisting of two sub-frames:

- Data sub-frame,
- Control sub-frame.

The control sub-frame is divided into MSH-CTRL-LEN transmission opportunities indicated in the ND (Network Descriptor). Each transmission opportunity comprises 7 OFDM symbols, so the length of the control sub-frame is fixed and equal to 7×MSH-CTRL-LEN OFDM symbols.

Generic Mac Header (6)	Mesh Subheader (2)	Payload (Optional)	CRC (4)

Figure 3. MAC PDU format

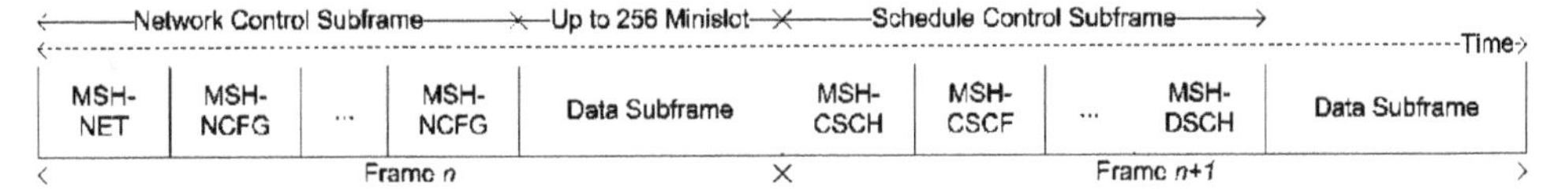

Figure 4. Frame structure for the mesh mode

Nodes can transmit based on the granted bandwidth and a transmission schedule which is worked out using a common distributed algorithm. The data sub-frame is used for this purpose which is divided into transmission opportunities comprising 256 mini-slots based on the standard. However, there may be fewer than 256 mini-slots depending on the frame size and the size of the control sub-frame. Frame duration which is indicated in ND is determined by MBS to avoid losing synchronization with the connecting nodes. MSH-CSCH-DATA-FRACTION indicated in ND specifies the fraction of data sub-frame which can be used for centralized scheduling. The remaining part of the data sub-frame is used for decentralized scheduling.

3.3. QoS Comparison between WiFi and WiMAX mesh mode

Providing QoS in IEEE 802.11e comes with a new coordination function called HCF. The HCF controlled channel access is for the parameterized QoS, which provides the QoS based on the contract between the AP and the corresponding QSTA(s). First, a traffic stream is established between the AP and an QSTA. A set of traffic characteristics and QoS requirement parameters are negotiated between the AP and QSTA and the traffic stream should be admitted by the AP. The QoS control field in the MAC frame format is a 16 bits field which facilitates the description of QoS requirements of application flows. Its TID (4 bits) identifies the TC (0-7) or the TS (8-15) to which the corresponding MSDU in the FB field belongs. The last eight bits are used usually by QAP to receive the queue size of QSTAs. After admission, the AP specifies the TXOP duration for the QSTA based on the traffic characteristics. So, the QoS is provided based on connections established between AP and QSTA(s).

Unlike WiFi, the QoS in the mesh mode of IEEE 802.16 is provided in a packet by packet basis. Each transmitted packet contains the mesh CID. Figure 5 shows the structure of mesh CID used in unicast messages. In order to enable differentiated handling of packets, the queuing and forwarding mechanisms deployed at individual nodes may make use of the values for the *Type, Reliability, Priority/Class,* and *Drop Precedence* fields. The *Type* field is used to distinguish between different categories of messages. This field may be used to prioritize the transmission of management messages transmitted in the data sub-frame (e.g., messages for uncoordinated distributed scheduling). The *Reliability* field is employed to specify unacknowledged transmitted packets (when ARQ is enabled). This allows the packet to be retransmitted for up to four times. The *Priority/Class* field allows the classification of the messages into eight priority classes. This can be used by the queuing and forwarding mechanisms at each node to differentiate the packet treatment for different classes. The *Drop Precedence* field indicates the likelihood of a packet being dropped during congestion.

Bits:2	1	3	2	8
Type	Reliability	Priority/Class	Drop Precedence	Xmt Link ID

Figure 5. Mesh CID format

4. System model

SSHC stationary end nodes (Figure 6) are able to communicate from one side with MTs and from the other side with MSSs or MBS via their PHY layers in both sides. In the following subsections we develop a genetic based system for scheduling the packets waiting in the SSHCs queues.

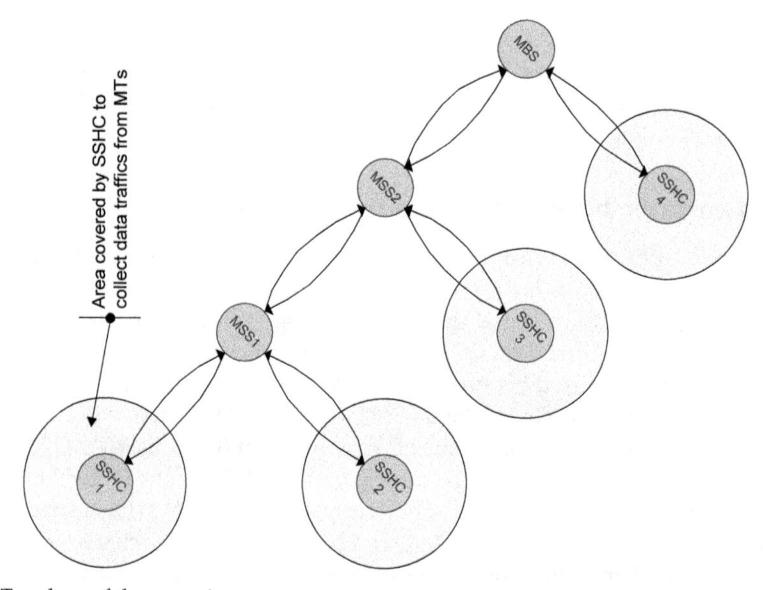

Figure 6. Topology of the scenario

4.1. Basic assumptions

In this chapter we consider a WMN with one MBS and some MSSs (Figure 6). We consider access traffic in the mesh network, so the routes of the traffic form a binary tree rooted at the MBS. MSSs relay numerous traffic types (data, voice or video) between their MTs and other

MSSs or the MBS. The MBS, MSSs, and MTs share the same frequency band. The routing tree that is made by the MBS is a binary tree [14], and we assume that it's known in advanced.

For better support of QoS, we consider MBS and MSSs use TDMA-based scheduling in their MAC layer; however, because of IEEE 802.11 deployment in most mobile devices (laptops and cell phones), MTs use contention-based medium access method. In a given TDMA frame (of the length for example 20ms), some MSSs are sending frames upward or downward the network, while the others are collecting (distributing) frames from (to) MTs. Since the tree rooted at the MBS is a binary one, each MSS has maximum of six logical links to its neighbor MSSs; three for sending and another three for receiving packets (Figure 1). As wireless transceivers are usually half duplex [8], they can't be used for reception and transmission at the same time. So there are six queues in each MSS, three of these are to store the outbound packets and three others are for inbound packets to queue for reception. However, in our model receiving queues are ignored as they are considered in sending nodes; hence at most three queues are considered for scheduling. It's worthy to note that we schedule only one queue at each leaf node and two queues at the MBS. Moreover we schedule only the links that have non-empty queues. Each queue is filled by MTs (shown in Figure 6) or the receiving links at that node. For example in Figure 1, the queue of e_2 can be filled by the queue of e_4.

Let, M be the set of all the stations (including MSSs and the MBS) in the system, indexed by m=1,2,…,M. We consider M>1; i.e., there is at least one MSS. In most of the mesh networks, the frame length is fixed and may not be changed; otherwise the whole system should be restarted [8]; hence the frame length is fixed at L milliseconds. Each transmission in the frame is along with some overheads, so the number of transmissions for each link should be limited to one per frame to minimize transmission overhead.

Let ψ be the set of all the links in the system. We take a subset I of ψ ($I \subset \psi$), in which the links have non empty queues. Each $i \in I$ has one queue per traffic type which are assumed to have unlimited sizes for the sake of simplicity.

Each queue has some restricted QoS traffic specifications; this means that each queue should be scheduled appropriately and get emptied in a desired time. Since the frame length is fixed and all of the links in the system should be scheduled at each frame (because of their restricted QoS requirements), there is a limited interval for each queue to get scheduled. Nevertheless, some of the nodes may not find enough transmission opportunity to evacuate all of their queues, causing the system not to be able to fulfill QoS constraints of delay sensitive traffic types. So, a scheduling method is strongly necessary to satisfy QoS requirements of voice and video traffic. On the other hand, bandwidth allocation to more stringent QoS traffic types may cause starvation for elastic traffic. As such, we define a threshold (k), to assure the elastic traffic types to be scheduled at each k frames.

Let $k_{m,i,j}$ be the length of the jth queue (filled by MTs or other MSSs), associated with ith outgoing link, related to mth MSS. So, $[k_{m,i,j}]$ is an $M \times I \times J$ matrix. Each queue should be scheduled to transmit its packets over the appropriate link based on the QoS requirements of its content.

$[k_{m,i,j}]$ is available at the MBS through some control messages. Assuming that a 64Kbps voice stream should be serviced in each frame, it generates a 160 bytes packet which is to be transmitted in each 20ms frame. In case of a video traffic stream, its packets should be scheduled every two frames [15].

Some of the main parameters of each traffic type in the system are as follows [16]:

- Delay Bound (*D*): Maximum amount of time allowed (including queuing delay) to transmit a frame across the wireless interface.
- Mean Data Rate (ρ): Average bit rate at the MAC layer required for the packet transmissions.
- Nominal Packet Size (*L*): Average packet size.
- Maximum Packet Size (*M*): Maximum packet size.
- Minimum Service Interval (*mSI*): Minimum interval between the start of successive service period.
- Maximum Service Interval (*MSI*): Maximum interval between the start of successive service period.
- Next, we describe the formal formulation of our system.

4.2. Problem formulation

Each queue is scheduled once in each frame. We assume that spatial reuse is deployed in the routing algorithm, so:

$$\sum_{m\in M}\sum_{i\in I}\sum_{j\in J} Tr_{m,i,j}^{F} \geq F, \quad \forall F = kL, \quad k = 1,2,3,\ldots \tag{1}$$

Where *Tr* is the data transferred from the queue *j* in the current frame, associated with *i*th outgoing link, related to *m*th MSS. *L* is the length of the frame and *F* is a parameter that specifies the scheduling duration which is a sub-multiple of *L*. The above inequality means that due to spatial reuse mechanism applied to the system, the number of transmissions may exceed the frame length.

At the end of each frame, the remaining number of packets in all of the queues should be minimized:

$$\min_K\left(\sum_{m\in M}\sum_{i\in I}\sum_{j\in J} K_{m.i.j}^{F} - \sum_{m\in M}\sum_{i\in I}\sum_{j\in J} Tr_{m,i,j}^{F}\right), \quad \forall F = kL \tag{2}$$

In the above minimization problem, *K* specifies the number of residual packets waiting in the queue for scheduling. We assume three different queues (e.g., CBR, VBR, and elastic traffic) for each link with higher priorities indexed by lower numbers. The following constraints are applied on queue depletions:

$$Tr_{m,i,1}^{F} > 0, \ \forall F = kL \tag{3}$$

$$Tr_{m,i,2}^{F} > 0,\ \forall F = kL \tag{4}$$

$$Tr_{m,i,3}^{F} > 0,\ \forall F = kL \tag{5}$$

$$Tr_{m,i,1}^{F} > Tr_{m,i,3}^{F}\ \wedge\ Tr_{m,i,2}^{F} > Tr_{m,i,3}^{F}\ \wedge\ Tr_{m,i,1}^{F} > Tr_{m,i,2}^{F} \quad \forall F = kL \tag{6}$$

Inequality (3) means that the first queue should be scheduled in every k frame. Two other inequalities state that their related queues may be scheduled in every k frame optionally. Inequality (6) demonstrates that the priority of the first queue is higher than the second one and the second queue is more prior than the third one.

The minimization problem (2) with its constraints (1), (3), (4), (5), and (6) are such that they can't be solved by simple mathematics; since the problem shown to be *NP*-complete [11]. Heuristic solutions might work in certain cases, but they fail to adapt to different network scenarios [2].

The above optimization problem can be bounded and reformulated such that speed convergence can be obtained as described in the following sentences. As the first queue is reserved for CBR traffic streams, the second queue is reserved for VBR traffic streams, and third one is reserved for elastic (ABR) traffic type, then the first queue should be serviced in each frame, and the second one should be services in every two frames [15]. After that, the remaining bandwidth (if any) is considered for the third queue. Available bandwidth should be fairly shared among the queues. For this purpose, we take advantage of a threshold (k) to force the scheduler to take a minimum percent of the available bandwidth for elastic data types. This causes a fair scheduling method for elastic traffic types and will be presented in the simulation results. Now we have the following optimization problem with its conditions. It can be seen from the minimization problem (7) that the number of queues per each link and the number of links per each node is bounded on 3 (as discussed earlier); so the search space is limited and convergence to the termination conditions will be faster than the previous problem (2).

$$\min_{K}\left(\sum_{m=1}^{M}\sum_{i=1}^{3}\sum_{j=1}^{3}K_{m,i,j}^{kL} - \sum_{m=1}^{M}\sum_{i=1}^{3}\sum_{j=1}^{3}Tr_{m,i,j}^{kL}\right),\quad k > 2 \tag{7}$$

Subject to:

$$\sum_{m=1}^{M}\sum_{i=1}^{3}\sum_{j=1}^{3}Tr_{m,i,j}^{L} \geq L \tag{8}$$

$$Tr_{m,i,1}^{L} > 0 \tag{9}$$

$$Tr_{m,i,2}^{2L} > 0 \tag{10}$$

$$Tr_{m,i,3}^{kL} > 0,\quad k > 2 \tag{11}$$

$$K_{m,i,1}^{L} > 0\ \vee\ K_{m,i,2}^{2L} > 0\ \vee\ K_{m,i,3}^{kL} > 0,\quad k > 2 \tag{12}$$

4.3. Admission control

The CBR queue should be emptied at the end of its schedule or the backlogged packets are to be dropped. So, the new connection should be rejected if the following equation is not satisfied:

$$K_{m,i,1}^{L} - Tr_{m.i,1}^{L} = 0 \quad m \in M, i \in I \tag{13}$$

While the VBR queue get filled in different intervals, we put a threshold on the top of its queue. If the number of the packets available in the queue is greater than this threshold, then any new call will be rejected. So:

$$K_{m,i,2}^{2L} - Tr_{m.i,2}^{2L} < \tau_1 \quad m \in M, i \in I \tag{14}$$

The elastic traffic queue can be filled every time a packet is generated, but may cause undesirable delay, so, we impose a threshold (τ_2) on the third queue as well. However τ_2 should be greater than τ_1, since elastic data types have lighter QoS constraints than VBR data types.

$$K_{m,i,3}^{kL} - Tr_{m.i,3}^{kL} < \tau_1 \quad k > 2, m \in M, i \in I \tag{15}$$

5. Application of genetic algorithm in scheduling of SSHCs queues

In the following subsections we first present an overview of genetic algorithm, and then we develop a GA-based scheduling mechanism for the problem.

5.1. Genetic algorithm: An overview

The genetic algorithm is a search heuristic that mimics the process of natural evolution. This heuristic is routinely used to generate useful solutions to optimization problems. Genetic algorithms belong to the larger class of evolutionary algorithms, which generate solutions to optimization problems using techniques inspired by natural evolution, such as inheritance, mutation, selection, and crossover [17], [18].

In a genetic algorithm, a population of strings (called chromosomes or the genotype of the genome), which encode candidate solutions (called individuals, creatures, or phenotypes) to an optimization problem, evolves toward better solutions. An initial population is created from a random selection of solutions (which are analogous to chromosomes). A value for fitness is assigned to each solution (chromosome) depending on how close it actually is to solve the problem and arrive to the answer of the problem. Those chromosomes with higher fitness values are more likely to reproduce offspring. The offspring is a product of the father and mother, whose composition consists of a combination of genes from them (this process is known as crossing over). This generational process is repeated until a termination condition has been reached. Common terminating conditions are:

- A solution is found that satisfies minimum criteria.
- Fixed number of generations reached.

- The highest ranking solution's fitness is reaching or has reached a plateau such that successive iterations no longer produce better results.
- Manual inspection
- Combinations of the above.

At each stage, crossover and mutation genetic operators may be applied to the new strings. Crossover is a genetic operator that combines two chromosomes (parents) to produce a new chromosome (offspring). The idea behind crossover is that the new chromosome may be better than both of the parents if it takes the best characteristics from each of the parents.

As an example, suppose there are two chromosomes 1 and 2 which are represented as a binary string, the most used way of encoding a chromosome, as the following:

Chromosome 1 1101100100110110
Chromosome 2 1101111000011110

Crossover selects genes from parent chromosomes and creates a new offspring. The simplest way how to do this is to choose randomly some crossover point and everything before this point copy from the first parent and everything after the crossover point copy from the second parent. I is the crossover point. The following shows this process:

Chromosome 1 11011 I 00100110110
Chromosome 2 11011 I 11000011110
Offspring 1 11011 I 11000011110
Offspring 2 11011 I 00100110110

After performing the crossover, mutation is used to maintain genetic diversity from one generation of population chromosomes to the next. It introduces some local modifications of the individuals in the current population on order to explore new possible solutions. For binary encoding of chromosome, we can switch a few randomly chosen bits from 1 to 0 or from 0 to 1. For our example, the mutation process is shown as the following:

Original offspring 1 1101111000011110
Original offspring 1 1101100100110110
Mutated offspring 1 1100111000011110
Mutated offspring 2 1101101100110110

The pseudo-code of a basic GA is summarized as follows:

1. Choose the initial population of individuals
2. Evaluate the fitness of each individual in that population
3. Repeat on this generation until termination: (time limit, sufficient fitness achieved, etc.)

i. Select the best-fit individuals for reproduction.
ii. Breed new individuals through crossover and mutation operations to give birth to offspring.
iii. Evaluate the individual fitness for new individuals.
iv. Replace least-fit population with new individuals.

5.2. A GA-based approach for the scheduling problem

In the previous section, the optimization problem which is needed to create a population was formally defined. This population is created by the MBS based on the fact that the MBS gathers queues' statistics of SSHCs through some control messages. Each chromosome of the population is such that every queue gets its service once per frame, so the scheduling overhead could be minimized. Each queue is scheduled as close to the beginning of the frame as possible, so that its transmission does not overlap with transmissions of its conflicting links. The scheduling period is set to two frames, since VBR traffic streams get their services every two frames.

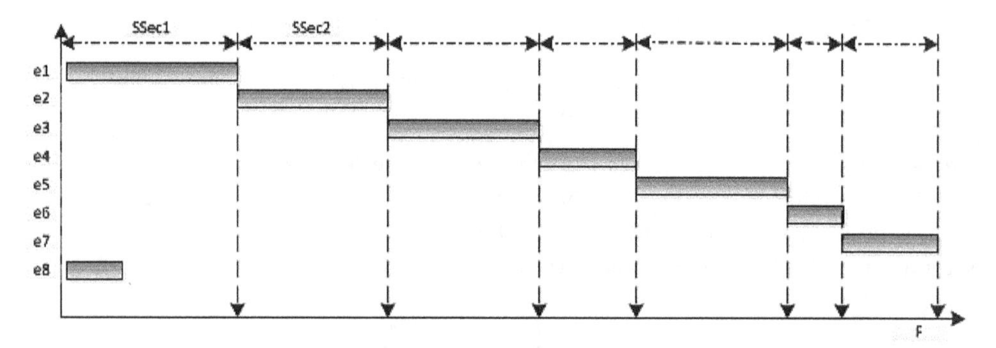

Figure 7. A typical frame (chromosome) with its SSecs (genes) associated with Figure 1

Each gene is defined as a scheduling section (SSec). A scheduling section is composed of one or more time slots in which a queue and its non-conflicting queues are scheduled to transmit in parallel. The first scheduling section is started at the beginning of the frame. All the scheduling sections are consecutive and non-overlapping.

The crossover operator is 0.5-uniform crossover [19]. Each SSec of one chromosome can be exchanged with equal-size SSec of another chromosome, with a constant probability of 0.5. Each scheduling section (gene) is subject to random mutation with a small independent probability. We use permutation encoding; hence each gene replaces with a duplicate of other equal-size genes (e.g., replace the SSec3 with the SSec5 in Figure 7).

Finally, we should use some QoS metrics of the network (that we want to optimize) to define the fitness function. It can be seen from the Eq. (7) that we are interested in depletion of all the queues in the scheduling period. On the other hand, when more queues get emptied, higher performance will be reached; hence, Eq. (7) can be explicated as Eq. (16).

$$\max_{Tr}\left(\sum_{m=1}^{M}\sum_{i=1}^{3}\sum_{j=1}^{3} Tr_{m,i,j}^{kL} - k \times L\right) \quad \forall l \in N, k > 2 \tag{16}$$

So we define the fitness function (*F.F.*) as follows:

$$F.F. = \frac{\sum_{m=1}^{M}\sum_{i=1}^{3}\sum_{j=1}^{3} Tr_{m,i,j}^{kL} - k \times L}{k \times L} \times 100 \tag{17}$$

6. Simulation results

We developed a TDMA-WMN system based on Orthogonal Frequency Division Multiplexing (OFDM) air interface that works in 5GHz frequency band using NS-2 simulator [20]. Basic OFDM parameters are listed in Table 1. OFDM symbol duration is about 14μs. TDMA frame duration (*L*) is set to 20ms. While TDMA-WMN uses BPSK-1/2 modulation technique, the underlying network (WLAN) uses 16QAM-1/2 modulation technique. Different modulation techniques have been used; because interference between MTs of different SSHCs should be avoided (Figure 6).

OFDM Parameters	Value	Scenario
Bandwidth		20MHz
Sampling rate	Depend on BW	23.04MHz
Useful time T_B	256.T	11.11μs
T_G/T_B	1/4,1/8,1/16,1/32	1/4
CP time T_G		2.78μs
Symbol time T_{sym}	T_G+T_B	13.89μs
Carriers N_{FFT}	256	
Data Carriers	192	

Table 1. Basic OFDM parameters

We define three types of traffic in the system: CBR, VBR, and ABR traffic streams. CBR traffic (e.g., voice over IP without silent suppression (G.711)), has constant packet size with constant packet interval. VBR traffic (e.g., H.263 video), has variable packet size with variable packet interval feature. At last, elastic traffic (e.g., FTP), can adjust its transmission rate gradually.

Voice and video traffic stream specifications are as follows:

a. G.711 voice (CBR traffic) which generates packets of 160 bytes with mean service interval of 20ms (64 Kb/s of average sending rate).
b. H.263 video (VBR traffic) which has been obtained from "Jurassic Park I" trace file, available in [21].

Traffic specifications of these two types of traffic are summarized in Table 2. For the sake of simplicity, we assume that elastic flows are generated using CBR traffic sources with packet size of 1000 bytes.

TSPEC Param.	G.711 Voice	H.263 Video (Park Jurassic I)
Mean Bit Rate (Kbps)	64	260
Delay Bound (ms)	20	20
Mean SDU Size (Byte)	160	4533.67
Maximum Burst Size (Byte)	160	11817
Minimum Service Interval (ms)	20	0
Maximum Service Interval (ms)	20	40

Table 2. Traffic specification parameters of traffic types

In order to evaluate the performance of the proposed scheduler and the admission control procedure, the topology of Figure 6 is considered as the scenario. All of the end nodes (SSHCs) are active, while the intermediate nodes (MSSs) pass only the traffic of the end nodes. SSHCs are configured to work in both WLAN and TDMA-WMN modes; while, MSSs work in TDMA-WMN mode. k is fixed at 4, since elastic traffic queues are scheduled every four frames. From one hand, this value is not too small that causes some drawbacks on delay sensitive traffic types and on the other hand it's not too large that leads to unfairness.

At first, we assume one VBR MT and one ABR MT and a number of CBR MTs which are gradually increased (Figure 8). It can be seen when there is no admission control mechanism, as the number of MTs exceeds 10, packets of the newly added MTs are dropped. The proposed admission control mechanism for this traffic type works well, since none of the packets has been dropped when it is applied.

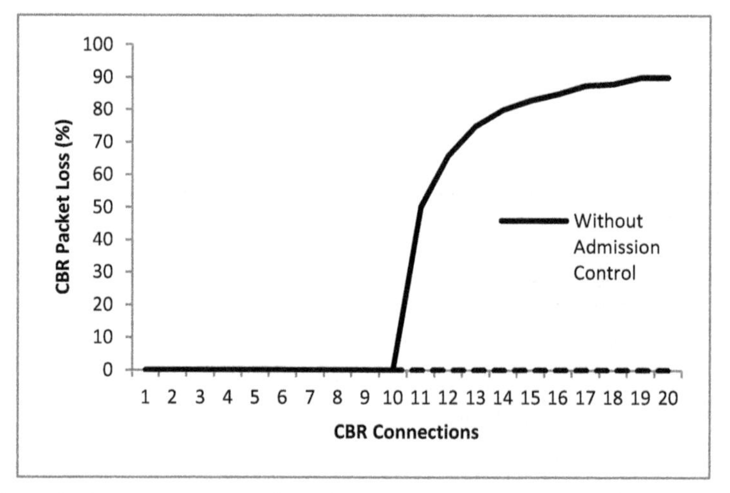

Figure 8. CBR packet loss versus increased number of CBR connection, while VBR and ABR connections are fixed at 1

In the next simulation, the number of CBR MTs and ABR MTs are fixed to one and the number of VBR MTs (Figure 9) is gradually increased. Since the packet size and the arrival time are variable in case of the packets of these traffic types, the number of admitted VBR MTs is less than the number of admitted CBR MTs. In this figure the packet loss is due to the threshold (τ_l) applied to the queue length. Here again the proposed admission control mechanism works well for this type of traffic.

For the last simulation, we removed all of the thresholds to see how many packets will be backlogged in the queues after scheduling the queues. For this purpose we monotonically increase the number of CBR, VBR, and ABR MTs in each SSHC. It can be seen from Figure 10 that the CBR queue is at its normal size, since almost all of its packets are serviced in appropriate time. However, after the second frame, all of the packets of elastic data type are queued, since there is no chance for them to be scheduled. Moreover, since all of the VBR packets do not receive the opportunity to be scheduled, the VBR queue length increases monotonically.

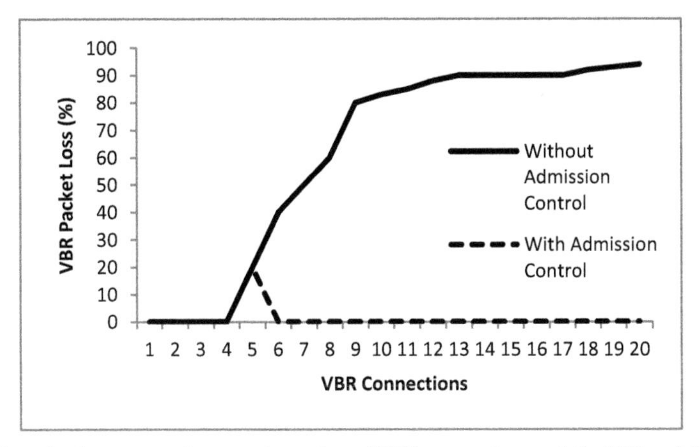

Figure 9. VBR packet loss versus increased number of VBR connections, while CBR and ABR connections are fixed at one

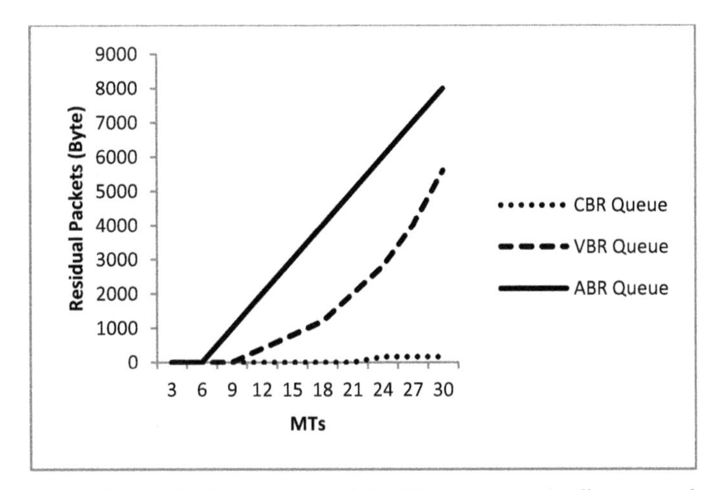

Figure 10. Residual packets in the three queues, while MTs are monotonically increased

Eliminating the thresholds causes the queue size of ABR and VBR traffic to increase, while by using these thresholds (Figure 8 and Figure 9) a fair bandwidth allocation can be reached.

7. Conclusion

In this chapter we considered an important aspect of TDMA-WMNs: Traffic flow requirements on scheduling the links. Moreover, we considered the underlying network which can affect the overall system performance despite previous research. We assumed three types of traffic with different QoS requirements and formulated a model describing the scheduling optimization problem which is to be solved to minimize queues in the system. We develop a genetic algorithm method to find the optimal schedule for each relaying node. Furthermore to be able to fulfill QoS requirements of established connections, we developed an admission control mechanism. Finally, the performance of the proposed GA algorithm along with the admission control procedure was evaluated by simulating a typical network scenario. Simulation results showed effectiveness of our admission control and scheduling mechanisms. In our next work, we introduce some new mechanisms including MIMO technique to the above-mentioned system and investigate its performance. Meanwhile, application of genetic algorithms in distributed scheduling of WMNs is investigated.

Author details

Vahid Sattari Naeini and Naser Movahhedinia
Department of Computer Engineering, University of Isfahan, Hezar Jerib Avenue, Isfahan, Iran

8. References

[1] X. Cheng, P. Mohapatra, S-J. Lee, and S. Banerjee, MARIA: Interference-Aware admission control and QoS routing in wireless mesh networks, ICC, 2008.

[2] L. Badia, A. Botta, and L. Lenzini, A genetic approach to joint routing and link scheduling for wireless mesh networks, Ad Hoc Networks, vol. 7, pp. 654-664, 2009.

[3] "IEEE 802.11e Medium access control (mac) quality of service (qos) enhancements," 2005.

[4] C. Cicconetti, L. Lenzini, E. Mingozzi, and G. Stea, An efficient cross layer scheduler for multimedia traffic in wireless local area networks with IEEE 802.11e HCCA, ACM SIGMOBILE Mobile Computing and Communications Review, vol. 11, pp. 31-46, July 2007.

[5] S. Liu, S. Feng, W. Ye, and H. Zhuang, Slot allocation algorithms in centralized scheduling scheme for IEEE 802.16 based wireless mesh networks, Computer Communications, vol. 32, pp. 943-953, 2009.

[6] W. Liao, S. P. Kedia, and A. K. Dubay, A centralized scheduling algorithm for WiMAX mesh network, NOMS, 2010, pp. 858-861.

[7] H.-Y. Wei, S. Ganguly, R. Izmailov, and Z. Haas, Interference-aware IEEE 802.16 WiMax mesh networks, VTC, 2005, pp. 3102-3106.

[8] P. Djukic and S. Valaee, Delay aware link scheduling for multi-hop TDMA wireless networks, IEEE/ACM Transactions on Networking, vol. 17, pp. 870- 883, June 2009.

[9] B. Han, W. Jia, and L. Lin, Performance evaluation of scheduling in IEEE 802.16 based wireless mesh networks, Computer Communications, vol. 30, pp. 782-792, Nov. 2007.

[10] D. Kim and A. Ganz, Fair and efficient multihop scheduling algorithm for IEEE 802.16 BWA systems, Broadnets, 2005, pp. 895-901.

[11] W. Wang, Y. Wang, X-Y. Li, W-Z, Song, O. Frieder, *Efficient Interference-Aware TDMA Link Scheduling for Static Wireless Networks*, MobiCom, 2006.

[12] Y-J. Shen and M-S. Wang, Broadcast scheduling in wireless sensor networks using fuzzy Hopfield neural network, Expert systems with applications, vol. 34, pp. 900-907, 2008.
[13] P. Fiengo, G. Giambene, and E. Trentin, Neural-based downlink scheduling algorithm for broadband wireless networks, Computer Communications, vol. 30, pp. 207-218, 2007.
[14] "IEEE Std 802.16-2004, IEEE standard for local and metropolitan area networks part 16: air interface for fixed broadband wireless access systems," 2004.
[15] X. Cheng, P. Mohapatra, S-J. Lee, and S. Banerjee, Performance evaluation of video streaming in multihop wireless mesh networks, NOSSDAV, 2008.
[16] I. Inan, F. Keceli, and E. Ayanoglu, An adaptive multimedia QoS scheduler for 802.11e wireless LANs, IEEE ICC, 2006.
[17] E. Ilavarasan and P. Thambidurai, Genetic Algorithm for Task Scheduling on Distributed Heterogeneous Computing System, International Review on Computers and Software, Vol. 1. n. 3, pp. 233-242, 2006.
[18] R. K. Jena, P. Srivastava, G. K. Sharma, A Review on Genetic Algorithm in Parallel & Distributed Environment, International Review on Computers and Software, Vol. 3. n. 5, pp. 532-544, 2008.
[19] S. N. Sivanandam, S. N. Deepa, Introduction to Genetic Algorithms (Springer-Verlag, Berlin Heidelberg, ISBN: 978-3-540-73189-4, 2008).
[20] NS2 Network Simulator, http://www.isi.edu/nsnam/ns/, [Accessed: March 2012].
[21] P. Seeling, M. Reisslein, and B. Kulapala, Network performance evaluation using frame size and quality traces of single-layer and two-layer video: a tutorial, IEEE Communications Surveys and Tutorials, vol.6, no. 2, pp. 58-78, 2004.
[22] "IEEE 802.11a, Part 11:Wireless LAN Medium Access Control (MAC) and Physical Layer (PHY) Specifications: High-Speed Physical Layer Extension in the 5 GHz Band, supplement to IEEE 802.11 Standard," 2000.
[23] J. Gross, M. Emmelmann, A. Punal, and A. Wolisz, Enhancing IEEE 802.11a/n with dynamic single-user OFDM adaptation, Performance Evaluation, vol. 66, pp.240-257, 2009.
[24] "IEEE 802.11n Enhancements for higher throughput, amendment 4 to ieee 802.11 part 11: Wireless lan medium access control (mac) and physical layer (phy) specifications," 2007.
[25] J. Zou and D. Zhao, Real-time CBR traffic scheduling in IEEE 802.16-based wireless mesh networks, Wireless Networks, vol.15, pp. 65-72, 2009.
[26] P. Djukic and S. Valaee, Scheduling algorithms for 802.16 mesh networks, WiMax/MobileFi: Advanced Research and Technology (Y. Xiao, ed.), Auerbach Publications, 2007, pp. 267-288.
[27] M. Rashid, E. Hossain, and V. K. Bhargava, Controlled channel access scheduling for guaranteed QoS in 802.11e-based WLANs, *IEEE Transactions on Wireless Communications*, vol. 7, no. 4, pp. 1287-1297, 2008.
[28] H. Cheng, X. Jia, H. Liu, Access Scheduling on the Control Channels in TDMA Wireless Mesh Networks, MSN, 2007.

[29] Y. Hou, K. Leung, A distributed scheduling framework for multi-user diversity gain and quality of service in wireless mesh networks, *IEEE Transactions on Wireless Communications*, vol. 8, no. 12, pp. 5904-5915, 2009.

[30] H. Cheng, N. Xiong, L. T. Yang, Y-S. Jeong, Distributed scheduling algorithms for channel access in TDMA wireless mesh networks, *The Journal of Supercomputing*, vol. 45, no. 1, pp. 105-128, 2008.

[31] "IEEE Standard for Information technology-Telecommunications and information exchange between systems-Local and metropolitan area networks-Specific requirements - Part 11: Wireless LAN Medium Access Control (MAC) and Physical Layer (PHY) Specifications," 2007.

[32] J. Luo, C. Rosenberg, A. Girard, Engineering Wireless Mesh Networks: Joint Scheduling, Routing, Power Control, and Rate Adaptation, IEEE/ACM Transactions on Networking, vol. 18, no. 5, pp. 1387-1400, Oct. 2010.

[33] J. Joseph, M. Princy, Analysis on Scheduling and Load Balancing Techniques in Wireless Mesh Networks, International Journal of Computer Applications, vol. 42, no. 12, 2012.

Chapter 3

Channel Assignment Using Topology Control Based on Power Control in Wireless Mesh Networks

Aizaz U. Chaudhry and Roshdy H.M. Hafez

Additional information is available at the end of the chapter

1. Introduction

In this section, an overview of Wireless Mesh Networks (WMNs) is presented, and some unique features which distinguish WMNs from Mobile Ad hoc Networks (MANETs) and Wireless Sensor Networks (WSNs) are listed. The main purpose of this chapter is discussed, and the contribution is presented along with the main features of this work.

1.1. Overview

The use of Wireless Local Area Networks (WLANs) has grown tremendously in the past few years due to their ease of deployment and maintenance. However, the access points in these WLANs have to be connected to the backbone network through wired media. Wireless Mesh Networks offer an attractive alternative for providing broadband wireless Internet connectivity by using a wireless backhaul network and eliminating the need for extensive cabling.

In traditional WLANs, each Access Point (AP) is connected to the wired network while only a subset of APs is connected to the wired network in WMNs. An AP that is connected to the wired network is called Gateway (GW); APs without wired connections are called Mesh Routers (MRs), and they connect to the GW through multiple hops. Like routers in a wired network, MRs in a WMN forward each other's traffic to establish and maintain their connectivity. MRs and GWs are similar in design, with the only difference that a GW is directly connected to the wired network, while a MR is not. Figure 1 shows a sample mesh network in a typical enterprise such as a university [1]. The following are some unique features that distinguish WMNs from MANETs and WSNs [1].

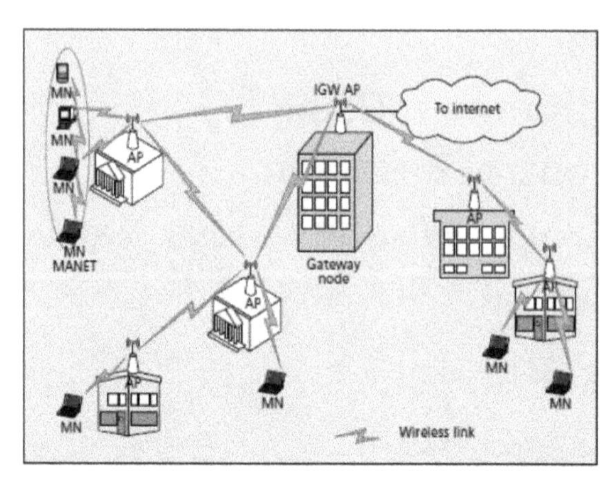

Figure 1. A sample enterprise WMN [1]

- **Mesh routers are static**

Mesh routers in a WMN are stationary; therefore, the route selection should focus on discovering links that interfere with as few nodes as possible to provide high end-to-end throughput.

- **Mesh routers have no power constraint**

In contrast to traditional wireless networks, such as MANETs and WSNs, where nodes are typically power-constrained, MRs have abundant power at their disposal.

- **Mesh routers have multiple radios**

With the reduced cost of radios, MRs can be equipped with multiple radios. Hence, simultaneous transmission and reception can be achieved using intelligent channel assignment to these radios.

- **The traffic model is different**

In MANETs, traffic can be from any peer Mobile Node (MN) to any other MN, while in WMNs, traffic is between MRs and the GW.

- **Traffic is concentrated along certain paths**

In MANETs, traffic distribution is generally assumed to be uniform, while in WMN, traffic is concentrated along the paths directed towards the GW.

- **Traffic volume is high**

MANETs have been designed essentially for enabling communication within a small group of people, while WMNs aim to provide high-bandwidth broadband connections to a large community, and thus should be able to accommodate a large number of users accessing the Internet. Due to high estimated traffic volume in WMNs, scalability and fault tolerance become important considerations in algorithm design.

1.2. Motivation

To the best of our knowledge, the proposed channel assignment algorithm is the first of its kind to use topology control based on power control for channel assignment in multi-radio multi-channel wireless mesh networks.

The main purpose of network topology control using power control is to minimize the interference between a MR and other MRs in the network by adjusting its transmission range using transmission power control. This leads to a better frequency reuse during channel assignment, which results in achieving the objective of significant improvement in the overall network throughput.

1.3. Contribution

Specifically, the contribution of this work is as follows: *"A new Topology-controlled Interference-aware Channel-assignment Algorithm (TICA) which intelligently assigns the available non-overlapping 802.11a frequency channels to the mesh routers with the objective of minimizing interference and, thereby, improving network throughput."* The main features of this work are as follows.

- A Topology Control Algorithm (TCA) named *Select x for less than x*; that builds the network topology by selecting the nearest neighbors for each node in the network, with the objective of minimizing interference among MRs and enhancing frequency reuse.
- A scheme that uses the minimum power as the link weight when building the Shortest Path Tree (SPT) with the required node degree with the objective of minimizing interference and enhancing frequency reuse.
- A Channel Assignment Algorithm (CAA), TICA, which assigns the available non-overlapping 802.11a frequency channels to the mesh nodes with the objective of improving the overall network throughput by minimizing interference between mesh nodes as well as ensuring connectivity between them. This work was first presented in [2].
- A centralized Failure Recovery Mechanism (FRM) for TICA which provides automatic and fast failure recovery by reorganizing the network in order to bypass the failed node and to restore connectivity. This work was first presented in [3].

1.4. Organization

The rest of the chapter is organized as follows. The related work on topology control and CAAs is presented in Section 2. Section 3 discusses the medium access issues encountered by IEEE 802.11 [4] single-radio single-channel nodes. The channel assignment problem is presented in this section with respect to multi-radio multi-channel wireless mesh network. Section 4 presents the network architecture for the proposed model and the proposed topology control algorithm and CAA, along with the details of their respective phases. The FRM of the proposed CAA is also presented in this section. Section 5 provides a performance evaluation of the proposed CAA. The network topologies used for performance evaluation are discussed.

The results of simulations for performance evaluation of the proposed CAA based on throughput analysis of a 36-node mesh network are presented in this section. Section 6 presents the conclusions, along with some directions for future work.

2. Background

This section discusses the effects of topology control on the operation of a network, gives the taxonomy of the topology control schemes for multihop wireless networks, and presents some related well-known topology control algorithms. The section also contains a taxonomical classification of the channel assignment schemes for wireless mesh networks, and discusses some related well-known channel assignment algorithms for each class.

2.1. Topology control schemes

The importance of Topology Control (TC) lies in the fact that it affects network spatial reuse and hence the traffic carrying capacity. Choosing a large transmit power results in excessive interference, while choosing a small transmit power results in a disconnected network [5]. Using TC through transmission power control, the network connectivity and hence the network topology is affected, interference levels are mitigated, which reduces the co-channel interference, and the opportunity of spatial channel re-use is enhanced [6].

2.1.1. Topology control in multi-radio WMNs

The connectivity graph in a multi-radio WMN is determined through topology control. So, the problem of TC in multi-radio WMNs involves the selection of transmission power for each radio interface of each mesh node in the network, so as to maintain the network connectivity with the use of minimum power.

2.1.2. Effect of topology control

The problem of TC is complex, since the choice of the transmit power fundamentally affects many aspects of the operation of the network [7].

a. Effect of TC on the Performance of the Network

TC via transmission power control has a multi-dimensional effect on the performance of the whole network.

- The transmit power levels determine the performance of medium access control, since the spatial channel reuse depends on the number of other nodes within the interference range.
- The choices of power levels affect the connectivity of the network, and consequently the ability to deliver a packet to its destination [8].
- The power level affects the throughput capacity of the network [9].
- Power control affects the network topology which affects the number of hops, and thus the end-to-end delay.

b. Effect of TC on the Performance of the MAC and Routing Protocols

In addition, the assumption of fixed power levels is so ingrained into the design of many protocols in the OSI stack that changing the power levels results in their malfunctioning.

- Changing power levels can create uni-directional links, which can happen when a node *i*'s power level is high enough for a node *j* to hear it, but not vice versa.
- Bi-directionality of links is implicitly assumed in many routing protocols.
- Medium access control protocols in IEEE 802.11 implicitly rely on bi-directionality assumption of links.

2.1.3. Taxonomy of topology control schemes

These schemes are mainly divided into two types [10].

a. Homogeneous

This is the basic type of TC, as all the nodes are assumed to use the same transmitting range. So, the topology control problem reduces to determining the minimum value of transmission range that ensures network connectivity. This minimum transmission range is also called the Critical Transmitting Range (CTR).

b. Non-homogeneous

In this type of TC, nodes are allowed to choose different transmitting ranges, provided they do not exceed the maximum range. Depending on the type of information that is used to compute the topology, non-homogeneous topology control is further classified into three categories.

i. Location-based schemes

In such schemes, exact node positions are known. If this information is used by a centralized authority to compute a set of transmitting range assignments which optimizes a certain measure such as the energy cost, this is the case of the Range Assignment Problem and its variants. The Local Minimum Spanning Tree (LMST) algorithm [5] and the Enhanced Local Minimum Shortest-Path Tree (ELMST) algorithm [6] are examples of location-based topology control schemes.

ii. Direction-based schemes

In such schemes, it is assumed that nodes do not know their position, but they can estimate the relative direction of each of their neighbors.

iii. Neighbor-based schemes

In such schemes, nodes are assumed to know only the ID of the neighbors, and are able to order them according to some criterion such as link quality.

2.2. Channel assignment schemes

Channel Assignment (CA) in a multi-radio WMN environment consists of assigning channels to the radios in order to achieve efficient channel utilization (i.e. minimize co-channel interference) and, simultaneously, to guarantee an adequate level of connectivity. The problem of optimally assigning channels in an arbitrary mesh topology has been proven to be NP-hard, based on its mapping to a graph-coloring problem [11]. Therefore, channel assignment schemes employ heuristic techniques to assign channels to radios belonging to mesh nodes. A taxonomical classification of various CA schemes for wireless mesh networks is as follows [12].

2.2.1. Fixed channel assignment schemes

Fixed assignment schemes assign channels to radios either permanently, or for intervals that are long with respect to the radio switching time. Such schemes can be further subdivided into two types.

a. Common Channel Assignment (CCA)

In CCA scheme [13], the radios of each node are all assigned the same set of channels. For example, if each node has two radios, then the same two channels are used at every node. The main benefit is that the connectivity of the network is the same as that of a single channel approach, while the use of multiple channels increases network throughput. However, it does not take into account the effect of interference on the channel assignment in a WMN.

b. Varying Channel Assignment (VCA)

In the VCA class of schemes, the radios of different nodes are assigned different sets of channels. However, the assignment of channels may lead to network partitions and/or topology changes, which may increase the length of routes between mesh nodes. Therefore, in such a scheme, channel assignment needs to be carried out carefully. The VCA approach is discussed in more detail by presenting algorithms that belong to this category.

i. Centralized Hyacinth (C-HYA)

C-HYA, a centralized channel assignment algorithm for multi-radio multi-channel WMNs, was proposed in [11]. Assuming that the offered traffic load is known, this algorithm assigns channels ensuring network connectivity and satisfying the bandwidth limitations of each link.

ii. Mesh-based Traffic and interference-aware Channel-assignment (MesTiC)

MesTiC, a fixed algorithm for centralized CA, was proposed in [14], and visits nodes once in the decreasing order of their rank. The rank of each node is computed on the basis of its link-traffic characteristics, topological properties and number of radios on a node.

2.2.2. *Dynamic channel assignment schemes*

In dynamic CA schemes, any radio can be assigned any channel but additionally, radios can frequently switch from one channel to another. Therefore, when nodes need to communicate with each other in such a scheme, a coordination mechanism is required to ensure that they are on a common channel.

a. Multi-channel Medium Access Control (MMAC)

MMAC [15] [16] is a link-layer multi-channel protocol for nodes with a single network interface. A node equipped with a single interface can only listen to one channel at a time. Therefore, in order to use multiple channels, the interface has to be switched between channels.

When nodes require to switch channels, a pair of nodes need to listen on the same channel at the time of communication and a channel coordination method is necessary, which is not required in TICA.

b. Distributed Hyacinth (D-HYA)

D-HYA, a dynamic and distributed channel assignment algorithm proposed in [17], can adapt to traffic load dynamically. The algorithm builds on a spanning tree network topology. The gateway node is the root of the spanning tree, and every mesh node belongs to that tree. Based on per-channel total load information, a WMN node determines the set of channels that are least used in its vicinity. As nodes higher up in the spanning trees need more relay bandwidth, they are given a higher priority in channel assignment. The priority of a WMN node is equal to its hop distance from the gateway.

The CA schemes, such as C-HYA, MesTiC and D-HYA, require the traffic load to be known before assigning channels, whereas TICA requires no such knowledge for channel assignment.

2.2.3. *Hybrid channel assignment schemes*

Hybrid channel assignment schemes combine both static and dynamic assignment properties by applying a fixed assignment for some radios and a dynamic assignment for other radios. The fixed radios can be assigned dedicated channels while the other radios can be switched dynamically among channels.

a. Hybrid Multi-Channel Protocol (HMCP)

HMCP [18] [19] is a link-layer multi-channel protocol for nodes with multiple radio interfaces. Out of the available interfaces at each node, X interfaces are assigned statically to X channels, and these interfaces are designated as "fixed interfaces." The fixed interfaces stay on the specified channels for long durations of time. The remaining interfaces can frequently switch between any of the remaining channels, based on the data traffic, and are designated as "switchable interfaces."

A co-ordination protocol is required to decide what channel to assign to the fixed interface, and also for enabling neighbors of a node X to know about the channels used by fixed interface of node X. Time synchronization and coordination between mesh nodes which is required in HMCP is not needed in TICA.

b. Breadth First Search - Channel Assignment (BFS-CA)

BFS-CA [20] is a centralized, interference-aware algorithm aimed at improving the capacity of the WMN backbone and at minimizing interference. This algorithm is based on an extension to the conflict graph concept called the Multi-radio Conflict Graph (MCG) where the vertices in the MCG represent edges between radios instead of edges between mesh routers.

BFS-CA requires certain number of MRs with certain number of radio interfaces to be placed at certain hops from the gateway, whereas TICA simply requires all MRs to have four data radios, does not require any careful router placement strategy, and works with any placement of routers as verified by a comprehensive performance evaluation.

3. Channel assignment problem

In this section, the medium access issues encountered by IEEE 802.11-based single-radio single-channel WMNs are presented, and a multiple-channel approach using multiple radios to overcome these problems is discussed. The key issue of channel assignment in Multi-Radio Multi-Channel (MRMC) WMNs is presented, along with its objectives and constraints.

3.1. IEEE 802.11 medium access issues

Since the WMN has to provide access to broadband Internet, it is expected to have higher bandwidth. Even though the physical layer can support very high bit rate, current MAC protocols are not able to utilize the entire bandwidth provided by the physical layer. The main reason for this poor performance is the suboptimal media access protocols, which were primarily designed for single-hop networks [1].

3.1.1. Hidden and exposed terminal problems

IEEE 802.11 Distributed Coordination Function (DCF) is one such widely accepted MAC protocol but, when used in a multihop network scenario, it results in poor performance and is therefore unacceptable. The reason is that some nodes remain starved due to hidden and exposed terminals in a multihop environment. Figure 2 illustrates these problems [1].

Node 2, which is outside the interference range of Node 3 and unaware of the ongoing transmission at Node 3, continues to send RTS to Node 1 causing collision. This is a case of the hidden terminal problem.

Node 4 is prevented from transmitting because of the neighboring transmission at Node 3. This is a case of the exposed terminal problem.

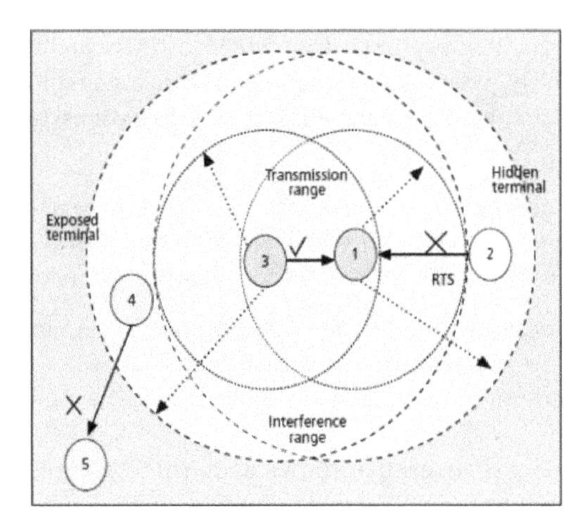

Figure 2. Hidden and exposed terminal problems [1]

3.2. Wireless mesh network architectures

3.2.1. Single-radio single-channel mesh network

A Single-Radio Single-Channel (SRSC) mesh architecture suffers from hidden and exposed terminal problems. Assigning orthogonal channels to the MRs within the interference range can help alleviate the hidden and exposed terminal problems, and assist in improving the overall capacity of the network. However, considering the traffic characteristics in a WMN, frequent channel switching may be required to communicate with neighboring nodes. In such scenarios, single-radio multi-channel MAC may not provide any significant performance gains due to high channel switching delay.

3.2.2. Multi-radio multi-channel mesh network

The use of a multi-channel approach using multiple radios overcomes the problems encountered in the previous architectures [1]. Two or more radios are employed for the backhaul link. The uplink and downlink backhaul radios operate at non-overlapping channels which eliminates the co-channel interference.

As each mesh router can be equipped with multiple radios, fixed channel assignment to these radios is a more viable solution. Efficient and intelligent channel assignment schemes have to be designed, as the number of channels is limited.

3.3. Channel assignment problem

In a typical WMN, the total number of radios is much higher than the number of available channels. Thus, many links between the mesh routers operate on the same set of channels and interference among transmissions on these channels decreases their utilization. Therefore, minimizing the effect of interference is required for the efficient reuse of the scarce radio spectrum. So, the key issue in a MRMC WMN architecture is the channel assignment problem, which involves assigning a channel to each radio of a MR in a way that minimizes interference on any given channel and guarantees connectivity between the mesh nodes [12].

Given the connectivity graph, the main challenge for CAA is to assign a channel to each radio in a way that minimizes interference between MRs and ensures connectivity between them. In order to achieve these goals, the CAA should satisfy the following requirements.

1. In order to communicate, a pair of nodes within transmission range of each other needs to have a common channel assigned to their end-point radios.
2. Links in direct interference range of each other should be assigned non-overlapping channels.
3. The number of distinct channels that can be assigned to a mesh router is bounded by the number of radios it has.
4. The total number of non-overlapping channels is fixed.
5. Since the traffic in a WMN is directed to and from the gateway, the traffic flows aggregate at routers close to the gateway. Therefore, priority in channel assignment should be given to links starting from the gateway based on the number of nodes that use a link to reach the gateway.

At first glance, the problem of assigning channels to links in a mesh network appears to be a graph-coloring problem. However, standard graph-coloring algorithms cannot satisfy all of its constraints, and it is NP-hard to find an optimal channel assignment to maximize the overall network throughput [11]. Also, the channel assignment problem for mesh networks is similar to the list coloring problem, which is NP-complete [21].

4. Topology control and channel assignment algorithms

In this section, the network architecture of the proposed model is presented. The section presents the proposed topology control and channel assignment algorithms for MRMC WMNs, and details on the working of different phases of the proposed algorithms are discussed. The procedure of fault recovery with the proposed CAA is also presented in this section.

4.1. Network architecture

In our proposed model, each mesh router is equipped with five radios which operate on IEEE 802.11a channels (5 GHz band). One of these radios is used for control purpose while other radios are used for data traffic.

The control radios of all mesh nodes operate on the same non-overlapping IEEE 802.11a channel. Out of the 12 available non-overlapping 802.11a channels, channel 12 is used as the common control channel. Each mesh router is equipped with 4 data radios in order to utilize the remaining 11 non-overlapping channels available in IEEE 802.11a frequency band. Each MR communicates with its transmission range (TR) neighbors using these data radios for data transmission. So, each MR can have a maximum of 4 TR neighbors with whom it can communicate for data transmission which implies that the Maximum Node Degree (MND) per node is four. The MND of 4 is selected in order to fully utilize the 11 available non-overlapping 802.11a frequency channels. Results have shown that with 12 available non-overlapping channels, the network throughput increases until a MND of four but saturates after that [11].

4.2. Topology control algorithm

The proposed Topology Control Algorithm (TCA) controls the network topology by selecting the nearest neighbors for each node in the network. The objective of the proposed TCA is to build a connectivity graph with a small node degree to mitigate the co-channel interference and enhance spatial channel reuse as well as preserve network connectivity with the use of minimal power, as less transmit power translates to less interference.

4.2.1. Gateway advertisement process

Initially, the gateway broadcasts a "Hello" message, using its control radio on the control channel, announcing itself as the gateway. Each mesh node that receives this Hello message over its control radio broadcasts it again and in this way, this Hello message is flooded throughout the mesh network. The Hello message contains a hop-count field that is incremented at each hop during its broadcast. So, a mesh node may receive multiple copies of the Hello message over its control radio. However, distance of a mesh node from the gateway is the shortest path length (shortest hop count) of the Hello message received by a mesh node through its control radio over different paths. In this way, each mesh node knows the next hop to reach the gateway using its control radio.

4.2.2. Assumptions

The proposed TCA assumes the following.

- Each node knows its location.
- Each node uses an omni-directional antenna for both transmission and reception.
- Each node is able to adjust its own transmission power.
- The maximum transmission power is the same for all nodes and hence, the maximum TR for any pair of nodes to communicate directly is also the same.

Note that all nodes start with the maximum transmission power, and that the initial topology graph created, when every node transmits with full power, is strongly connected.

4.2.3. Phases of TCA

The proposed TCA consists of the following five phases.

a. Exchange of Information Between Nodes

In the first exchange, each node broadcasts a HELLO message at maximum transmission power containing its node ID and the node position.

b. Building the Maximum Power Neighbor Table (MPNT)

From the information in the received HELLO messages, each node arranges its neighboring nodes in the ascending order of their distance. The result is the Maximum Power Neighbor Table (MPNT). Then, each node sends its MPNT along with its position and node ID to the gateway node using its control radio over the control channel.

c. Building the Direct Neighbor Table (DNT)

For each node in the network, the gateway builds a Direct Neighbor Table (DNT). Based on information in the MPNT of node v and the MPNTs of its neighbors, if

(a) node w is in the MPNT of node v, and
(b) node w is closer to any other node y in the MPNT of node w than to node v, then gateway eliminates node w from the MPNT of node v.

If, after removing nodes from the MPNT of node v, the remaining number of nodes in the MPNT of node v is less than "x," then the gateway selects "x" nearest nodes as neighbors of node v which results in the DNT. However, if after removing nodes from the MPNT of node v, the remaining number of nodes is greater than or equal to "x," then the result is the DNT.

This algorithm is called *Select x for less than x* TCA where x is a positive integer. The *Select x for less than x* TCA ensures that each node has at least x neighbors, as shown in Figure 3.

d. Converting into Bi-directional Links

For each node in the network, the gateway converts the uni-directional links in the DNT of a node into bi-directional links. For each uni-directional link, this is done by adding a reverse link in the DNT of the neighboring node. This converts the DNT into Bi-directional DNT. This results in the Final Neighbor Table (FNT).

e. Calculating the Minimum Power Required

For each node in the network, the gateway calculates the minimum power, P_{min}, required to reach each of the nodes in the FNT of a node, using the appropriate propagation model.

4.2.4. Propagation models

The free space model is used for short distances and the two ray ground reflection model is used for longer distances, depending on the value of the Euclidean distance in relation to the cross-over distance. The cross-over distance is calculated by [22]

$$Cross\_over\_dist = \frac{4\pi h_t h_r}{\lambda}, \tag{1}$$

where h_t and h_r are the antenna heights of the transmitter and receiver, respectively. If the distance between two nodes is less than the cross over distance, i.e. $d(u,v) < Cross\_over\_dist$, Free Space propagation model is used, whereas if $d(u,v) > Cross\_over\_dist$, Two-ray propagation model is used. The minimum power for the free-space propagation model is calculated by [22]

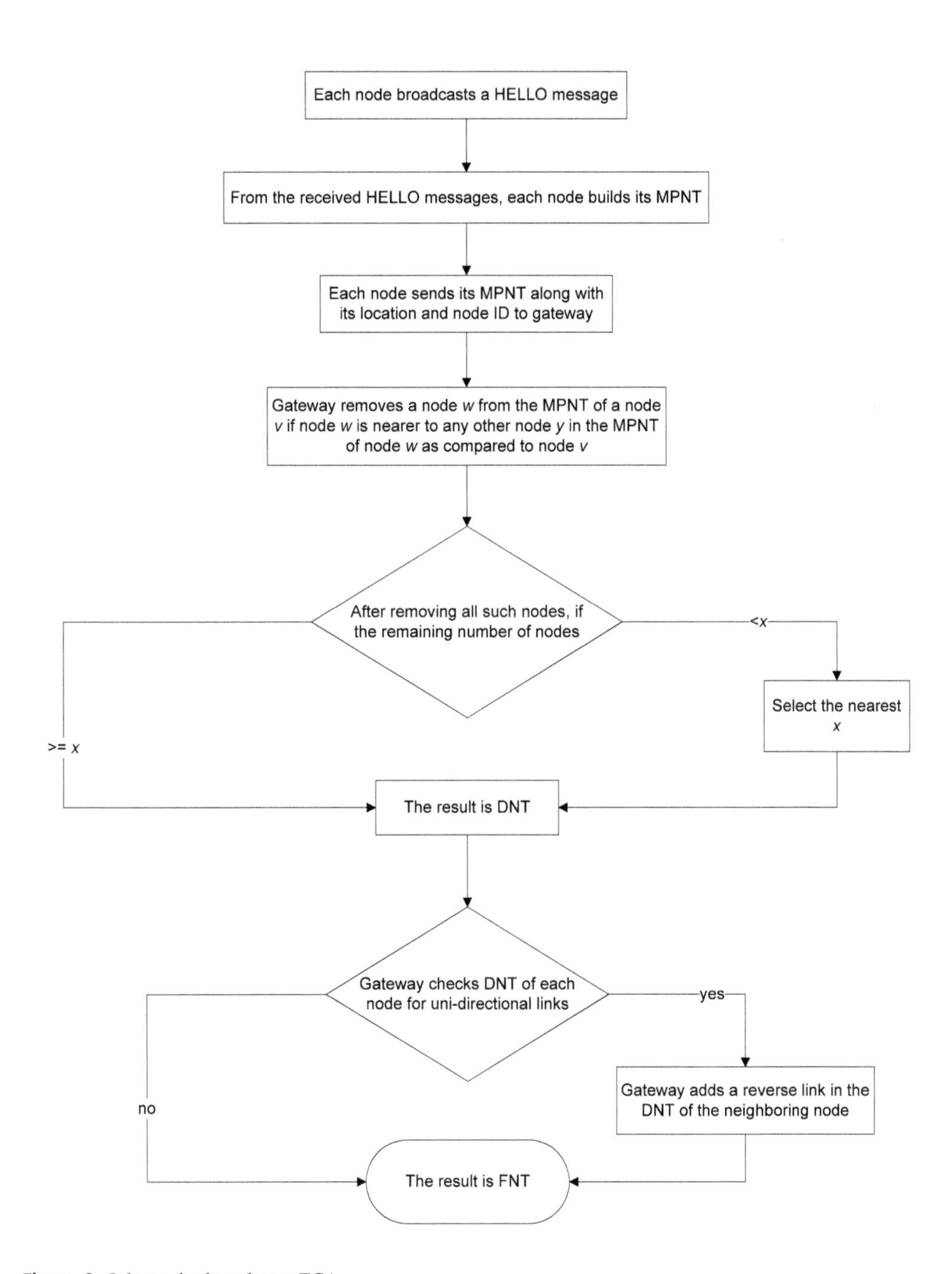

Figure 3. Select x for less than x TCA

$$P_{\min} = \frac{RxThresh(4\pi d)^2}{G_t G_r \lambda^2}. \quad (2)$$

The minimum power for the two-ray propagation model is calculated by [22]

$$P_{\min} = \frac{RxThresh(d)^4}{G_t G_r h_t^2 h_r^2}, \quad (3)$$

where G_t and G_r are transmitter and receiver antenna gains respectively, and *RxThresh* is power threshold required by radio interface of receiving node to correctly understand the message.

4.3. Channel assignment algorithm

4.3.1. Interference-range edge coloring

If K be the number of available colors (channels), then for $K \geq 4$, the distance-2 edge coloring problem, also known as strong edge coloring problem, is NP-complete [23]. A distance-2 edge coloring of a graph G is an assignment of colors to edges so that any two edges within distance 2 of each other have distinct colors. Two edges of G are within distance 2 of each other if either they are adjacent, as shown in Figure 4a or there is some other edge that is adjacent to both of them, as shown in Figure 4b. The distance-2 edge coloring has been used in [24] for channel assignment, where the authors have described the interference model as two-hop interference model. In this model, two edges interfere with each other if they are within two-hop distance. In other words, two edges e_1 and e_2 cannot transmit simultaneously on the same channel if they are sharing a node or are adjacent to a common edge.

To minimize co-channel interference in a wireless mesh network, it is necessary to assign channels to links such that links within interference range of each other are assigned different channels (colors). This problem can be termed as *interference-range edge coloring,* and the corresponding interference model can be called *interference-range interference model.* In a grid topology where links are of equal length, the interference-range edge coloring is similar to distance-2 edge coloring, as shown in Figure 5a. The channel assigned to link l_1 cannot be assigned to links l_2 and l_3 as they are within the interference range of link l_1. Note that l_2 and l_3 are also within two-hop distance of l_1.

However, in a random topology where links are of different lengths due to the random nature of the topology, the interference-range edge coloring can be harder than distance-2 edge coloring as shown in Figure 5b. In this case, the channel assigned to link l_1 cannot be assigned to links l_2, l_3 and l_4 as they are within the interference range of link l_1. Note that l_2, l_3 and l_4 are within three-hop distance of l_1.

In the proposed network model, the number of available channels (colors) is 11 which means that $K = 11$. Based on its similarity to distance-2 edge coloring problem which is NP-complete for $K \geq 4$, the interference-range edge coloring problem is, therefore, also NP-complete. Therefore, we propose an approximate algorithm for channel assignment. The proposed channel assignment algorithm, TICA, is shown in Figure 6 and has the following phases.

Figure 4. Two edges at distance-2 of each other [23]

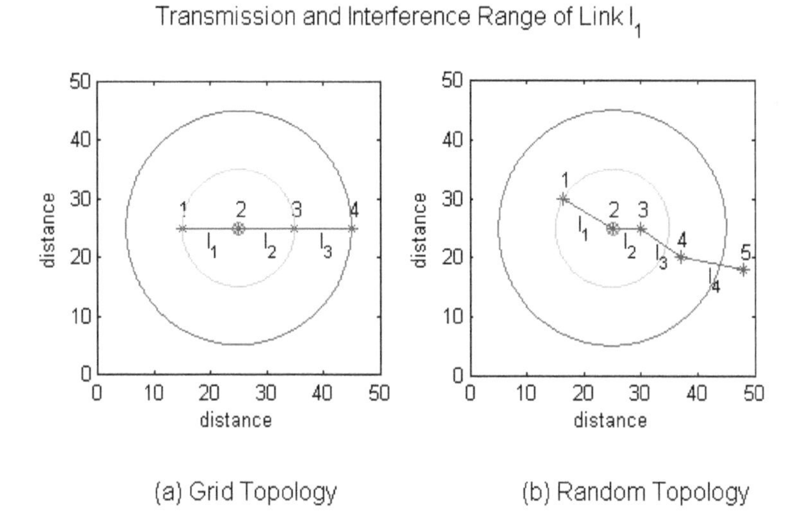

Figure 5. Interference-range edge coloring

4.3.2. *Phases of TICA*

a. Topology Control

In order to create the network connectivity graph with the aim of reducing the interference between MRs, network topology is controlled using the topology control algorithm.

All nodes send their MPNTs to the gateway using their control radio. Note that in order to send its MPNT to the gateway, each mesh node knows the next hop to reach the gateway using its control radio via the "gateway advertisement process." The gateway starts with the *Select 1 for less than 1* TCA and builds the FNTs of all nodes.

b. Connectivity Graph

Based on the FNTs of all nodes, the gateway builds the connectivity graph and checks the resulting network for connectivity. A topology is said to be connected if the gateway can reach any node in the connectivity graph directly or through intermediate hops.

If the resulting network is not connected, the gateway moves to the next higher TCA by incrementing x in the *Select x for less than x* TCA, builds the connectivity graph and checks the resulting network for connectivity. The gateway keeps on moving to a higher TCA until it finds that the network resulting from the connectivity graph is connected.

c. Minimum Power-based Shortest Path Tree with a MND of Four

After ensuring that the connectivity graph is connected, the gateway builds the Shortest Path Tree (SPT), using Dijkstra's algorithm [25], based on the connectivity graph. The metric for path selection is minimum power.

The node degree is defined as the number of TR neighbors of a node. The number of TR neighbors of a mesh router is bounded by the number of its radios and each node has four data radios. So, if any node in the shortest path tree has more than four links, the gateway selects those four links for that node which have the minimum weight and sets the weight of all other links to infinity. In other words, the gateway ensures that each node can have a maximum of four TR neighbors and builds a Minimum Power-based SPT (MPSPT) with a MND of 4 per node. The gateway checks the resulting MPSPT graph for connectivity. If the resulting MPSPT graph is not connected, the gateway moves to a higher TCA.

Once the MPSPT graph is determined, the gateway has to assign channels to links of the MPSPT. Now, the objective is to assign channels to the links of the MPSPT such that the interference between simultaneous transmissions on links operating on the same channel is minimized and the overall network throughput is maximized.

d. Link Ranking

In order to assign channels to the links of the MPSPT graph, each link is assigned a ranking by the gateway. The ranking associated with each link is derived from the number of nodes that use a link to reach the gateway node. If l is link and n is node using link l to reach the gateway, then rank of link l, i.e. r_l, is given by

$$r_l = \sum_{n=1}^{N} I_{n,l}, \tag{4}$$

where N is the total number of nodes in the network. $I_{n,l}$ is 1 if node n is using link l and 0 otherwise.

In the case of two or more links that have the same rank, the link whose power of the farthest node to the gateway is smaller is given priority in channel assignment. If there are some links that still have the same rank, the link with smaller node IDs is given priority in channel assignment.

e. Channel Assignment

The gateway assigns a channel to each link in the order of its rank, and it begins with assigning the 11 available non-overlapping channels to the 11 highest-ranked links such that Channel 1 is assigned to the highest-ranked link. For the 12th-ranked link and onwards, the gateway checks the channel assignment of all links within the interference range of both nodes that constitute that link.

i. Non-conflicting Channel

Out of the 11 available channels, channels which are not assigned to any link within the interference range of both nodes that constitute the 12th-ranked link are termed as non-conflicting channels. If the gateway finds one or more non-conflicting channels, it assigns that channel from the unassigned non-conflicting channels to the 12th-ranked link which has the highest channel number.

ii. Least Interfering Channel

If the gateway cannot find any channel among the 11 available channels that is not assigned to any link within the interference range of both nodes that constitute the 12th-ranked link, it selects the least interfering channel and assigns it to that link. A Least Interfering Channel (LIC) is a channel which causes minimum interference within the interference range of both nodes that constitute the 12th-ranked link.

iii. Interference Level

In order to find out the LIC, the gateway builds the interference level (IL) for all the 11 channels. The LIC is the channel with the minimum IL, which means that assigning this channel to the 12th-ranked link would result in minimum interference in the network.

In order to build the IL for Channel One, the gateway finds all links within the interference range of each of the two nodes that constitute the 12th-ranked link that use Channel One, and calculates IL of each link based on its rank and distance from a node of the 12th-ranked link. It sums up individual ILs of all links that use Channel One within the interference range of each of the two nodes that constitute the 12th-ranked link, to find out total IL for Channel One. This is done by

$$(IL)_i = \sum_m \left(\frac{r_m}{R}\right)\left(\frac{1}{d_m^{\alpha}}\right), \quad (5)$$

where i is the channel that has value between 1 and 11,
$(IL)_i$ is interference level of channel i,
m is a link using channel i that is within the interference range of a node of the 12th-ranked link,
r is the rank of link m,
R is the maximum rank assigned to a link,
d is distance from a node of link m to a node of the 12th-ranked link, and
α is the path loss exponent and is 2 or 4, depending on cross over distance.

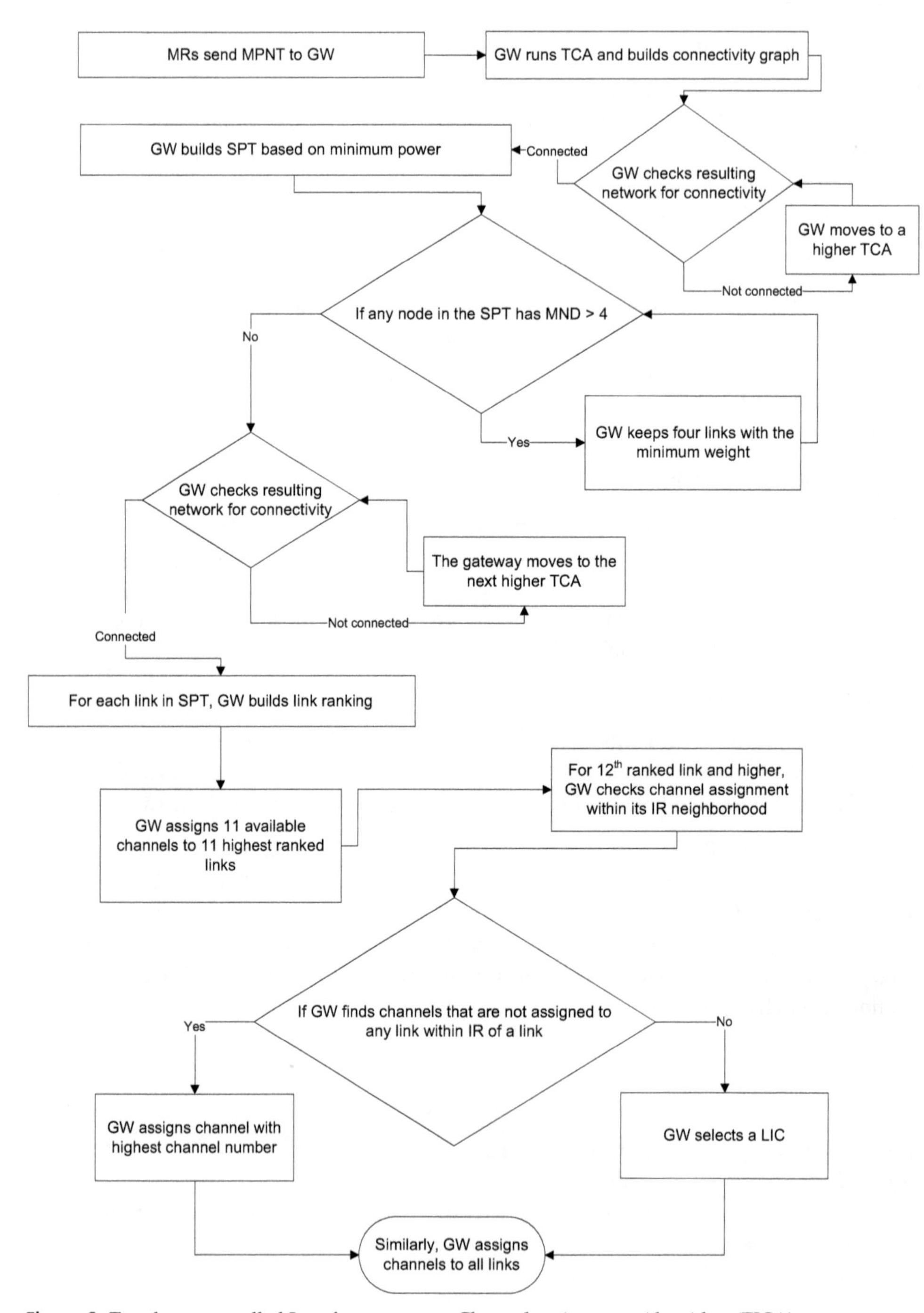

Figure 6. Topology-controlled Interference-aware Channel-assignment Algorithm (TICA)

If a link is emanating from either of the two nodes that constitute the 12th-ranked link and a channel has been assigned to that link, then the IL for this channel is set to infinity.

The LIC is the channel with the minimum interference level and is selected by

$$(IL)_{LIC} = \min\left[(IL)_1, (IL)_2, \ldots\ldots\ldots, (IL)_{11}\right]. \tag{6}$$

Similarly, the gateway assigns channels to all the links of the MPSPT.

iv. Channel Assignment and Routing Message

Using its control radio, the gateway then sends each mesh node the Channel Assignment and Routing Message (CARM). For each channel assigned to a mesh router, CARM contains the channel number and the neighbor node to communicate with, using this channel. The CARM also contains the next hop to reach the gateway for data traffic. Based on the channel assigned to a mesh router to communicate with a neighbor and its distance to that neighbor, the mesh router applies power control and adjusts its transmission power accordingly by using the appropriate propagation model.

4.3.3. Failure recovery mechanism of TICA

When a node fails, nodes in its sub-tree lose their connectivity to the gateway and hence, the Internet through the wired world. TICA supports automatic and fast failure recovery and reorganizes the network to bypass the failed node and to restore the connectivity. In case of node failure, the FRM of TICA, which is shown in Figure 7, is initiated by the gateway.

All nodes send periodic "keep-alive" messages to the gateway on the control channel using their control radios. The keep-alive message from a node tells the gateway that the node is active. If the gateway does not receive three consecutive keep-alives from a node z, then it concludes that node z has failed and is no longer active. The gateway then deletes the MPNT for this node and deletes node z from the MPNT of all its neighboring nodes. Note that the gateway has MPNTs of all nodes, as all nodes sent their MPNTs to the gateway during the setup phase. The gateway builds the FNTs for all nodes using the *Select x for less than x* TCA.

Based on the FNTs of all nodes, the gateway builds the connectivity graph, the MPSPT with a MND of four, the link ranking for the links of the MPSPT and assigns the channels to all links of the MPSPT. The gateway then sends the new CARM to all nodes in the network on the control channel.

5. Performance evaluation

In this section, the performance evaluation of the proposed channel assignment algorithm is provided. Different topologies used for performance evaluation are presented. The performance of the proposed channel assignment algorithm for MRMC WMNs is compared against a "Single-Radio Single-Channel" (SRSC) scheme and a "Common Channel

Assignment" (CCA) scheme for multi-radio mesh nodes. The detailed results of simulations for performance evaluation of TICA based on throughput analysis of a 36-node network are presented, and analyzed. The features' comparison of TICA with related well known channel assignment schemes is also given.

5.1. Simulation environment

For the performance evaluation via throughput analysis, NS2 (version 2.30) [26] simulation tool is used. However, MATLAB [27] is used to generate the power controlled topology, the MPSPT graph, the link ranking of the MPSPT and the channel assignment for the links of the MPSPT.

Multi-interface wireless mesh nodes are created in NS2 by modifying the built-in IEEE 802.11 node model in NS2, using the procedure given in [28]. Based on the channel assignment by the gateway, the radio interfaces are configured for each node and the transmission power for each radio of each mesh node is set accordingly. All the mesh nodes at the periphery of the network send traffic to the gateway. Each of these nodes generates an 8 Mbps Constant Bit Rate (CBR) traffic stream consisting of 1024 byte packets, and sends data to the gateway node at the same time. They do not stop transmitting until the end of the simulation. So, this is a scenario in which multiple flows within the mesh network interfere with each other.

All radios are IEEE 802.11a radios and support 12 channels. The first 11 non-overlapping channels are used by the data radios, whereas the 12th channel is used by the control radio on each node. If the distance between the nodes is less than the cross-over distance, free space propagation model is used; if the distance between the nodes is greater than the cross-over distance, two-ray propagation model is used. As per [4], the minimum receiver sensitivity (*RxThresh*) is set to -65 dBm (3.16227×10^{-10} Watts) in order to achieve a maximum data rate of 54 Mbps supported by IEEE 802.11a.

In order to achieve a strongly connected topology, the maximum transmission power for all the radios is set to 27 dBm. The maximum power transmission range is 164 meters and the maximum power interference range is 328 meters. RTS/CTS is disabled. Note that in the CCA and SRSC schemes, the mesh nodes do not control their power, transmit with the same maximum power (27 dBm) and use AODV (Ad hoc On-Demand Distance Vector) [29] routing protocol as their routing agent.

5.2. Network topologies

Three types of topologies are used in the evaluation. Each topology consists of 36 mesh nodes distributed in an area of 500x500 meters.

Topology 1 is a grid topology; Topology 2 is a randomly generated topology while in Topology 3, called controlled random, the physical terrain is divided into a number of cells and a mesh node is placed randomly within each cell.

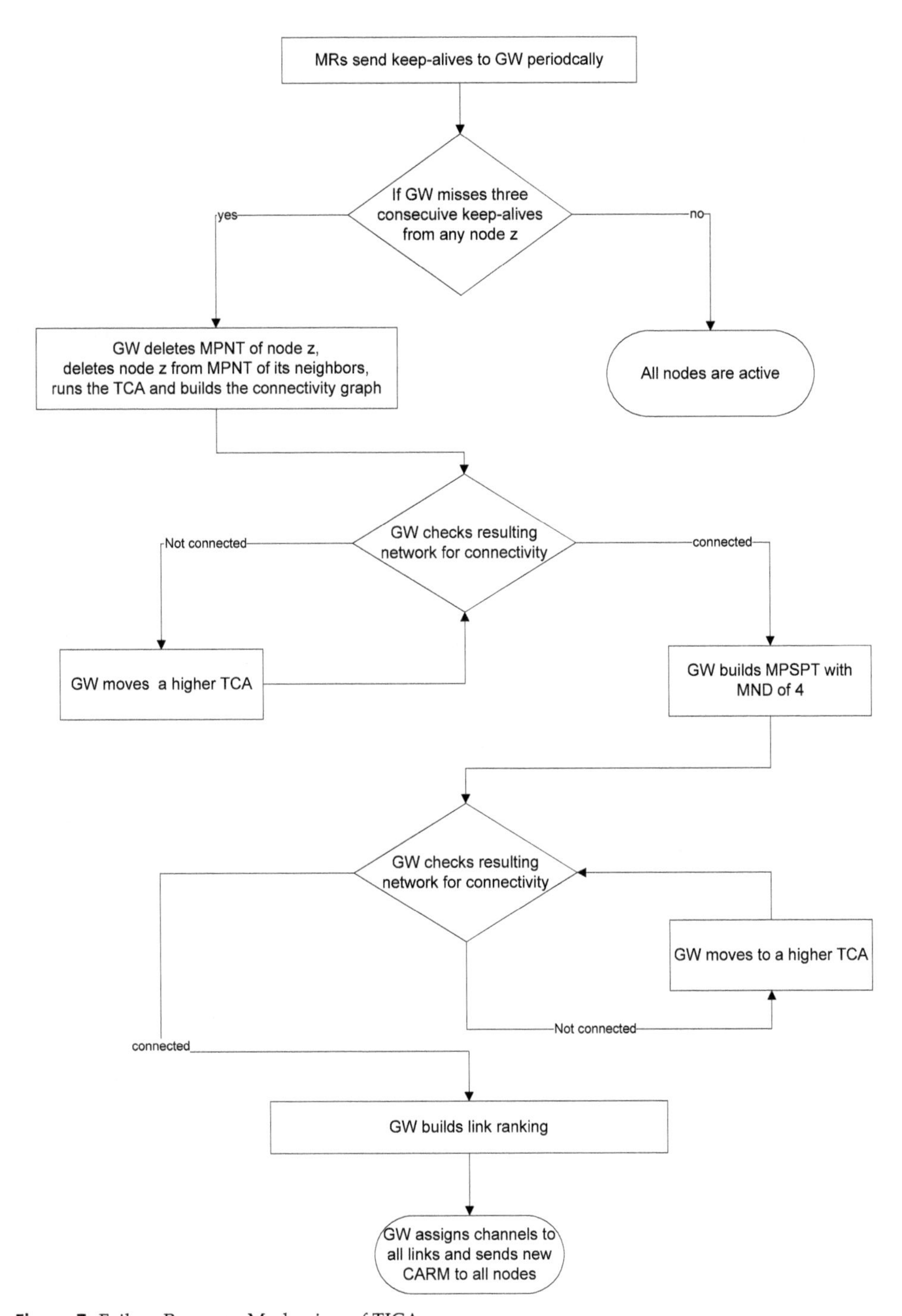

Figure 7. Failure Recovery Mechanism of TICA

Grid Topology (GT) is used to evaluate TICA in a densely populated topology. Random Topology (RT) is used to evaluate TICA in an unplanned deployment of randomly and uniformly distributed mesh nodes. Controlled Random Topology (CRT) is used to reflect real-world deployments where mesh routers are uniformly distributed for maximum coverage.

5.3. Simulation results based on throughput analysis

5.3.1. Simulation parameters

The physical layer and MAC layer settings of the node which are used during the simulation are shown in Tables 1 and 2, respectively. Note that out of the 12 available non-overlapping 802.11a channels, 11 channels are used for data traffic and channel 12 is used for control. Based on the channel assignment by the gateway node, IEEE 802.11a channels are assigned to the links between the mesh nodes and transmission power for each radio of each mesh node is set accordingly.

Physical Layer Parameter	**Setting**
Antenna Type	Omni Antenna
TX/RX Antenna Height (meters)	3
Gain of TX/RX Antenna	1
Packet Capture Threshold (SIR) (dB)	10
Packet Reception Threshold (watts)	3.16227e-10
Carrier Sense Threshold (watts)	7.90569e-11
System Loss Factor	1

Table 1. Physical layer node configuration in NS2

As mentioned earlier, the maximum transmission power for all the radios is 27 dBm. In the CCA and SRSC schemes, MRs do not control their power, transmit with the same maximum power (27 dBm), and use AODV (Ad hoc On-Demand Distance Vector) routing protocol as their routing agent.

MAC Layer Parameter	**Setting**
Minimum Contention Window	15
Maximum Contention Window	1023
Slot Time (micro seconds)	9
SIFS period (micro seconds)	16
Preamble Length (bits)	96
PLCP Header Length (bits)	24
PLCP Data Rate (Mbps)	6
Basic Rate (Mbps)	6
Data Rate (Mbps)	54
RTS/CTS Threshold (bytes)	10192 (disabled)

Table 2. MAC layer node configuration in NS2

All the mesh nodes at the periphery of the network send traffic to the gateway. Each of these nodes generates an 8 Mbps Constant Bit Rate (CBR) traffic stream consisting of 1024 byte packets and sends data to the gateway node at the same time. They do not stop transmitting until the end of the simulation, which is 600 seconds (10 minutes).

5.3.2. *Simulation results*

The Average Throughput (AT) in Mega bits per second at the gateway node is calculated using the following formula:

$$AT = \frac{TPR \times 8 \times 1024}{(TrafficStopTime - TrafficStartTime) \times 1 \times 10^6} \tag{7}$$

Note that *TPR* is the Total Packets Received in (7).

a. Random Topology

Figure 8 shows a graphical comparison of the average throughput of all schemes for ten random topologies. The results in this figure clearly indicate that the proposed algorithm TICA significantly outperforms other schemes for all different random topologies.

b. Controlled Random Topology

Figure 9 shows a graphical comparison of average throughput of all schemes for ten controlled random topologies. The results in this figure clearly indicate that the proposed algorithm TICA significantly outperforms other schemes for all different controlled random topologies.

The placement of the nodes and hence, the length of links in the MPSPT of a topology affects the interference range and hence, the channel assignment. In random and controlled random topologies, the random placement of nodes results in variation in the length of links in the MPSPT. This results in LICs, which may cause significant interference in the network and degrade the overall throughput.

c. Throughput Comparison for the three topologies

Figure 10 shows the comparison of average throughput of all schemes for the three topologies (random, controlled random and grid) for a network of 36 nodes.

Note that for random and controlled random topologies in Figure 10, the average throughput is the average over ten different random and controlled random topologies, respectively. Figure 10 shows that as compared to the CCA scheme, the throughput improvement with TICA is 3 times for random topology, 11 times for controlled random topology and 12 times for grid topology. In comparison to the SRSC scheme, the throughput improvement with TICA is 8 times for random topology, 95 times for controlled random topology and 133 times for grid topology.

The results in this figure clearly indicate that the proposed algorithm, TICA, significantly outperforms other schemes for the three topologies.

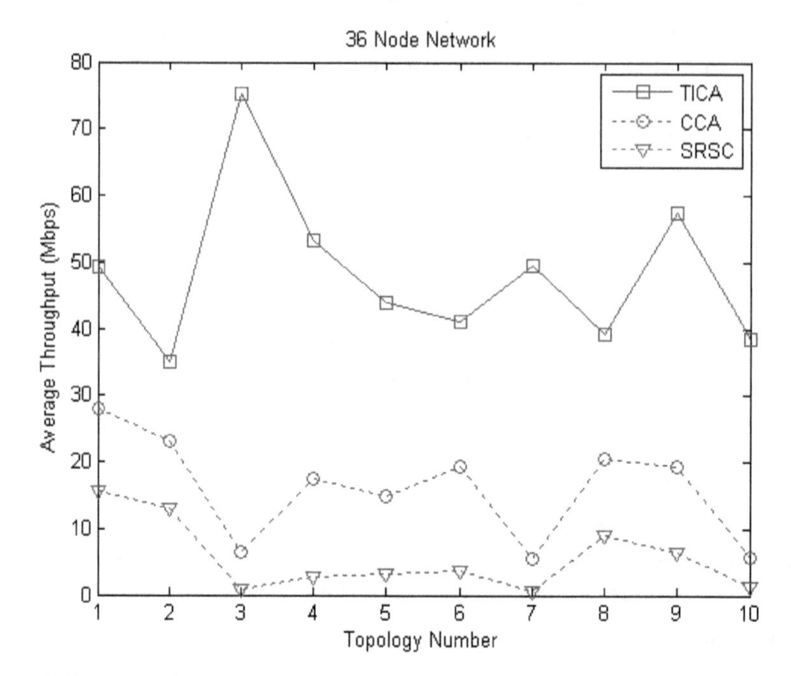

Figure 8. AT for ten random topologies

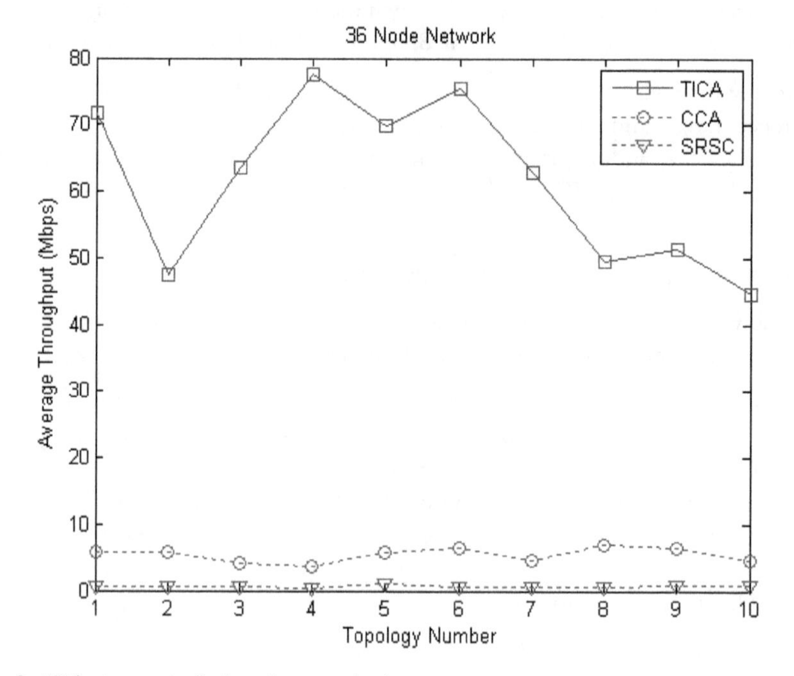

Figure 9. AT for ten controlled random topologies

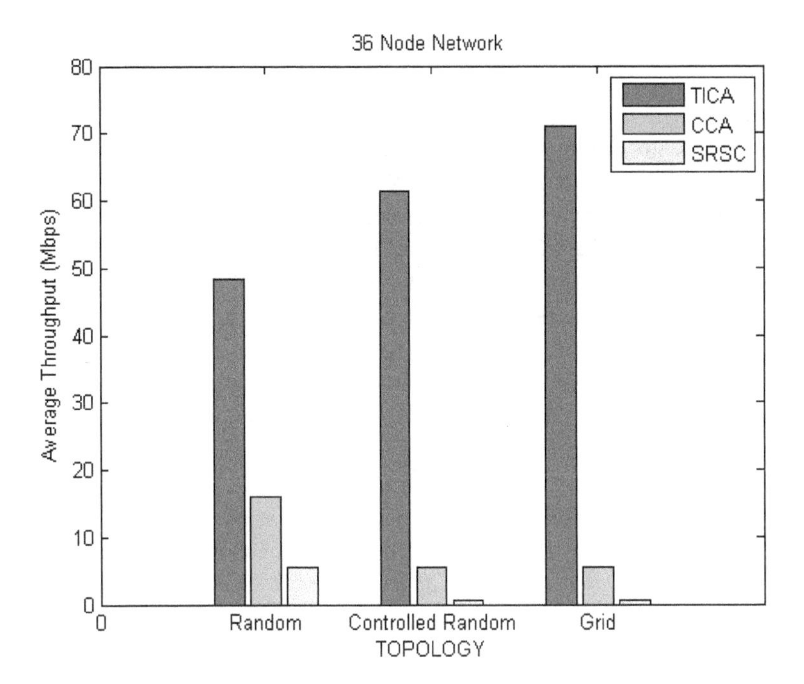

Figure 10. AT of all schemes for the three topologies

5.4. Features' comparison of related CA schemes

The important features of the related CA schemes are summarized in Table 3. They include channel switching, topology control, power control, knowledge of traffic load, connectivity, fault tolerance and CA control.

The most significant difference between TICA and existing CA schemes is that TICA uses topology control based on power control to build the topology for CA with the objective of minimizing the interference between the MRs whereas no other CA scheme has used topology control based on power control for CA.

Another significant difference between TICA and most other existing CA schemes is that TICA performs routing in addition to CA whereas most other CA schemes rely on routing protocols and complicated routing metrics for route selection.

Unlike TICA and D-HYA, other well-known CA schemes do not possess fault tolerance and have not provided any mechanism for recovery after a node failure.

6. Conclusion and future work

The chapter finally concludes in this section along with some directions for future work.

6.1. Conclusions

- A new topology control algorithm for multi-radio WMNs, *Select x for less than x*, is proposed. It controls the network topology by selecting the nearest neighbors for each node in the network.
- A new channel assignment algorithm for MRMC IEEE 802.11a-based WMNs, TICA, is proposed, which is based on topology control. As verified by a comprehensive performance evaluation, the improvement in the overall network throughput with TICA is significantly higher than the CCA scheme but is much higher than SRSC scheme in all three topologies. This is due to the fact that topology control based on power control results in an efficient frequency reuse during CA, which leads to an overall improvement in the network throughput.

  In the CCA and SRSC schemes, which are two commonly-used benchmark schemes, MRs do not control their power, transmit with maximum power, and use AODV routing protocol as their routing agent. TICA, on the other hand, uses topology control by selecting the nearest neighbors for each mesh node in the network, which reduces the interference among the mesh routers. It uses the minimum power as the link weight when building the SPT with the required node degree as less power translates to less interference among the mesh routers. It uses a scheme to build the IL for all the frequency channels if all the available frequency channels have already been assigned within the interference range of a link. It assigns a LIC to a link if all the available frequency channels have already been assigned within the interference range of that link. LIC is the channel with the minimum IL, which means that assigning this channel to the link would result in minimum interference in the network. In TICA, from the information in the received CARM, each mesh router applies power control and adjusts its transmission power accordingly, based on the channel assigned to that mesh router to communicate with a neighbor and its distance to that neighbor.
- As per the specification of the IEEE 802.11s standard for wireless mesh networks, the mesh routing protocol has to reside in the MAC layer. TICA not only performs channel assignment but performs routing as well by providing each node with the next hop information to reach the gateway node and hence, conforms to the requirements of IEEE 802.11s.
- A centralized Failure Recovery Mechanism for TICA is proposed, which supports automatic and fast failure recovery. In case of node failure, after detecting the failed node, deleting the MPNT of the failed node and deleting it from the MPNT of its neighbors, the gateway runs the TCA and builds the MPSPT. Based on the new MPSPT, the gateway calculates the link rankings and the channel assignments. The new channel assignments are communicated to the mesh nodes by the gateway via a new CARM using the control radio over the control channel.
- The proposed CA algorithm, TICA, does not require modifications to the MAC protocol and therefore, can work with existing IEEE 802.11a-based interface hardware.

6.2. Future work

Following are some aspects of this work which can be extended in future.

- Since TICA uses the new approach of building the interference level for all the frequency channels, it can be enhanced to model and account for the interference from co-located wireless networks.
- Other algorithms for building the tree topology, such as Prim's algorithm [30] and Kruskal's algorithm [31], can be used for building the minimum power based tree, which may lead to a further enhancement in the overall network throughput.

Scheme / Feature	TICA	C-HYA [11]	BFS-CA [20]	MesTiC [14]	D-HYA [17]	HMCP [18]
CA control	Centralized	Centralized	Centralized	Centralized	Distributed	Distributed
Knowledge of traffic load	Not required	Required	Not required	Required	Required	Not required
Channel switching	Required for failure recovery	Not required	Periodic channel switching required	Not required	Required as CA changes with traffic	Per-packet channel switching required
Topology	Topology controlled using TCA	Fixed	Fixed	Fixed	Topology defined by spanning tree	Topology changing due to channel switching
Power control	Yes	No	No	No	No	No
Connectivity	Ensured by common control radio, ensured by the CA scheme	Ensured by the CA scheme	Ensured by common control radio	Ensured by common control radio	Ensured by the CA scheme	Ensured by channel switching
Routing	Performed by the CA scheme	No	No	No	Performed by the CA scheme	No
Failure recovery	Yes	No	No	No	Yes	No

Table 3. Features' comparison of related CA schemes

- Other schemes for building the tree topology with the required node degree, such as the one proposed in [32], can be used which may lead to a better performance.
- The propagation model used is free-space model or two-ray model depending upon the cross-over distance. The performance of the proposed algorithm may be tested under more realistic propagation models such as Shadowing, Rayleigh-fading.
- The carrier sensing range is generally assumed to be twice the transmission range. However, carrier sensing range is a tunable parameter and an optimally tuned carrier sensing range can improve the network throughput in wireless mesh networks by enhancing the spatial frequency reuse and reducing collisions [33]. The performance of the proposed algorithm can be improved by controlling the carrier sensing range.
- The phenomenon of topology control based on power control impacts the per-node fairness of medium access based on CSMA/CA and hence the per-flow end-to-end throughput fairness [34]. The proposed algorithm can be extended to ensure per-node and hence per-flow fairness.
- All the traffic in a WMN is directed towards the gateway. The traffic bottleneck at the gateway is the main reason of the capacity limitation of a WMN. The use of multiple gateways can increase the capacity of the WMN by preventing the formation of traffic bottlenecks [35]. The proposed algorithm can be enhanced by extending it for multiple gateways.

Author details

Aizaz U. Chaudhry and Roshdy H.M. Hafez
Department of Systems and Computer Engineering, Carleton University, Ottawa, Canada

List of acronyms and abbreviations

AODV	Ad hoc On-Demand Distance Vector
AP	Access Point
AT	Average Throughput
BFS-CA	Breadth First Search – Channel Assignment
CA	Channel Assignment
CAA	Channel Assignment Algorithm
CARM	Channel Assignment and Routing Message
CBR	Constant Bit Rate
CCA	Common Channel Assignment
C-HYA	Centralized Hyacinth
CRT	Controlled Random Topology
CSMA/CA	Carrier Sense Multiple Access with Collision Avoidance
D-HYA	Distributed Hyacinth
DNT	Direct Neighbor Table
FNT	Final Neighbor Table

FRM	Failure Recovery Mechanism
GT	Grid Topology
GW	Gateway
HMCP	Hybrid Multi-Channel Protocol
IL	Interference Level
LIC	Least Interfering Channel
MAC	Medium Access Control
MANET	Mobile Ad hoc Network
MCG	Multi-Radio Conflict Graph
MesTiC	Mesh-based Traffic and interference-aware Channel-assignment
MMAC	Multi-channel Medium Access Control
MND	Maximum Node Degree
MPNT	Maximum Power Neighbor Table
MPSPT	Minimum Power-based Shortest Path Tree
MR	Mesh Router
MRMC	Multi-Radio Multi-Channel
NS2	Network Simulator Version 2
PHY	Physical Layer
RT	Random Topology
SPT	Shortest Path Tree
SRSC	Single-Radio Single-Channel
TC	Topology Control
TCA	Topology Control Algorithm
TICA	Topology-controlled Interference-aware Channel-assignment Algorithm
VCA	Varying Channel Assignment
WLAN	Wireless Local Area Network
WMN	Wireless Mesh Network
WSN	Wireless Sensor Network

7. References

[1] N. Nandiraju, D. Nandiraju, L. Santhanam, B. He, J. Wang, and D.P. Agrawal, "Wireless Mesh Networks: Current Challenges and Future Directions of Web-in-the-Sky," *IEEE Wireless Communications Magazine*, Vol. 14(4), pp. 79–89, August 2007.

[2] A. U. Chaudhry, R. H. M. Hafez, O. Aboul-Magd, and S. A. Mahmoud, "Throughput Improvement in Multi-Radio Multi-Channel 802.11a-based Wireless Mesh Networks," *Proceedings of IEEE Globecom 2010*, December 2010.

[3] A. U. Chaudhry, R. H. M. Hafez, O. Aboul-Magd, and S. A. Mahmoud, "Fault-Tolerant and Scalable Channel Assignment for Multi-Radio Multi-Channel IEEE 802.11a-based Wireless Mesh Networks," *Proceedings of IEEE Globecom 2010 Work-*

shop on Mobile Computing and Emerging Communication Networks (MCECN' 2010), December 2010.

[4] "IEEE Standard for Information Technology-Telecommunications and Information Exchange between Systems-Local and Metropolitan Area Networks-Specific Requirements - Part 11: Wireless LAN MAC and PHY Specifications," *IEEE Std 802.11-2007 (Revision of IEEE Std 802.11-1999)*, June 2007.

[5] N. Li, J. Hou, and L. Sha, "Design and Analysis of an MST-Based Topology Control Algorithm," *Proceedings of IEEE International Conference on Computer Communications*, June 2003.

[6] F. O. Aron, T. O. Olwal, A. Kurien, and Y. Hamam, "Network Preservation Through a Topology Control Algorithm for Wireless Mesh Networks," *Proceedings of 2nd IASTED Africa Conference on Modelling and Simulation*, September 2008.

[7] V. Kawadia and P.R. Kumar, "Principles and Protocols for Power Control in Wireless Ad Hoc Networks", *IEEE Journal on Selected Areas in Communications: Special Issues in Wireless Ad Hoc Networks*, 2005.

[8] P. Gupta and P. R. Kumar, "Critical Power for Asymptotic Connectivity in Wireless Network," *Stochastic Analysis, Control, Optimization and Applications*: A Volume in Honor of W.H. Fleming, W. M. McEneany, G. Yin, and Q. Zhan, Eds. Birkhauser, Boston, pp. 547–566, 1998.

[9] P. Gupta and P. R. Kumar, "The Capacity of Wireless Networks," *IEEE Transactions on Information Theory*, Vol. IT-46, pp. 388–404, 2000.

[10] P. Santi, "Topology Control in Wireless Ad Hoc and Sensor Networks," *ACM Computing Survey*, 37(2), pp. 164–194, 2005.

[11] A. Raniwala, K. Gopalan, and T. Chiueh, "Centralized Channel Assignment and Routing Algorithms for Multi-channel Wireless Mesh Networks," *ACM Mobile Computing and Communications Review*, Vol. 8(2), pp. 50 – 65, 2004.

[12] E. Hossain and K. Leung, "*Wireless Mesh Networks: Architectures and Protocols*," Springer US, 2007.

[13] R. Draves, J. Padhye, and B. Zill, "Routing in Multi-Radio, Multi-Hop Wireless Mesh Networks," *Proceedings of ACM International Conference on Mobile Communications and Networking*, 2004.

[14] H. Skalli, S. Ghosh, S. K. Das, L. Lenzini, and M. Conti, "Channel Assignment Strategies for Multiradio Wireless Mesh Networks: Issues and Solutions," *IEEE Comm. Magazine, Special Issue on "Wireless Mesh Networks"*, pp. 86-93, November 2007.

[15] P. Kyasanur, J. So, C. Chereddi, and N. Vaidya, "Multi-Channel Mesh Networks: Challenges and Protocols," *IEEE Wireless Communications Magazine*, Vol. 13(2), pp.30-36, April 2006.

[16] J. So and N. H. Vaidya, "Multi-Channel MAC for Ad Hoc Networks: Handling Multi-Channel Hidden Terminals Using a Single Transceiver," *Proceedings of ACM International Symposium on Mobile Ad Hoc Networking and Computing*, 2004.

[17] A. Raniwala and T.-C. Chiueh, "Architecture and Algorithms for an IEEE 802.11-Based Multi-Channel Wireless Mesh Network," *Proceedings of IEEE International Conference on Computer Communications*, 2005.

[18] P. Kyasanur and N. Vaidya, "Routing and Interface Assignment in Multi-Channel Multi-Interface Wireless Networks," *Proceedings of IEEE Wireless Communications and Networking Conference*, New Orleans, LA, March 2005.

[19] P. Kyasanur and N. H. Vaidya, "Routing and Link-layer Protocols for Multi-Channel Multi-Interface Ad Hoc Wireless Networks," *ACM Mobile Computing and Communications Review*, Vol. 10(1), pp. 31– 43, January 2006.

[20] K. Ramachandran, E. Belding, K. Almeroth, and M. Buddhikot, "Interference-Aware Channel Assignment in Multi-Radio Wireless Mesh Networks," *Proceedings of IEEE International Conference on Computer Communications*, 2006.

[21] K. Rosen, "*Discrete Mathematics and its Applications*," McGraw Hill, 1999.

[22] T. Rappaport, "*Wireless Communications: Principles and Practice*," 2nd Ed. Prentice Hall, Upper Saddle River, NJ, 2002.

[23] M. Mahdian, "On the Computational Complexity of Strong Edge Coloring," *Discrete Applied Mathematics*, Vol. 118, pp. 239–248, 2002.

[24] M. Shin, S. Lee, and Y.-A. Kim, "Distributed Channel Assignment for Multi-Radio Wireless Networks," *Proceedings of IEEE International Conference on Mobile Ad hoc and Sensor Systems*, pp. 417–426, Oct. 8–11, 2006.

[25] E.W. Dijkstra, "A Note on Two Problems in Connection with Graphs," *Numerische Mathematik*, Vol. 1, pp. 269-271, 1959.

[26] The VINT Project, "*Network Simulator (NS) – version 2*," available at http://www.isi.edu/nsnam/ns.

[27] The MathWorks Inc., "*MATLAB® Reference Guide*," Natick, MA, USA, 1992.

[28] R. A. Calvo and J. P. Campo, "Adding Multiple Interface Support in NS-2," University of Cantabria, Jan. 2007, available at http:/personales.unican.es/aguerocr/files/ucMultiIfacesSupport.pdf.

[29] Ad hoc On-Demand Distance Vector (AODV) Routing Protocol, RFC3561, available at http://www.ietf.org/rfc/rfc3561.txt.

[30] R. Prim, "Shortest Connection Networks and Some Generalizations," *The Bell System Technical Journal*, Vol. 36, pp. 1389-1401, 1957.

[31] J. Kruskal, "On the Shortest Spanning Tree for a Graph and Traveling Salesman Problem," *Proceedings of American Mathematics Society*, Vol. 7, pp. 48-50, 1956.

[32] S. C. Narula and C. A. Ho, "Degree-Constrained Minimum Spanning Tree," *Computers and Operations Research*, Vol. 7, pp. 239–249, 1980.

[33] J. Deng, B. Liang, and P. K. Varshney, "Tuning the Carrier Sensing Range of IEEE 802.11 MAC," *Proceedings of IEEE Global Communications Conference*, Dec. 2004.

[34] H. Y. Hsieh and R. Sivakumar, "Improving Fairness and Throughput in Multi-Hop Wireless Networks," *Proceedings of IEEE International Conference on Networks*, pp.569-578, 2001.

[35] S. Lakshmanan, R. Sivakumar, and K. Sundaresan, "Multi-Gateway Association in Wireless Mesh Networks," *Ad Hoc Networks*, Vol. 7(3), pp. 622–637, May 2009.

Chapter 4

Partially Overlapping Channel Assignments in Wireless Mesh Networks

Fawaz Bokhari and Gergely Záruba

Additional information is available at the end of the chapter

1. Introduction

The concept of wireless mesh networks (WMN) has emerged as a promising technology for the provision of affordable and low-cost solutions for a wide range of applications such as broadband wireless internet access in developing regions with no or limited wired infrastructure, security surveillance, and emergency networking, One concrete example is WMNs for public safety teams like firefighters who can still be connected with the help of mesh nodes mounted on street poles even if all infrastructure communications fail. The main reason for this vast acceptance of mesh networks in the industry and academia is because of their self-maintenance feature and the low cost of wireless routers. In addition, the self-forming features of WMNs make the deployment of a mesh network easy thereby enabling large-scale networks. Mesh networks which are of most commercial interests are characterized as fixed backbone WMNs where mesh nodes (routers or access points) are generally static and are mostly supplied by a permanent power source. Such a wireless mesh network architecture is illustrated in Figure 1, consisting of mesh routers, clients, and gateway nodes. Mesh routers (MR) communicate with peers in a multi hop fashion such that packets are mostly transmitted over multiple wireless links (hops). Therefore, nodes forward packets to other nodes that are on the route but may not be within direct transmission range of each other. Routers which are connected to the outside world are called gateway nodes (GWN). These GWNs carry traffic in and out of the mesh network. The collection of such routers and gateway nodes connected together in a multi hop fashion form the basis for an infrastructure WMN (also called backbone mesh). Moreover, the multi hop packet transmission in an infrastructure WMN extends the area of wireless broadband coverage without wiring the network; thus WMNs can be used as extensions to cellular networks, ad hoc networks (MANET), sensor and vehicular networks, IEEE 802.11 WLANs (Wi-Fi), and IEEE 802.16 based broadband wireless (WiMax) networks [1].

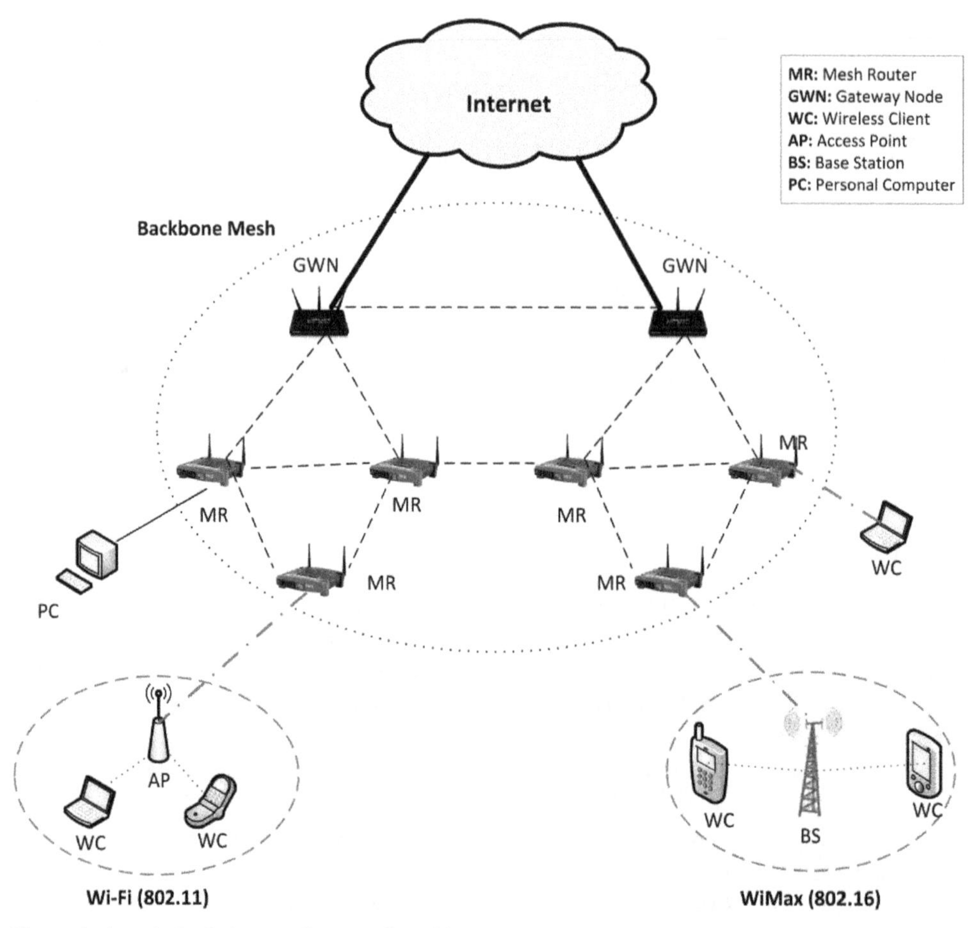

Figure 1. A typical wireless mesh network architecture.

WMNs can be classified based on the number of radios on each mesh router. In single-radio mesh networks, each node is equipped with only one radio. In multi-radio mesh networks, multiple radios are installed on each mesh node in the backbone mesh. Depending upon the radio to channel configuration (also called interface to channel assignment), mesh networks can be further classified into single-radio single-channel (SRSC), single-radio multi-channel (SRMC), and multi-radio multi-channel (MRMC) wireless mesh networks. (Note, that we did not list multi-radio single-channel WMNs as that would mean that nodes are equipped with multiple radios but all of the radios in the network are configured on the same single channel defeating any purpose of multi-radios.) In a SRSC-WMN, as the name suggests, all nodes are configured to use the same wireless channel. This ensures network connectivity; however, capacity of the network is greatly affected as all nodes are competing to access the same channel. Therefore, interference minimization is the major issue in such networks. SRMC-WMNs can achieve parallel transmissions by assigning different orthogonal channels

(OCs) to radios belonging to different nodes, thus improving network capacity. However, such networks severely suffer from network disconnections due to having a single radio at each node possibly configured at different channels. In, MRMC-WMNs, with the availability of off-the-shelf, low cost, IEEE 802.11 based networking hardware, it is possible to incorporate multiple radio interfaces operating on different radio channels on a single mesh router. This enables a potentially large improvement in the capacity of the WMN (compared to all the previous forms of mesh networks) [20].

Wireless mesh networks, particularly infrastructure WMNs, have some unique characteristics that set them apart from other wireless networks, such as MANETs and sensor networks. For example, nodes (at least relay nodes) in a typical infrastructure mesh network are generally static and have no significant constraints on power consumption, as opposed to MANETs, where nodes have limited energy and are mostly mobile. Similarly, due to the shared nature of the wireless medium, nodes compete with each other for channel access when they transmit on the same channel resulting in possible interference among the nodes. Unlike MANETs, where the general traffic model describes traffic flows between any pair of mobile nodes, in WMNs data flows are typically between mesh nodes and GWNs. In general, in WMNs certain paths and nodes are much more likely to be saturated as the distribution of flows over nodes is less uniform compared to that in MANETs. Therefore, load balancing is of utmost importance to avoid hot spots and to increase network utilization.

In a typical multi radio mesh network, the total number of radios within the network is usually significantly higher than the number of available channels in the network (e.g. only 11 channels are available in the U.S.A. for IEEE 802.11b/g). This forces many links to operate on the same (set of) channels, resulting in possible interference among transmissions. The existence of such interference if not accounted for, can affect the capacity of the network. Therefore, understanding and mitigating interference has become one of the fundamental issues in WMNs; recently a number of channel assignment (CA) solutions have been proposed to address this problem [5, 10-13, 15-20, 33-35].

The problem of channel assignment (frequency assignment) has been widely studied in cellular networks [2]. However, with the proliferation of IEEE 802.11 based technologies in the wireless arena (WLANs, sensor networks, WMNs), the need for channel assignment solutions outside of cellular networks has surfaced. CA algorithms are usually designed based on the peculiar characteristics of individual networks; since the differences in characteristics are vast, CA algorithms for WMNs must be significantly different from those of cellular networks. For example, base stations in a cellular network are typically connected by cables, whereas mesh nodes in a mesh network are connected wirelessly (and usually on the same channels as are used for providing service). This brings up several interference issues in mesh networks between mesh nodes which are not found in cellular networks between base stations (as in cellular networks BSs are not competing for the shared medium as they have dedicated bandwidth for intra-BS communication). The bottleneck in cellular networks is from the base stations to the client devices, whereas, in WMNs, the bottleneck is usually inside the mesh backbone, typically along the route from the mesh routers to the gateway nodes. In addition nearby BSs are usually configured on completely orthogonal channels (OCs) to avoid interference; this is rarely possible in backbone meshes, as the nodal density of a typical WMN can be high and the number of available orthogonal channels is limited. Most existing deployed mesh networks are IEEE 802.11 technology based; among the standards of IEEE 802.11, the most widely used are IEEE 802.11b/g, which support up to 14 channels in the unlicensed Industrial, Scientific, and Medical (ISM) radio bands at the nominal 2.4 GHz carrier frequency [32]. Out of these 14 channels, only 11 channels are available for use in the U.S.A., 13 channels are open in EU, while Japan has made all of them available. Figure 2. shows the 2.4 GHz ISM band's division into 11 IEEE 802.11b/g channels in the U.S.A.; the channel numbers have a one-to-one relationship with the corresponding center frequency of that channel. (For example, channel 6 operates at 2.437GHz.) Each channel's bandwidth is 22 MHz and each channel's center frequency is separated from the next channel's by 5MHz. Therefore, in general, a channel overlaps with 4 of its neighboring channels leaving only three non-overlapping (orthogonal) channels, i.e., channels 1, 6, and 11 as depicted in Figure 2. Similarly, IEEE 802.11a operates in 5GHz ISM band and provides 12 orthogonal channels, but since it operates in a higher frequency band, it has a shorter range as opposed to 802.11b/g (higher frequencies in general have higher inabilities to penetrate walls and obstructions). Recently, the IEEE 802.11n standard was proposed which operates in both the 2.4GHz and 5 GHz bands and provides legacy support to devices operating based on previous standards (b/g). It provides data rates of up to 600Mbps using Multiple Input, Multiple Output (MIMO) technology with Orthogonal Frequency Division Multiplexing (OFDM).

Most existing research on CA algorithms in WMNs has been focused on assigning orthogonal (non-overlapping) channels [33-35] to links belonging to neighboring nodes in order to minimize the interference in the network. Since, links operating on orthogonal channels do not interfere at all, multiple parallel transmissions can be possible resulting in overall network throughput improvement. The number of non-overlapping channels in commodity wireless platforms such as 802.11b/g is very small (again, only three orthogonal channels out of total 11 channels) while nodal density in a typical MRMC-WMN is high. This realization has recently drawn significant attention to the study of partially overlapped channels (POC) for channel assignment [5]. The basic idea is to make all channels available to nodes for channel selection as a result of which, partially overlapped channels may be employed. This could enable multiple concurrent transmissions on radios configured on POCs and therefore could increase network capacity assuming that the interference is lessened in POCs compared to completely overlapping channels.

Previously, an algorithm for channel assignment based solely on orthogonal channels had to deal with only co-channel interference. However, one of the major issues in designing efficient channel assignment schemes using POCs is the adjacent channel interference, which is the interference between two neighbors configured on adjacent (partially overlapping) channels. The effect of such adjacent channel interference has a direct relationship with the geographical location of these two nodes, i.e., the farther two nodes are apart, the less interference is created on adjacent channels. Nonetheless, the assignment of orthogonal and non-orthogonal channels in high density mesh networks needs to be carefully coordinated; the key issue lies in the fact that the interference between adjacent channels has to be considered. This needs to be done intelligently so that channel capacity is maximized, otherwise the shared nature of wireless medium can lead to serious performance degradation of the whole mesh network. Thus, recently POCs for channel assignment in wireless networks has received some attention [5, 10-13, 15-20].

Within the scope of this chapter, we focus on the problem of channel assignment using partially overlapping channels in the context of both single- and multi-radio WMNs. The rest of the chapter is organized as follows. Section 2 describes different types of interferences that may exist in a typical WMN. Section 3 demonstrates the benefits of using partially overlapping channels for channel assignment in WMNs with the help of experiments performed on a real testbed. In Section 4, we provide a comprehensive review of some of the recent well-known channel assignment schemes exploiting POCs in WMNs and classify these POC-based CA schemes according to their most prominent attributes together with the objectives and limitations of each of the approaches. In Section 5, we discuss open issues and challenges in the design of partially overlapping channel assignment schemes, followed by the chapter's conclusion in Section 6.

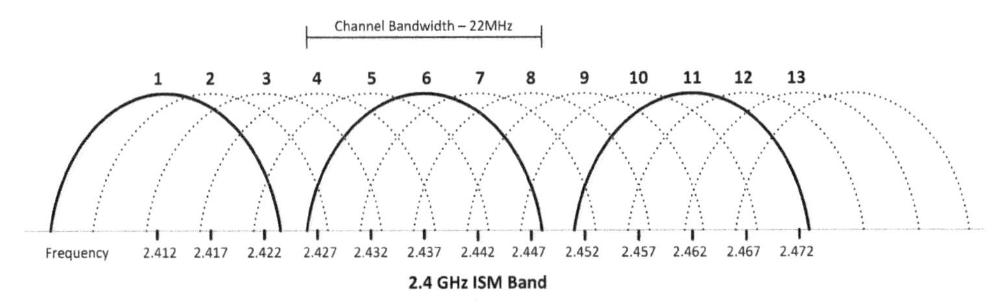

Figure 2. IEEE 802.11b/g channels, showing the three orthogonal channels in bold

2. Interference in Wireless Networks

In a typical WMN, flows on links belonging to different nodes compete with each other to access the wireless medium. This results in possible interference among the nodes therefore severely affecting network performance. Multiple types of interferences exist in WMNs depending on flow characteristics and on interface to channel configurations. We first explain what the different types of flow interferences are particularly in infrastructure WMNs. We will also present another interference classification in mesh networks based on the configuration of the channels to radios and also on the number of radios installed in nodes.

2.1. Flow based interference

2.1.1. Inter-flow Interference

This type of interference occurs when neighboring nodes carrying different flows compete for channel access when they transmit on the same channel as depicted in Figure 3(a). This effectively means that whenever a node is involved in a transmission; its neighboring nodes should not communicate at the same time.

2.1.2. *Intra-flow Interference*

Nodes on the path of a same flow compete with each other for channel access when they transmit on the same channel. This is referred to as intra-flow interference and is shown in Figure 3(b).

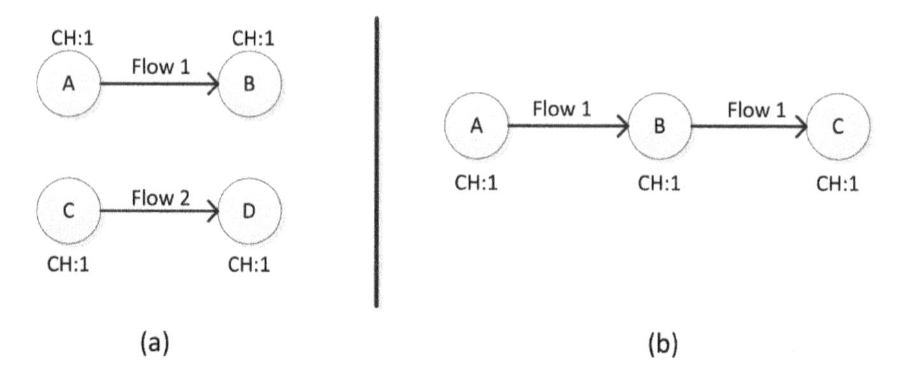

Figure 3. Flow based interference. (a) Inter-flow interference (b) Intra-flow interference

2.2. Interference based on interface to channel configuration

A wireless mesh network utilizing both orthogonal and non-orthogonal channels may suffer from interferences which can be characterized as follows.

2.2.1. *Co-channel Interference (CCI)*

Co-channel interference is the most common type of interference that exists in almost all wireless networks (depicted in Figure 4-a). It refers to the fact that radios belonging to two nodes, operating on the same channel would interfere with each other, if they are within the interference range of each other. This effectively means that parallel communications from two separate in-range nodes is not possible.

2.2.2. *Adjacent Channel Interference (ACI)*

We talk about adjacent channel interference when radios on two nearby nodes are configured to partially overlapping channels. For example, in Figure 4(b), a radio on node *A* is configured on channel-4 while another radio at neighboring node C is configured on channel-1; then the transmission from either node would experience some sort of partial interference. This type of interference also restricts parallel communication depending upon the channel separation and the physical distance between the two nodes.

2.2.3. *Self-Interference (SI)*

Self-interference is defined as a transmission from a node interfering with one of its own transmissions. This is related to situations when nodes are equipped with multiple radios in a mesh network. Parallel communication cannot be achieved using multiple radios installed on a node, unless they are configured on **completely orthogonal** channels as shown in Figure 4(c).

All of the above types of interferences have to be considered when designing channel assignment algorithms to exploit the full potential of the available wireless spectrum. Therefore, the first step in developing mechanisms which take advantage of the partial overlap is to build a model that captures the channel overlap in a quantitative fashion.

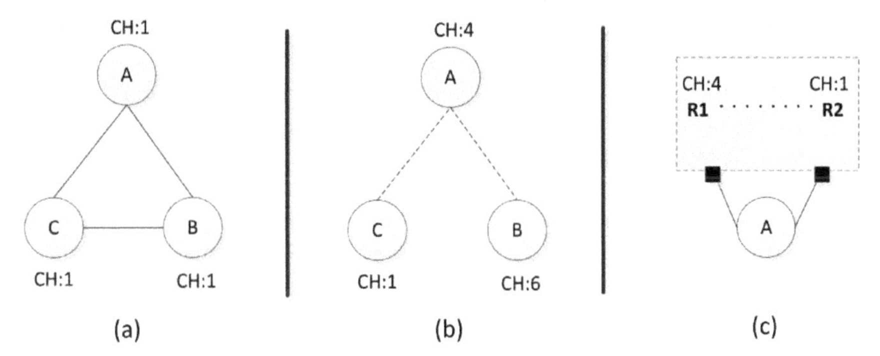

Figure 4. Types of interferences (a) co-channel interference (b) adjacent channel interference (c) self-interference

3. Benefits of using Partially Overlapped Channels

In this section, we will discuss the benefits of using POCs in WMNs. First, we will explain what the different scenarios are, where the use of partial overlap among channels will be useful. We follow that by a quick testbed experiment to demonstrate the effectiveness of using POCs in WMNs.

Mishra, et al., in [6] have performed detailed experiments to demonstrate the effectiveness of using partial overlap among channels in WMNs. The authors have measured the signal to noise ratio (SNR) of two communicating nodes configured on adjacent channels and mapped them onto a normalized [0,1] scale with 0 representing the minimum signal received. Their results are shown in Table I.

Channel	1	2	3	4	5	6	7	8	9	10	11
Normalized SNR (I-factor)	0	0.22	0.60	0.72	0.77	1.0	0.96	0.77	0.66	0.39	0

Table 1. SNR of transmission made on channel 6 as received on channels 1 ... 11.

A typical bandwidth of an IEEE 802.11b channel which uses direct sequence spread spectrum (DSSS) is 44MHz. It is distributed equally on each side of the center frequency of that channel i.e. 22MHz on each side. A transmit spectrum mask (band pass filter) is applied to the signal at the transmitting station (with a typical example shown in Figure 5) which is basically used by the transmitter to limit the output power on nearby frequencies. As it can be seen in the figure, the mask is set to 0dB at the center frequency where signals are passed without any attenuation. However, at frequencies beyond 11MHz on either side of the center frequency, the signal's power is attenuated by as much as 30dB and at 22MHz as much as 50dB. The receiver also uses a band pass filter centered around the nominal transmission frequency of the channel. Three scenarios are discussed in [6] where the use of partial overlap among the channels can be useful in the context of wireless mesh networks:

- *Multi-channel communication:* The first scenario is when a node can communicate with two of its neighboring nodes configured on orthogonal channels (OCs) by operating on a partial overlapping channel. Basically, for a little reduction in throughput, one can use partially overlapping channels and this can give flexibility in topology construction while reducing the extra overhead in channel switching to enable communication.
- *Throughput improvement:* The second scenario is when nodes in a mesh network have only one radio and therefore, they can be configured to only one channel at a time. There is a possibility of network disconnection while assigning different channels to nodes in the network. Channels with partial overlap can be assigned to nodes in such a manner that improves the overall network throughput capacity. In this way, the assignment of partially overlapping channels has to be intelligent enough to utilize the maximum bandwidth available and therefore can result in significant throughput improvements.
- *Channel re-use:* Shorter ranges for frequency reuse can be obtained if two interfering links are assigned partially overlapping channels rather than orthogonal channels. It is possible to significantly improve the overall channel re-use (i.e., by reducing the distance between nodes using POCs) by careful assignment of channels which will result in higher peak throughputs.

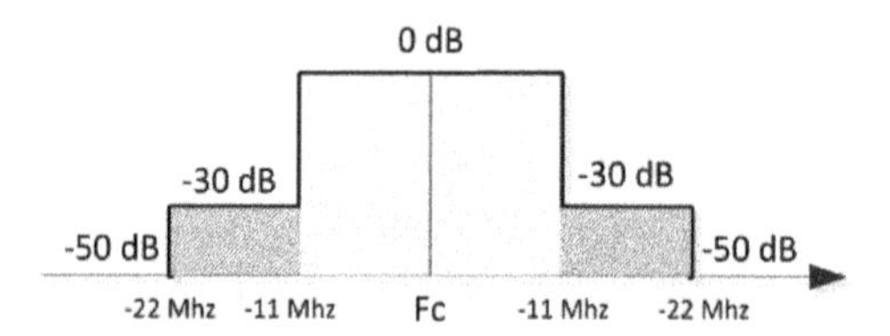

Figure 5. A typical IEEE 802.11b transmit spectrum mask

Later, in [3], the same authors have shown the advantage of using POCs in two different types of networks, i.e., WLANs and WMNs. In a WLAN setup, nearby access points can be assigned POCs such that the signal attenuation due to the overlap degrades to a tolerable level. In other words, the interference range of APs is reduced as perceived by neighboring APs operating on a partially overlapping channel. This provides efficient spatial re-use of channels and more APs can operate concurrently providing better service to clients. Similarly, in a single radio WMN environment, throughput can be improved when nodes can be configured to overlapping channels in order to avoid network disconnection and also to avoid any channel switching overhead.

3.1. Experimental evaluation

Next we will show results from experiments performed on a real testbed in order to evaluate the benefits of using partially overlapping channels in mesh networks. Our experimental testbed consists of four Linksys WRT54GLv1.1 wireless routers, each equipped with one radio. We installed the Freifunk firmware [28] on these routers for more freedom in our experiments. We created two point-to-point networks between two router pairs and thus formed two links each consisting of two routers as shown in Figure 6. Link-1 belongs to Pair-1 and Link-2 belongs to Pair-2. Each radio on Link-1 is fixed on channel 6; we varied the channels of Pair-2 from 1 to 6. The distance between nodes belonging to the same link is kept constant throughout the experiment. Pair-1 nodes have fixed locations while Pair-2 is moved to various distances from Pair-1 ranging from 5 to 30 meters (but Pair-2 nodes are kept equidistance to each other during the experiments). UDP and TCP traffic is generated on both links lasting for 10 seconds. The throughput on Link-2 is measured and the results are averaged over several runs. Three different IEEE 802.11b defined data rates are used for conducting the experiments, i.e. 2Mbps, 5.5Mbps and 11Mbps.

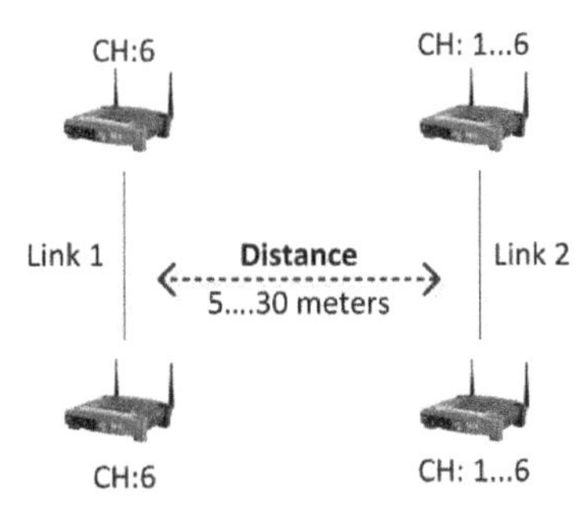

Figure 6. POC measurement testbed

Figure 7(a), (b), and (c) show the UDP throughput on Link-2 with different channel separations for the three data rates. It can be seen that as the distance between the two interfering links is increased, the throughput increases due to the reduced amount of interference. In this setup we did not see any further improvements when nodes were more than 30 meters apart. However, the same maximum throughput can be achieved at significantly lower distances with increased channel separation between the two links. For example, at about 20 meters, Link-2 achieves the maximum benchmark throughput, when the channel separation between the two links is three. For data rates 5.5Mbps and 11Mbps, we notice similar results; however, maximum throughput can be achieved by eliminating interference at a much lower distances i.e. about 15 meters, when the channel separation is three as compared to 30 meters, when both the channels are separated by only one. Figures 8 (a), (b), and (c) show the comparable results when TCP traffic is used on all the three 802.11b data rates.

From these results, we can extrapolate the interference ranges of nodes with varying channel separations and at different data rates; this comprehension is shown in Figure 9. Each point in the graph represents the minimum distance that is required between the two links in order for them to experience no interference and achieve maximum throughput when they are on particular partially overlapping channels (with a given channel separation). We can observe that the interference ranges are decreasing with increasing channel separation and increasing data rates. From these measurements, we can empirically conclude that the interference range of nodes operating on POCs is significantly less than the range when they are on the same channel. (Similar experiments have been performed before in [3, 5-7, 16]; however, those experiments were done either on wireless card equipped computers or a computer attached to an access point. We believe, that our setup is easier to reproduce and is more representative for a WMN and thus provides a better understanding of POCs in mesh networks.)

Therefore, there is a tradeoff between efficient utilization of the wireless spectrum and a slight decrease in the throughput. An intelligent assignment of partially overlapping channels can decrease the impact of interference, eventually resulting in more efficient utilization of the spectrum.

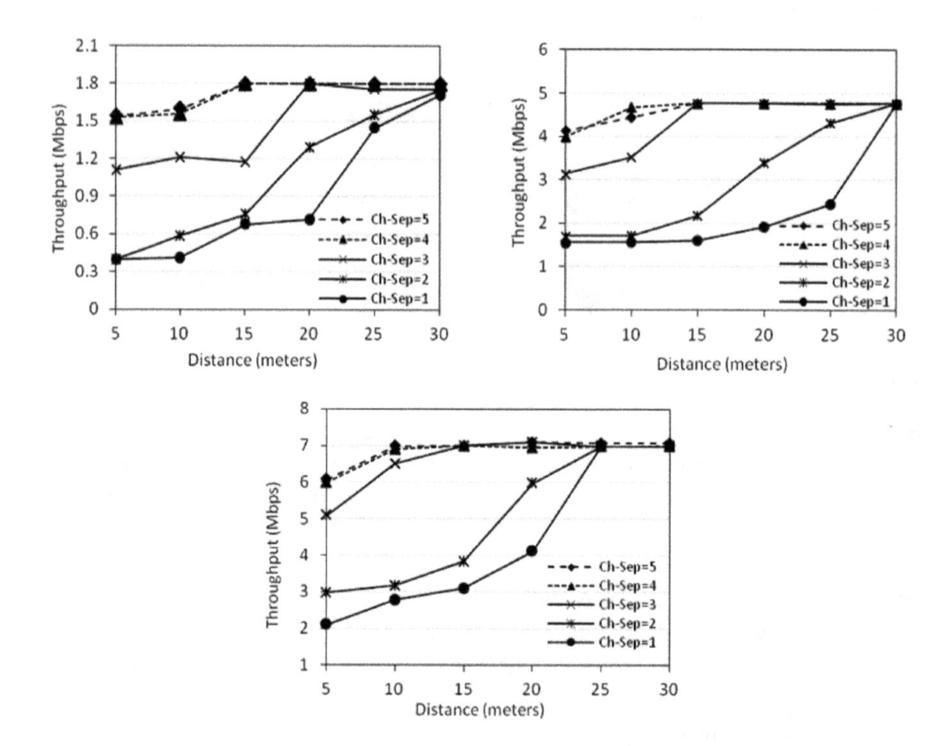

Figure 7. UDP throughput of two interfering links as a function of channel separation. (a) 2 Mbps (b) 5.5 Mbps (c) 11 Mbps

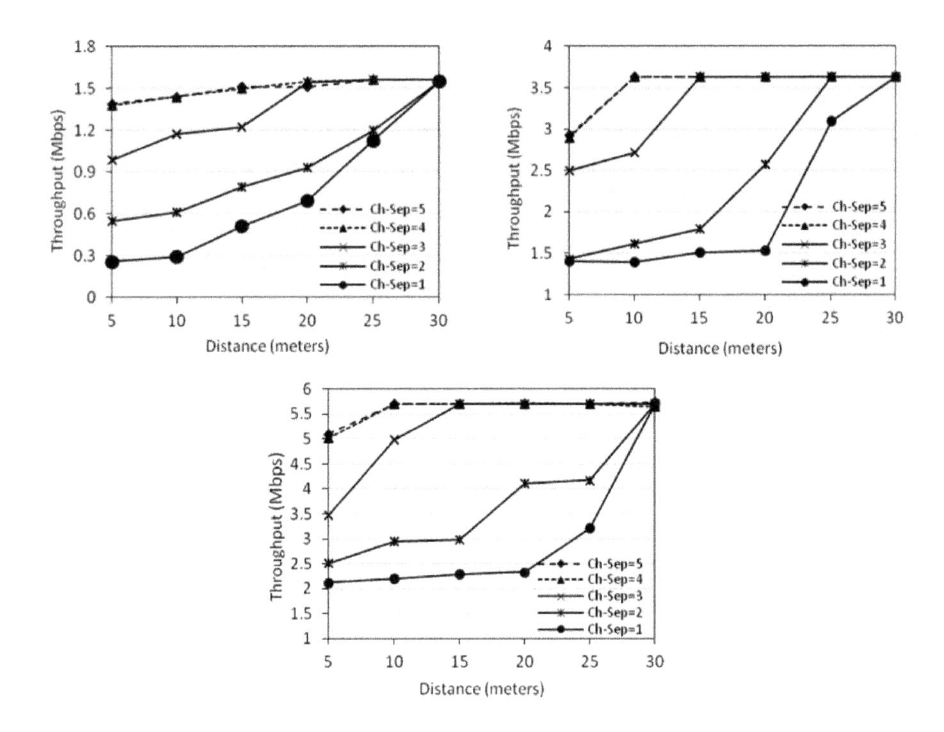

Figure 8. TCP throughput of two interfering links as a function of channel separation. (a) 2 Mbps (b) 5.5 Mbps (c) 11 Mbps

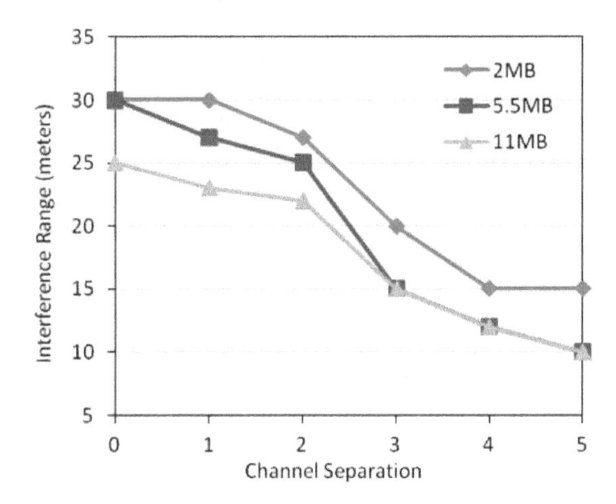

Figure 9. Interference range as a function of data rates

4. Classification of POCA Schemes in Wireless Mesh Networks

Partially overlapping channel assignment (POCA) schemes can be classified based on different criteria and approaches. The criteria that we have used for classification is the *interference model,* which is defined as the technique for capturing interference of radios belonging to nodes operating on partially overlapping channels in a WMN. Figure 10 presents the classification on which the rest of the section is based. Note that our classification based on interference model may not create disjoint categories and thus, a particular scheme may have significant overlaps with another scheme belonging to a different category.

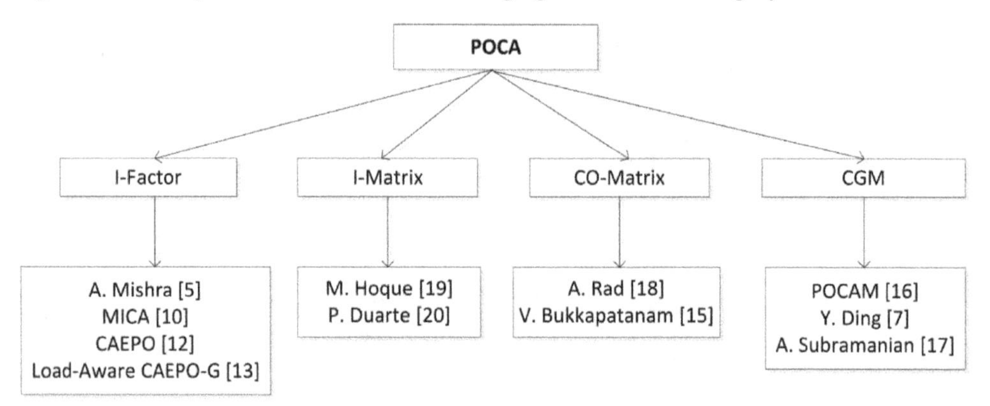

Figure 10. Classification of partially overlapping channel assignment algorithms based on the interference model employed.

4.1. Interference factor model (I-Factor)

4.1.1. Revised Channel Assignment Schemes for Wireless Networks

One of the first models to capture partial interference in wireless networks was presented by A. Mishra, et al. in [5]. They have extensively studied the practicality of using POCs in WLANs and WMNs. Through analytical formulation they have shown the benefits of POCs in terms of how they increase network capacity and improve channel-reuse. In order to model the interference generated by nodes operating on channels with partial overlaps, they have proposed a novel concept called interference factor (I-factor) capturing the extent of overlap between two communicating nodes. They define I-factor as:

$$IF_{(t,r)}(\delta) = \int_{-\infty}^{+\infty} S_t(f)B_r(f-\delta)df$$

where t and r are indices of the transmitting and receiving nodes, and δ denotes the difference of the frequencies of the transmitting and receiving nodes. In other words, parameter δ represents the amount of overlap between the two frequencies and is defined as a continuous variable. $S_t(f)$ is the transmitter's signal's power distribution and $B_r(f)$ denotes the frequency response of the receiver's band pass filter. In lay man terms: if we measure the area of intersection between a transmitter's signal spectrum and receiver's band-pass filter, we can calculate how much overlap there is between these signals; this is defined as the interference factor (I-factor). Since, IEEE 802.11 standards operate on a set of discrete channels, the continuous variable δ can be discretized as follows: $\delta = 5|i-j|$ (in MHz).

The authors of [5] have also revised two existing channel assignment algorithms in the context of WLANs and WMNs and have applied the I-factor model to these algorithms. First, an existing algorithm [22] was modified which is a centralized greedy-style approach for CA in WLANs using only orthogonal channels with the objective of increasing overall spectrum utilization. The algorithm employs an indicator variable to model the interference in WLANs and the authors have modified this indicator variable to capture not only the orthogonal channels (which was previously the case) but also channels with partial overlap (using their I-factor model). The actual channel assignment problem is formulated as a conflict set coloring problem where a conflict is present when clients belonging to a particular AP experience interference from neighboring clients (which are attached to their respective APs). The objective function is a min-max formulation to capture the total interference experienced by each client. The algorithm starts with a random permutation on how channels are assigned to APs; this is followed by the computation of the objective function. The best channel with minimum interference among the available channels is chosen and the process repeats for each AP. The modification lies in the interference calculation function to incorporate POCs into the algorithm. Interferences among channels with partial overlaps are calculated based on the I-factor interference model either empirically or analytically; this enables the possibility of assigning all available channels to the WLAN.

Still in [5], another CA algorithm which was designed for wireless mesh networks using only orthogonal channels [21] was modified to include POCs. It is a joint channel assignment, routing and link scheduling approach and a mathematical formulation in the form of a linear program (LP) is presented. The formulation also includes an indicator variable to model interference in the network. The authors have modified the link scheduling part of the joint mathematical formulation to change the conflict links' constraints to include the I-factor model (partial interference). They have evaluated the performance of this modified LP to show improved throughput in WMNs. The revised algorithms demonstrate that careful use of POCs can lead to significant improvements in spectrum utilization and application performance. They have performed extensive simulations to show that the use of POCs can improve network throughput (the extent of which depends on the nodal density of the network).

4.1.2. Channel Assignment Exploiting Partially Overlapping Channels (CAEPO)

The authors in [12] have proposed a POC channel assignment scheme called CAEPO. The main contribution of their work is the design of a traffic-aware metric that captures the degree of overlap among the channels when measuring interference. It is a hybrid distributed channel assignment protocol, where each node collects information locally and hence performs the channel assignment locally. The proposed I-factor based metric captures the interference experienced by nodes operating on channels with partial interference. Each node measures the interference according to the degree of overlap between channels and scales it to the traffic load experienced by its neighboring node (this information is maintained by each node). Each node does this for all of its neighbors and combines the results to determine the total interference it is "suffering" due to its neighboring nodes. Thus, the interference metric at node *i* is calculated as:

$$Interference[i] = \sum_{j \in N(i)} f[i][j] * B(j)$$

where *B(j)* is the proportion of the busy time of a neighboring node *j*, and *N(i)* is the set of neighbors of node *i*; *f[i][j]* captures the extent of overlap a node operating on a particular channel has from its neighboring nodes configured on another channel. This is based on the extent of the channel separation between the channels used by the two nodes (taken from [23]).

More precisely, CAEPO works as follows: each node in the network is equipped with two interfaces; the first interface is configured to a fixed channel while the other interface can be dynamically switched between channels. The algorithm starts with each node assigning a fixed channel to its fixed interface and a default channel to its switchable interface using the interference estimation metric with the initial value of *B(j)*=1. Then, this channel assignment information, together with the interference measurements are relayed to all neighbors. After this initial channel assignment, each node periodically calculates the interference using the interference metric described above and if the fixed interface channel needs to be changed, then that information is relayed on the default channel of the switchable interface. Similarly, when a node has data to send, it switches its dynamic interface to the fixed channel of the receiver node's interface. Performance evaluations of CAEPO show improved network performance when all 11 channels of IEEE 802.11b are used.

4.1.3. Load-Aware Channel Assignment Exploiting Partially Overlapping Channels (Load-Aware CAEPO-G)

The authors of [13] present an extension to the previously discussed CAEPO [12] to make it traffic load-aware in addition to being interference-aware. A grouping algorithm is also proposed with the goal of achieving better aggregate network throughput. In the grouping algorithm, each node sends periodic hello messages; based on a node's weight (which is determined by how many hello messages it has received so far from its one-hop neighbors) the node may become a group leader. There can only be one group leader in the one-hop vicinity of any particular node. New nodes can join the group by sending a *join message* and similarly existing nodes can leave the group by sending a *quit message* to the group leader. Once the group leaders have been assigned (grouping is done), channels are assigned to links similarly to that in [12], with only one major difference: any update of the channel (i.e., channel switching) has to be initiated by the group leader. If a node "feels a need" to switch to a new, less contentious channel, it will send a "channel switch" request to its corresponding group leader who if agrees relays the information onwards to the other members in the group. Because of the addition of a new grouping algorithm and the load-aware feature, load-aware CAEPO-G achieves much better performance than the original CAEPO.

4.1.4. Minimum Interference for Channel Allocation (MICA)

In [10], the authors have introduced the concept of *node orthogonality:* two nodes, operating over adjacent and partially overlapping channels, are considered orthogonal if they are sufficiently physically apart. A novel interference model is proposed that captures the adjacent channel interference and also takes into account the physical distance of the two nodes configured on POCs. The proposed interference factor $I_c(i,j)$ is defined as follows:

$$I_c(i,j) = 1 - \frac{\min\{d_{i,j}, D_i(c_i,c_j)\}}{D_i(c_i,c_j)}$$

where $D_i(c_i,c_j)$ is the adjacent channel interference range between channels i and j, extracted from the physical model of the I-factor described in [3-6]. $D_i(c_i,c_j)$ captures both the channel separation and physical distance among the nodes to model the interference due to POCs. The proposed interference factor $I_c(i,j)$ can be used to define *node orthogonality* by stating that two nodes are orthogonal if and only if their interference factor value is equal to 0.

Given a particular channel assignment, a weighted interference graph can be constructed with weights on the edges measured by the interference factor $I_c(i,j)$; Figure 11 shows an example. Here, it is assumed that the data rate and the transmit power for all the APs are the same.

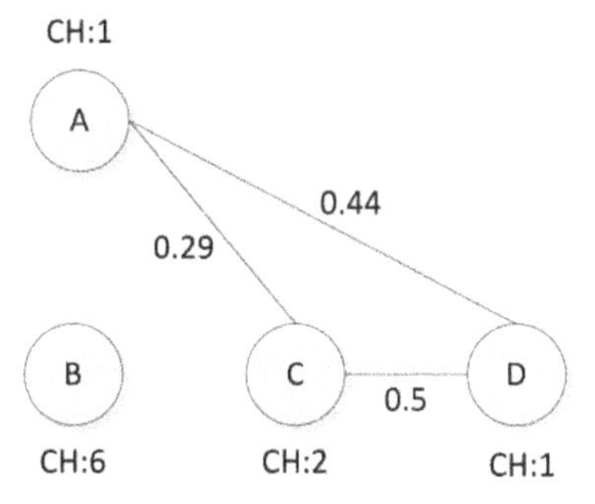

Figure 11. Construction of a weighted interference graph

Using the weighted interference graph model, a minimum weighted interference optimization problem is formulated with the objective of minimizing the sum of weights in the interference graph. A centralized heuristic is proposed called minimum interference for channel allocation (MICA) to obtain a near-optimal solution which relaxes the formulated minimum interference problem in order to find fractional interference in polynomial time and eventually to assign POCs to APs (after rounding off the fractional solution to the nearest integer).

In addition to the above approaches, there have been other research efforts in designing MAC protocols that exploit POCs in wireless networks. One such scheme is presented in [11] in which some of the challenges that may be faced when using overlapping channels in the design of a MAC protocol are discussed. Analytical models are designed to capture partial interference at the MAC layer in order to improve channel utilization and to enhance network capacity. Based on the model, an efficient medium access scheme with collision avoidance mechanism is developed which increases network throughput (exploiting multiple channel transmissions).

The authors of [14] study the use of POCs for data aggregation in sensor networks. In a typical sensor network, the job of each sensor node is to collect the data, aggregate it and send it back to the sink for further processing. Arguably, reducing latency of data aggregation is therefore one of the fundamental issues in sensor networks. This is also called the minimum latency scheduling (MLS) problem in which a conflict free transmission schedule is designed with the objective of minimizing the overall data transmission latency. The concept of POCs is used in order to reduce the data aggregation latency; a joint tree construction, channel assignment and scheduling algorithm is proposed to solve the MLS problem. The basic idea is to compute a partially overlapping channel assignment algorithm for the sensor network, and then construct a data aggregation tree for the whole network followed by finally designing a link schedule so that the data aggregation latency is minimized.

Table II provides a side-by-side comparison for the above four POCA schemes based on their objectives, the procedures that are used in obtaining a partially overlapping channel assignment algorithm and their limitations.

4.2. Interference matrix model (I-Matrix)

The second type of interference model we consider for POCA schemes was originally presented in [19]. The model is called I-Matrix, and is designed to measure the adjacent channel interference (ACI) among different POCs on adjacent nodes as well as self-interference (SI) among different radios on a single node. I-Matrix captures the interference that a channel belonging to a particular radio experiences due to all other possible channels (10 channels in the case of 802.11b). The proposed interference model (I-Matrix) is made up of three components, namely the interference factor, the interference vector, and the I-Matrix itself. The interference factor is derived from the I-factor of [5] and is the ratio of the interference range and the physical distance between two radios configured on adjacent channels ($f_{i,j} = IR(\delta)/d$). In other words, the interference factor captures both the physical distance and the channel separation between nodes. This means that even if the respective channels of two nodes are overlapping, but their physical distance is greater than the interference range (demonstrated by $IR(\delta)$ and taken from [8, 24]), the value of $f_{i,j}$ will be zero. The interference factor is computed for all the channels with respect to a particular channel and put in a vector called the interference vector as shown in Table III. Similarly, each node combines all the interference vectors it has calculated for each channel and constructs the I-Matrix as outlined in Table IV.

Technique	Objective	Methodology	Limitations
A. Mishra [5]	Maximization of the total throughput (maximizes simultaneous link activations)	Routing, channel assignment, and link flow scheduling; performed stepwise until optimal CA and routing solution is found	Complexity; ignoring switching overhead; SI not considered
MICA [10]	Minimization of the sum of the weighted interference in an interference graph	Approximate algorithm for channel allocation using integer linear programming (ILP) formulation.	Offline solution; only designed for single radio networks; SI not considered
CAEPO [12]	Minimization of the network interference	Heuristic distributed load-aware algorithm. Channel assignment based on traffic-aware interference estimation and packet loss ratio metrics	Simplistic interference model; SI not considered; scalability issues
Load-Aware CAEPO-G [13]	Minimization of the network interference	Extension of [12] with the addition of self-interference factor and a grouping algorithm to make CA scalable.	Simplistic interference model

Table 2. Comparison of POCA schemes based on the I-factor model.

CH	d_i	Interference Factor experienced at channels										
		1	2	3	4	5	6	7	8	9	10	11
6	d_6	0	$f_{6,2}$	$f_{6,3}$	$f_{6,4}$	$f_{6,5}$	∞	$f_{6,7}$	$f_{6,8}$	$f_{6,9}$	$f_{6,10}$	0

Table 3. Interference vector for channel 6.

CH	d_i	Interference Factor experienced at channels										
		1	2	3	4	5	6	7	8	9	10	11
1	d_1	∞	$f_{1,2}$	$f_{1,3}$	$f_{1,4}$	$f_{1,5}$	0	0	0	0	0	0
2	d_2	$f_{2,1}$	∞	$f_{2,3}$	$f_{2,4}$	$f_{2,5}$	$f_{2,6}$	0	0	0	0	0
⋮	⋮	⋮	⋮	⋮	⋮	⋮	⋮	⋮	⋮	⋮	⋮	⋮
11	d_{11}	0	0	0	0	0	0	$f_{11,7}$	$f_{11,8}$	$f_{11,9}$	$f_{11,10}$	∞

Table 4. I-Matrix

[19] also proposes a heuristic channel assignment algorithm exploiting POCs based on the I-Matrix model. The algorithm assigns channels to the maximum number of links with the objective of minimizing network interference. The algorithm starts with an input describing the number of links that need to have channel assignments. The links are then assigned to their respective nodes and those nodes are sorted in descending order of their degrees. For each node, its incident link is assigned a channel which has the minimum interference calculated from the I-Matrix; accordingly after the channel assignment, the interference vectors of the corresponding channel are updated. This in turn forces the node to update the I-Matrix with the new channel's interference measurements against all other channels. [19] shows that using POCs can improve network capacity by as much as 15% compared to when only non-overlapping channels are used.

4.2.1. Channel assignment based on I-Matrix model

In [20], the authors have extended the work of [19] by trying to remove some of the limitations in the proposed I-Matrix interference model and the channel assignment algorithm. More precisely, the CA algorithm in [19] sorts the links in descending order based on nodal degrees; however, this is not practical in multi hop WMNs as most of the traffic is targeted to gateway nodes. Therefore, the descending order should be based on the traffic load, implying that the busiest link should be assigned the channel first, i.e., gateway links should be first (thus being in accordance with typical WMN traffic characteristics). Another, shortcoming of [19] pointed out is that it suffers from the network partitioning problem, in the sense that some of the links may remain unassigned because the CA algorithm only assigns POCs and never assigns the same channel (as it tries to completely avoid the co-channel interference). To overcome this limitation, the I-Matrix model is modified to consider co-channel interference by adding a co-channel column to the matrix. This ensures network connectivity (because now the links can be assigned the same channels).

The algorithm of [20] consists of two phases. In the first phase, instead of the number of links as the input, links with traffic load information are provided as input and they are sorted in descending order of the traffic they carry. Then a suitable channel with the minimum interference is extracted from the I-Matrix. The second phase guarantees network connectivity in which the algorithm looks for those nodes that do not have a path to the gateway and if such nodes are found, their radios can be configured to the same channel on which one of their neighbor node's radio is already configured on. This ensures full network connectivity at the cost of co-channel interference. They have shown through experiments that the existence of such co-channel interference does not strongly influence the network performance (as such formerly disconnected nodes are likely to be at the peripheral of the network).

Table V summarizes the I-Matrix POCA schemes. It states the objective of each algorithm, the procedures used in obtaining a partially overlapping channel assignment algorithm, and the limitations of each scheme.

Technique	Objective	Methodology	Limitations
M. Hoque [19]	Maximization of network capacity	Greedy heuristic channel assignment algorithm based on I-Matrix interference model, links are visited in descending order of the node degrees	Simplistic interference model, network can be disconnected, CCI is not considered, topology is not preserved
P. Duarte [20]	Minimization of network interference	Extended [19] to incorporate traffic load into I-Matrix for channel assignment, ensures network connectivity, links are visited in descending order of the traffic load	Simplistic interference model

Table 5. Summary of the two I-Matrix based approaches.

4.3. Channel Overlapping Matrix Model (CO-Matrix)

In order to model orthogonal and non-orthogonal channels, a novel interference model called *channel overlapping matrix* is proposed in [18]. Consider a MRMC-WMN consisting of N routers, each equipped with I radios and C available frequency channels. For any two routers $a,b \in N$, a channel assignment vector x_{ab} of size $C\ x\ 1$ can be defined which defines the channel on which the two routers are communicating (that particular element in the matrix becomes 1). Similarly, a vector of size $I\ x1$ defines an interface assignment vector y_{ab}, which tells which radio belonging to a particular router a is used to communicate with router b (by changing the value of that element in the vector to 1). To model the partial overlap among channels, a $C\ x\ C$ channel overlapping matrix W was proposed whose m[th] row, r[th] column entry can be calculated as:

$$W_{mn} = \frac{\int_{-\infty}^{+\infty} \mathrm{F_m(w)F_n(w)dw}}{\int_{-\infty}^{+\infty} \mathrm{F_m^2(w)dw}}$$

where $F_m(w)$ denotes the power spectral density (PSD) function of the band-pass filter for channel m and consequently the same for channel n. Based on this channel overlap matrix, the authors have formulated a linear mixed-integer program consisting of few integer variables in order to solve a joint channel assignment, interface assignment and scheduling problem when the whole spectrum of the IEEE 802.11 frequencies is to be used.

4.3.1. Channel Assignment based on Channel Overlapping Matrix Model

Another channel assignment algorithm based on the channel overlapping matrix was proposed in [15]. Here, a joint channel assignment and flow allocation problem in MRMC WMNs is considered. [15] formulates this joint problem into a mixed integer linear program with the objectives of maximizing aggregate end-to-end throughput while minimizing queuing delays in the network (given that the traffic characteristics are known). In order to model the partially overlapping channels, the I-factor of [5] is used, capturing the overlap between two different nodes configured on two different channels. Based on the I-factor a $C \times C$ symmetric channel overlapping matrix O is proposed:

$$o_{ij} = \begin{cases} 1 & i = j \\ \frac{I(5|i-j|)}{I(0)} & i \neq j \end{cases}$$

where o_{ij} represents an entry in the ith row and j^{th} column of the matrix O. To model the impact of interference, a physical model is employed [25].

Table VI provide a side-by-side comparison of the two algorithms surveyed above based on their objectives, the methodology used to assign POCs, and their limitations.

Technique	Objective	Methodology	Limitations
A. Rad [18]	Minimization of the maximum link utilization	Joint CA, interface assignment and flow scheduling algorithm based on channel overlapping matrix to model POCs / linear mixed-integer program formulation	SI is not considered, extensive computational complexity
V. Bukkapatanam [15]	Maximization of the aggregate end-to-end flow allocations	Joint CA and flow allocation algorithm based on CO matrix / mixed integer linear program formulation	extensive computational complexity, offline solution; no bounds on completion

Table 6. Comparison of the two CO-Matrix based POCA schemes.

4.4. Conflict Graph based Model (CGM)

4.4.1. Channel Assignment with Partially Overlapped Channels

A weighted conflict graph model is proposed in [7] to more accurately model interference among nodes operating on overlapping channels. In order to measure the partial interference, a metric called interference factor (IF) is defined:

$$IF = \frac{t_1' + t_2'}{t_1 + t_2}$$

where t_1 and t_2 are the throughputs of two links (link-1 and link-2) each belonging to a pair of nodes which are placed at various locations to measure interference when the other link is idle. Similarly, t'_1 and t'_2 are the corresponding link throughputs when both links are active. As it can be seen from the formula, a higher *IF* value indicates lower interference. Experimental studies of [7] measured link interference (*IF*) and found out that for a particular channel separation, the interference between two links degrades quickly (higher IF factor) even with a slight increase in distance. From this *IF* metric, the interference range of two links separated by a fixed number of channels can be extracted. Multiple interference ranges are calculated for all five possible channel separations under different IEEE 802.11b bitrates (i.e., 2Mbps, 5.5Mbps, and 11Mbps).

The concept of interference range is then applied to formulate the channel assignment problem into a weighted conflict graph model where the edges in the conflict graph are labeled by the minimum channel separation that two interfering links must have in order to have a conflict free communication. This weighted graph serves as an input to select the edges having minimum weights, eventually minimizing the overall network interference. A greedy partially overlapping channel assignment algorithm is proposed to solve the weighted conflict graph problem. The algorithm consists of two parts, namely *select* and *assign*. During *select*, the link with the minimum expected interference among all available links is selected. In the *assign* phase, a channel is assigned to this link with the minimum interference to all previously assigned channels. These steps are repeated until all links are covered, i.e., all links are assigned channels. In addition, the authors in [7] have also designed a novel genetic algorithm for channel assignment which produces slightly better results compared to the greedy algorithm for solving the assignment using the conflict graph. In order to map the partially overlapping channel assignment algorithm, a channel assigned to a single link is considered as a DNA sequence and the channel assignments of the all the links are mapped to an individual. In a typical genetic algorithm, a generation consists of a set of individuals; therefore, in this case, it will be a series of channel assignment solutions. An example of this mapping of the channel assignment problem to a genetic algorithm is shown in Figure 12 [7].

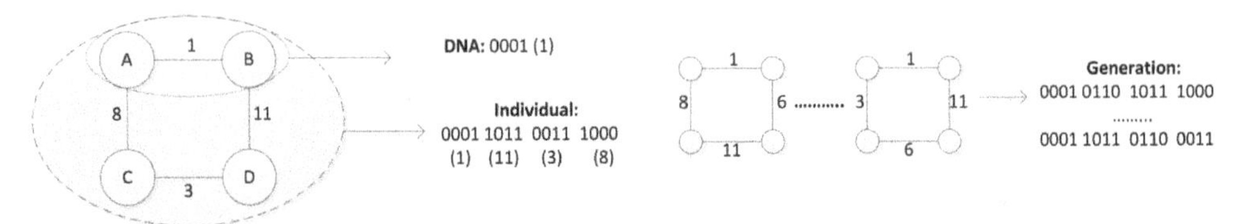

Figure 12. Example of POCA using a genetic algorithm [7]

The procedure for encoding the channel assignment scheme into an individual in a genetic algorithm requires first to sort the links, convert them to fixed length binary strings (a DNA sequence), and then to concatenate the binary strings together to form a single individual. The fitness function is defined as the inverse of the total interference in the network. The algorithm starts with randomly generating N channel assignment schemes (individuals).

The selection strategy selects two individuals (from the *N* sized population) by using the roulette wheel selection method and then choosing the better one of them according to the tournament selection strategy. These two strategies are commonly referred to as the stochastic selection strategy. After the selection stage, a reproduction step is performed in which one-point crossover and two-point crossover and mutation is applied to the selected two individuals. Both the greedy and genetic channel assignment algorithms are evaluated on various sets of topologies. The greedy algorithm is faster but the genetic algorithm provides better results and thus can generate better channel assignment schemes which eventually result in improved network capacity.

4.4.2. Partially Overlapped Channel Assignment (POCAM)

In [16], a new partially overlapped channel assignment for multi-radio multi-channel wireless mesh networks called POCAM is proposed, where the interference model stems from measurements of commercial radios using real testbeds. An extensive set of testbed experiments were performed to analyze the effect of partial interference and self-interference in WMNs. Through these tests it is shown that the self-interference issue is worse than it is usually assumed as it still needs to be considered even if the two radios on the same node are configured on non-overlapping channels. The proposed POCAM algorithm consists of two steps and incorporates the traffic load distribution. First, a transformation of the partially overlapped channel assignment problem into a weighted conflict graph (WCG) is performed followed by calculation on that weighted conflict graph. The WCG is a graph $G = (V,E)$ where V represents the number of nodes in a WMN. For each edge in E, edge weights are assigned based on a table in [7] capturing interference ranges against each channel separation. The WCG is constructed with links represented as vertices in the conflict graph and there is a weighted edge between two vertices in the conflict graph if those two links interfere. The WCG formulation becomes a constraint satisfaction problem (CSP) which is an NP hard problem. CSPs are usually solved by applying backtracking search algorithms [27], thus [7] shows a design of three heuristics specially tailored for WMN characteristics.

4.4.3. Minimum Interference Channel Assignment

The authors in [17] propose a centralized channel assignment algorithm based on the tabu-search heuristic [26] which is used to find quasi-optimal solution for a graph coloring problem. The objective of the channel assignment algorithm is to minimize the overall network interference by assigning channels to links in a WMN. Network interference is captured as a graph coloring problem by assigning colors (channels) to the vertices of a conflict graph using K colors while maintaining interface constraints. The interface constraints limit the number of different channels assigned to interfaces belonging to a single node by the number of interfaces on that node. The proposed tabu-search based channel assignment algorithm consists of two phases. In the first phase, the algorithm starts with a random solution by assigning random colors to each vertex in the conflict graph, followed by a series of solutions which are created with the objective of minimizing overall network interference by assigning colors to vertices such that the conflicts is minimized. In each iteration, a tabu list of the colors (channels) that have already been assigned is maintained to avoid their assignment a second time and to achieve fast convergence. This phase terminates after a certain number of iterations (solutions). In the second phase, the interface constraints are satisfied by a merge operation in which, those nodes who have been assigned more distinct colors (channels) to links than how many radios they have, have their colors merged to bring them to be equal to their number of radios. To ensure network connectivity by this merge operation, the just changed color is propagated to all the other links that were assigned the old color to repeat the merge operation on them (those links must be part of the common node).

A distributed greedy heuristic channel assignment algorithm based on Max K-cut is also proposed by the authors [17]. Given a conflict graph, the max K-cut problem deals with dividing vertices into K partitions to maximize the number of edges that lie in different partitions. Two formulations of their proposed channel assignment problem are provided, one is a semi-definite programming (SDP) formulation and the other is a linear programming formulation in order to obtain tighter lower bounds on optimal network interference. The linear programming formulation is modified to capture partial interference that exist when overlapping channels are being used and in order to make the formulation compatible to POCs. The SDP formulation however turns out to be too complex and therefore, it is not been evaluated.

Mishra et al., in [4] formulate the channel assignment problem as a weighted variant of the graph coloring problem incorporating realistic channel interference based on the I-factor model. The channel assignment problem is formulated as a weighted graph coloring problem with APs representing vertices in the graph and potential interference among them is represented by an edge between the vertices in the weighted graph. The weight on each edge depicts the significance of using different colors for the vertices that are connected by that edge. The weights are defined as the number of clients attached to an AP, scaled by the degree of interference between the chosen channels (I-factor). Therefore, the goal of the weighted graph coloring solution is to minimize the objective function. A higher weight translates to higher amounts of partial overlap between the channels; the algorithm attempts to assign different channels or channels with higher spatial difference to the edges in the graph. An edge weight of zero means that there is no interference among the clients of the corresponding APs. It is proved that the proposed weighted graph coloring problem is NP-hard, therefore, two distributed channel assignment techniques are proposed with the objective of minimizing the overall network interference. The first technique tries to minimize each individual AP's interference and does not require any inter-AP communication. It consists of two steps; i.e. an initialization and an optimization step. The initialization step starts with assigning the same channel to all the APs. In the optimization step (which is incremental in nature), each AP performs the greedy optimization trying to minimize its local maximum interference by taking the maximum weight edge (which eventually minimizes the objective function). The algorithm stops when the network achieves an acceptable "coloring" configuration. The second channel assignment algorithm requires collaboration among APs and is intended to minimize interference by reducing the number of clients that are experiencing interference. Simulations and testbed experiments show that the proposed channel assignment algorithms achieve 45.5% reduction in interference when the network is sparse. The algorithms are scalable and provide better performance than existing channel assignment algorithms.

A heuristic-based channel assignment and link scheduling algorithm is proposed in [9] to enhance network capacity by exploiting partially overlapping channels in WMNs. Since, finding optimal channel assignment and link scheduling together for a given network is NP-hard, heuristic based policies are summoned to provide a sub-optimal solution. The problem is divided into two parts; first channel assignment is performed and then based on that an optimal link scheduling is explored. For the channel allocation, a genetic algorithm [28] is used. The authors have also studied some of the factors that influence the performance of POCs in channel assignment in a wireless mesh network (such as node density and node distribution).

All of the above three POCA schemes make use of graph-theory to model partial overlap among nodes in MRMC-WMNs except [7] which is designed for single radio WMNs. The approaches then apply a heuristic for channel assignment. Table VII provides a side-by-side comparison of the three POCA schemes based on their objectives, methodology, limitations.

Technique	Objective	Methodology	Limitations
POCAM [16]	Minimization of network interference	Weighted conflict graph, constraint satisfaction problem, heuristic based backtracking search algorithm	Simplistic interference model, no SI is considered
Y. Ding [7]	Minimization of network interference	Weighted conflict graph, graph coloring, greedy CA algorithm, genetic algorithm based on partially overlapped channel assignment	SI is not considered, extensive computational complexity, edge weight assignment is difficult, does not consider traffic load
A. Subramanian [17]	Minimization of network interference	Conflict graph, Max K-cut, SDP and ILP formulation, tabu based CA and heuristic based greedy CA algorithm	Extensive computational complexity, SI is not considered, ignores switching overhead

Table 7. Comparison on the objective, methodology and limitations of POCA schemes based on CGM.

4.5. Summary of all POCA approaches

In this section, we provided a survey of existing POCA schemes in WMNs and summarized them based on their objectives, methodologies and limitations. Table VIII presents an overall summary of all the POCA approaches examined; the table shows the comparison of these schemes based on the following six questions:

- *Implementation:* Is the proposed POCA centralized or distributed?
- *Multi-radio support:* Is the POCA scheme designed for multi-radio WMNs?
- *Interference:* What type of interference does the proposed POCA capture?
- *Routing dependency:* Is the POCA dependent on a particular routing algorithm?
- *Channel switching frequency:* How frequently are the channels switched?
- *Connectivity:* Does the algorithm ensure network connectivity?

5. Open issues in POCA design

In spite of a reasonable amount of research in the late literature, there are still some challenges and open issues that need to be addressed in designing efficient channel assignment schemes exploiting POCs, particularly in WMNs. Below, we outline what we believe some of these challenges and open issues are.

5.1. Capturing self interference

As explained in Section 2.2, self-interference restricts parallel communication originating from a node having more than one radio unless these radios operate on completely orthogonal channels (OCs). Since, there are only three OCs for IEEE 802.11b/g in the 2.4GHz band, there is a need to further investigate how the self-interference issue can be better addressed. Few CA schemes have addressed self-interference in multi radio MWNs and we believe that there is room for improvement.

5.2. Modeling interference of POCs

More robust and efficient modeling schemes are required to intelligently capture the interference experienced by neighboring nodes operating on POCs in MRMC-WMNs. Although existing approaches do partially capture one or two types of interferences in a WMN, they are not complete solutions (they do not capture all the different types of interferences realistically). Furthermore issues arising from geographical positions of neighboring nodes and the availability of variable data rates still pose major challenges for POCA algorithms.

5.3. Lack of simulation tools

Most existing simulators [29-31] still do not support underlying physical models and easy POC evaluation scripting to capture partial interference between adjacent nodes in WMNs. However, we believe the reason for the lack of this feature is because the concept of POCs in CA schemes is relatively new and is still progressing and evolving to its maturity.

5.4. Multi-rate capability

To the best of our knowledge, there is no partially overlapping channel assignment algorithm that has been proposed to explore the multi-rate capability of IEEE 802.11 based hardware in MRMC-WMNs. Almost all the aforementioned works have assumed a fixed transmission rate (homogeneous links) which make the problem of channel assignment simple, whereas a POCA scheme with adaptive rates could potentially achieve significantly better performance.

Characteristics	Implementation	Multi-radio Support	Interference	Routing Dependency	Channel Switching Frequency	Focus on Connectivity
A. Mishra [5]	Centralized	No	ACI	No	Dynamic	No
MICA [10]	Centralized	No	ACI	No	Fixed	Yes
CAEPO [12]	Distributed	No	ACI	Yes	Hybrid	Yes
Load-Aware CAEPO-G [13]	Distributed	Yes	ACI	Yes	Hybrid	Yes
M. Hoque [19]	Centralized	Yes	ACI and SI	No	Dynamic	No
P. Duarte [20]	Centralized	Yes	ACI, SI and CCI	No	Dynamic	Yes
A. Rad [18]	Centralized	Yes	ACI and CCI	Joint	Fixed	Yes
V. Bukkapatanam [15]	Centralized	Yes	ACI	Joint	Fixed	Yes
POCAM [16]	Centralized	Yes	ACI and SI	No	Hybrid	Yes
Y. Ding [7]	Centralized	No	ACI	No	Dynamic	No
A. Subramanian [17]	Both	Yes	ACI	No	Dynamic	Yes

Table 8. Summary of characteristics of all POCA approaches in wireless mesh networks.

6. Conclusions

In this chapter, we have discussed the problem of assigning channels with partial overlaps to radios in single- and multi-radio WMNs. We have characterized different types of interferences that may exist in a WMN depending on the flow characteristics and on the particular configuration of interfaces to channel assignments. We then presented IEEE 802.11 standard constraints on communications and evaluated the benefits of using partially overlapped channels (POCs) for the design of efficient channel assignment schemes with the help of experiments performed on a real testbed. Our, and previous experiments demonstrated that the use of POCs: i) improves network capacity by enabling more parallel communications and ii) provides more efficient utilization of the available spectrum. We have also provided a survey of some of the existing POC assignment schemes in WMNs and have classified them based on the interference models that they employ. Finally, we discussed some of the challenges and open issues in designing efficient channel assignment schemes utilizing both orthogonal and non-orthogonal channels in WMNs.

Author details

Fawaz Bokhari and Gergely Záruba

Department of Computer Science and Engineering, The University of Texas at Arlington, Texas, USA

7. References

[1] Ian F. Akyildiz, Xudong Wang, and Weilin Wang, "Wireless mesh networks: a survey," Computer Networks and ISDN Systems, v.47 n.4, pp. 445-487, March 2005.
[2] I. Katzela and M. Naghshineh, "Channel assignment schemes for cellular mobile telecommunication systems," IEEE Personal Communications, vol. 3, pp. 10–31, 1996.
[3] A. Mishra, E. Rozner, S. Banerjee, and W. Arbaugh, "Exploiting partially overlapping channels in wireless networks: Turning a peril into an advantage," *ACM SIGCOMM*, pp. 29–29, 2005.
[4] A. Mishra, S. Banerjee, and W. Arbaugh, "Weighted coloring based channel assignment for wlans," *ACM SIGMOBILE Mobile Computing and Communications Review*, no. 3, pp. 19–31, July 2005.
[5] A. Mishra, V. Shrivastava, S. Banerjee, and W. Arbaugh, "Partially overlapped channels not considered harmful," *ACM SIGMetrics/Performance*, pp. 63–74, 2006.
[6] A. Mishra, E. Rozner, S. Banerjee, and W. Arbaugh "Using partially overlapped channels in wireless meshes," *Wimesh*, Santa Clara, September 2005.
[7] Y. Ding, Y. Huang, G. Zeng, and L. Xiao, "Channel assignment with partially overlapping channels in wireless mesh networks," *WICON*, 2008.
[8] Z. Feng and Y. Yang, "How much improvement can we get from partially overlapped channels?," *IEEE WCNC*, pp. 2957–2962, 2008.
[9] H. Liu, H. Yu, X. Liu, C.-N. Chuah, and P. Mohapatra, "Scheduling multiple partially overlapped channels in wireless mesh networks," *IEEE ICC*, pp. 3817–3822, 2007.
[10] Yong Cui, Wei Li, and Xiuzhen Cheng, "Partially overlapping channel assignment based on "node orthogonality" for 802.11 wireless networks," *IEEE INFOCOM 2011*, Shanghai, April 2011.
[11] K. Shih, C. Chang, D. Deng, and H. Chen, "Improving channel utilization by exploiting partially overlapping channels in wireless ad hoc networks," *IEEE GLOBECOM*, Miami, December 2010.
[12] Y. Liu, R. Venkatesan, and C. Li, "Channel assignment exploiting partially overlapping channels for wireless mesh networks," *IEEE GLOBECOM*, Honolulu, USA, November 2009.
[13] Yuting Liu, R. Venkatesan, and Cheng Li. "Load-Aware channel assignment exploiting partially overlapping channels for wireless mesh networks," *IEEE GLOBECOM*, Miami, December 2010.
[14] B. Wang, X. Jia, "Reduing data aggregation by using partially overlapped channels in sensor networks," *IEEE GLOBECOM*, Honolulu, USA, November 2009.
[15] V. Bukkapatanam, A. Franklin, and C. Murthy, "Using partially overlapped channels for end-to-end flow allocation and channel assignment in wireless mesh networks," *IEEE ICC*, Dresden, Germany, June 2009.
[16] D. Wang, P. Lv, Y. Chen, and M. Xu, "POCAM: Partially overlapped channel assignment in multi-radio multi-channel wireless mesh networks," *IEEE ISCIT '11.*, Hangzhou, China, October 2011.
[17] A.P. Subramanian, H. Gupta, and S.R. Das, "Minimum interference channel assignment in multi-radio wireless mesh networks," IEEE Transactions on Mobile Computing, vol. 7, no. 11, 2008.

[18] A. Mohsenian Rad, V.W.S. Wong, "Partially overlapped channel assignment for Multi-channel wireless mesh network," *IEEE ICC*, pp. 3770-2775, 2007.
[19] M. Hoque, X. Hong, and F. Afroz, "Multiple radio channel assignment utilizing partially overlapped channels," *IEEE GLOBECOM*, Honolulu, USA, November 2009.
[20] P.B.F. Duarte, Z.M. Fadlullah, K. Hashimoto, N. Kato, "Partially overlapped channel assignment on wireless mesh network backbone," *IEEE GLOBECOM*, Miami, December 2010.
[21] M. Alicherry, R. Bhatia, and L. Li, "Joint channel assignment and routing for throughput optimization in multi-radio wireless mesh networks," *ACM MobiCom*, 2005.
[22] A. Mishra, V. Brik, S. Banerjee, A. Srinivasan, and W. Arbaugh, "A client-driven approach for channel management in wireless lans," *IEEE INFOCOM*, 2006.
[23] M. Burton, "Channel overlap calculations for 802.11b networks," White Paper, Cirond Technologies Inc., November 2002.
[24] Zhenhua Feng and Yaling Yang, "Characterizing the impact of partially overlapped channel on the performance of wireless networks," *IEEE GLOBECOM*, 2008.
[25] P. Gupta and P. Kumar, "The capacity of wireless networks," *IEEE Transactions on Information Theory*, vol. 46, no. 2, pp. 388-404, 2000.
[26] A. Hertz, and D. de Werra, "Using tabu search techniques for graph coloring," *Computing*, vol. 39, no. 4, 1987
[27] S.J. Russell and P. Norvig, "Artificial intelligence: a modern approach," *Prentice hall*, 2010.
[28] Freifunk firmware, http://ff-firmware.sourceforge.net/
[29] Fall and Varadhan, "NS notes and documentation," in The VINT Project, UC berkely, LBL, USC/ISI, and Xerox PARC, 1997.
[30] X. Zeng, R. Bagrodia, and M. Gerla, "GloMoSim: a library for parallel simulation of large-scale wireless networks," *Proceedings of the 1998 Workshop on Parallel and Distributed Simulation*, pp. 154-161, 1998.
[31] X. Chang, "Network simulations with OPNET," *Proceedings of the 1999 Winter Simulation Conference*. 1999.
[32] IEEE Std 802.11-2007, "IEEE standard for information technology — Telecommunications and information exchange between systems — Local and metropolitan area networks-specific requirements — Part 11: Wireless LAN medium access control (MAC) and physical layer (PHY) specifications," LAN/MAN Standards Committee, New York, NY, USA, pp. C1–1184, June 2007. [Online]. Available: http://dx.doi.org/10.1109/IEEESTD.2007.373646
[33] K.N. Ramachandran, E.M. Belding, K.C. Almeroth, and M.M. Buddhikot. "Interference-aware channel assignment in multi-radio wireless mesh networks," *IEEE INFOCOM*, 2006.
[34] C.L. Barrett, G. Istrate, V.S.A. Kumar, M.V. Marathe, S. Thite, S. Thulasidasan, "Strong edge coloring for channel assignment in wireless radio networks," *Proceedings of IEEE PerCom Workshops*, 2006.
[35] A.K. Das, R. Vijayakumar, S. Roy,"WLC30-4: static channel assignment in multi-radio multi-channel 802.11 wireless mesh networks: issues, metrics and algorithms," *IEEE GLOBECOM* , November 2006.

Chapter 5

Channel Assignment Schemes Optimization for Multi-Interface Wireless Mesh Networks Based on Link Load

Stefan Pollak and Vladimir Wieser

Additional information is available at the end of the chapter

1. Introduction

In recent years, wireless mesh networks (WMNs) were deployed as a type of next generation wireless broadband networks. WMNs provide wireless broadband accessibility to extend the Internet connectivity to the last mile and improve the network coverage. WMN consists of a set of mesh routers and mesh clients (Fig. 1). *Mesh routers* are usually stationary and form multi-hop wireless backbone network (i.e. mesh routers are interconnected with each other via wireless medium). Some or all of the mesh routers also serve as access points for mobile users (*mesh clients*) under their coverage. Usually one or more mesh routers have direct connections to wired network and serve as Internet gateways for the rest of the network. These nodes are called *mesh gateways*. Compared to traditional wireless LANs, the main feature of WMNs is their multi-hop wireless backbone capability (Conti et al., 2007).

Traditionally, wireless networks are equipped with only one IEEE 802.11 radio interface. However, a single-interface inherently restricts the whole network by using only one single channel (Fig. 3a). In order to communicate successfully, two neighboring routers have to build a logical link which operates on a common channel. Due to that, all wireless nodes have to use only one radio interface, all logical links in network must use the same channel. If two neighboring links operate on the same channel and transfer data simultaneously, then they definitely interfere with each other. The network capacity and the performance may degrade significantly because of the interference (Gupta & Kumar, 2000). The key factor for reducing the effect of interference is the using of non-overlapping channels (standard IEEE 802.11b/g provides 3 and standard IEEE 802.11a up to 12 non-overlapping channels) (Ramachandran et al., 2006). In practice, IEEE 802.11b/g defines 11 communication channels (number of communication channels varies due to regulations of different countries) but only 3 of them are non-overlapping (Fig.2).

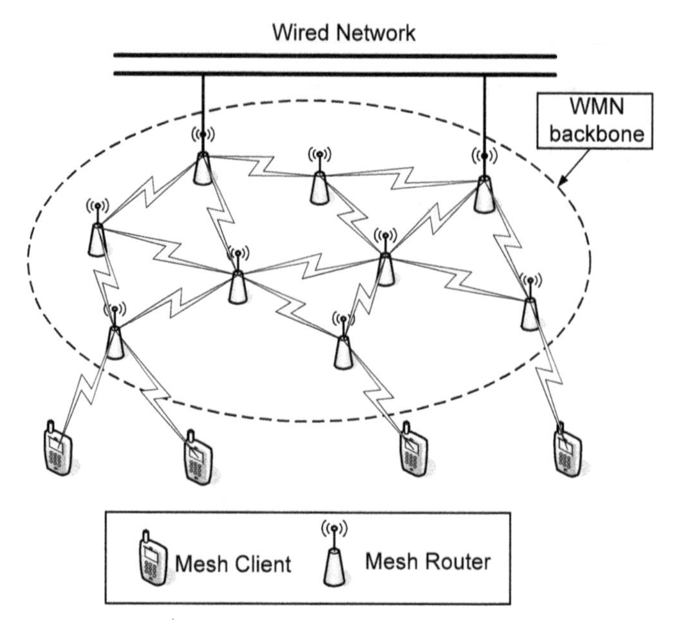

Figure 1. WMN architecture

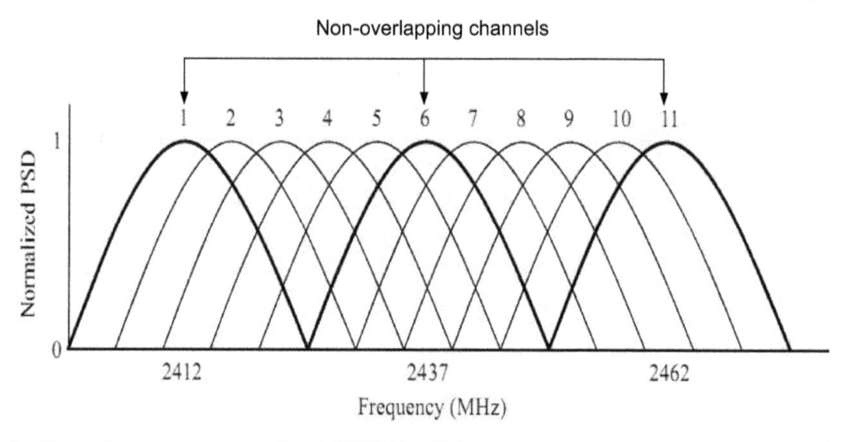

Figure 2. Channel spectrum occupation in IEEE 802.11b/g

Using multiple non-overlapping channels in single interface network disconnects the subset of nodes using one channel from other nodes that are not using the same channel (Fig. 3b). For this reason this approach generally requires MAC layer modification and per packet channel switching capability for radio interfaces (Marina & Das, 2005). Before every data transmission a channel selection mechanism evaluates the available channels and selects a channel to transmit. There are also some problems introduced with channel switching mechanism. These problems include multi-channel hidden terminal problem, broadcast problem, deafness problem and channel deadlock problem (Raniwala et al., 2004).

One of the most promising approaches lies in using multiple radio interfaces and multiple non-overlapping channels (Fig. 3c). This solution is better than previous one, because of providing the effective usage of given frequency spectrum (Conti et al., 2007). This architecture overcomes deficiencies of single interface solution. It allows using of multiple interfaces per node to allow the simultaneous transmission and reception on different radio interfaces tuned to different channels, which can essentially improve network capacity. However, the number of radio interfaces is always much higher than the number of effective channels, which causes an existence of many different links between mesh routers operating on the same channel. For this reason, the suitable channel assignment method is needed to maintain the connectivity between mesh nodes and to minimize the effect of interference (Raniwala et al., 2004).

The channel assignment (CA) in a multi-interface WMN consists of a task to assign channels to the radio interfaces by such a way to achieve efficient channel utilization and to minimize the interference. The problem of optimally assigning channels in an arbitrary mesh topology has been proved to be NP-hard (non-deterministic polynomial-time hard) based on its mapping to a graph-coloring problem. Therefore, channel assignment schemes predominantly employ heuristic techniques to assign channels to radio interfaces belonging to WMN nodes.

The channel assignment algorithms can be divided into three main categories: fixed, dynamic and hybrid, depending on the frequency with which it is modified by the channel assignment scheme. In a fixed scheme, the CA is almost constant, while in a dynamic one it is continuously updated to improve performance. A hybrid scheme applies a fixed scheme for some radio interfaces and dynamic one for the others (Yulong Chen et al., 2010).

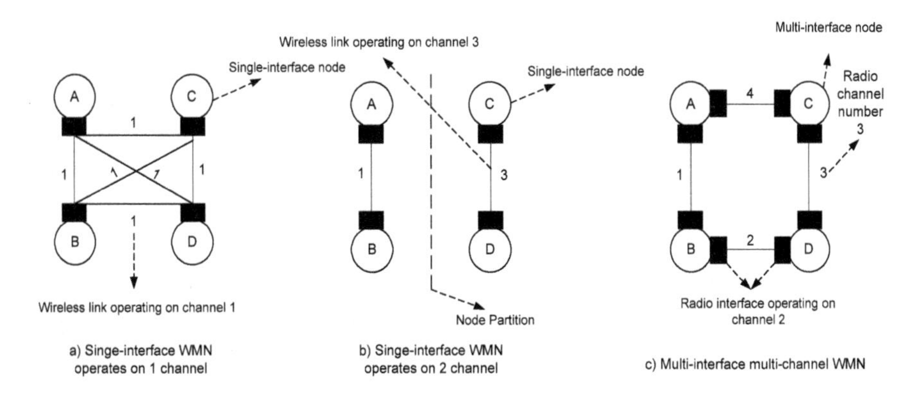

Figure 3. Different types of WMNs

The main objective of this chapter is to give to reader the compact information about problems connected with optimal using of radio interfaces and radio channels in wireless mesh networks. The optimal using is computed from several different points of view, e.g. network topology, number of data flows, number of nodes by comparison of selected QoS parameters. In the second part of the chapter, the new proposed centralized channel assignment concept called First Random Channel Assignment algorithm (FRCA) is compared with two other channel assignment techniques (CCA, LACA) by the same QoS parameters.

The rest of this chapter is organized as follows. In section 2, the related work is summarized. In section 3, the methods and simulation results to find the optimal number of radio interfaces per node are introduced. In section 4 the mathematical background and graph based mathematical model is described and in the next section different types of channel assignment methods based on links load are analyzed. Section 6 concludes the chapter.

2. Related work

There exist a large number of studies which address the channel assignment problem in wireless mesh networks. Several works have proposed MAC protocols for utilizing multiple channels (So & Vaidya, 2004, Gong & Midkiff, 2005), but these multi-channel protocols require changes to existing standards and therefore cannot be deployed by using existing hardware. In (Adya et al., 2004) was proposed a link-layer solution for transmitting data over multiple radio interfaces, but this approach is designed for scenario where the number of radio interfaces is equal to the number of channels. In (Gupta & Kumar, 2000) the performance of multi-channel ad-hoc networks was studied, where each channel was assigned to an interface. In (Draves et al., 2004) several methods for increasing the performance in single-channel per interface were proposed. The most studies is focused only to one problem - to find the efficient channel assignment method, but did not suggest the optimal number of radio interfaces per node. In (Husnain et al., 2004) were compared different static centralized algorithms, but for evaluation of optimal number of radio interfaces was used only one parameter - total interference (number of links in conflict graph). (Raniwala et al., 2004) proposed centralized channel assignment and routing method, where results about number of radio interfaces were shown but only for network cross-section goodput. In (Chi Moon Oh et al., 2008) the study of optimal number of radio interfaces was created but only for grid network, using simple channel assignment method and for one QoS parameter (throughput).

3. The study of optimal number of radio interfaces

In this section several simulations were created to find the optimal number of radio interfaces for static WMN. In this study we focus only to one problem - to find the optimal number of radio interfaces for different conditions therefore, for channel assignment we used simple CCA approach (section 5.1).

Nowadays the availability of the cheap off-the-shelf commodity hardware also makes multi-radio solutions economically attractive. This condition provides the using much more radio interfaces per node, which shows the investigating of optimal number of interfaces as a reasonable argument.

We have included in our simulations several QoS parameters, data flows, number of nodes and network topologies to find the optimum number of radio interfaces for services which required the real time transmission (e.g. video conference).

3.1. Simulation environment

A simulation WMN model was developed in NS-2 network simulator, with additional function to support multi-channel and multi-interface solution (Calvo & Campo, 2007). Each mesh node used the number of interfaces between 1 to 8 and the same number of channels. Two different network topologies were created. The first one was grid topology, which consisted of 25 static wireless mesh nodes placed in an area of 1000 x 1000 meters. Transmission range for each node was set to 200 meters (Fig.4a). The second topology consists of 25 nodes, which were randomly placed in an area of 1000 x 1000 meters (Fig.4b). For simulation evaluations, ten random topologies and computed average values of chosen QoS parameters were studied. We have used the WMN with 25 nodes, because of the typical number of mesh nodes in WMN (25 to 30) (Skalli et al., 2006). For traffic generation, 5 CBR (Constant Bit Rate) flows were used and the packet size was set to 512 bytes. The same radio default parameters as in (ns-2, 2008) were used, except that we set the channel data rate to 11 Mbit/s. Simulation parameters are summarized in Table 1.

Parameter	Value
Test Area	1000x1000 m
Mac protocol	IEEE 802.11
Propagation model	Two ray ground
Routing protocol	AODV
Antenna type	Omni-directional
Traffic type	CBR
Packet size	512 bytes
Simulation time	100 seconds

Table 1. Simulation parameters

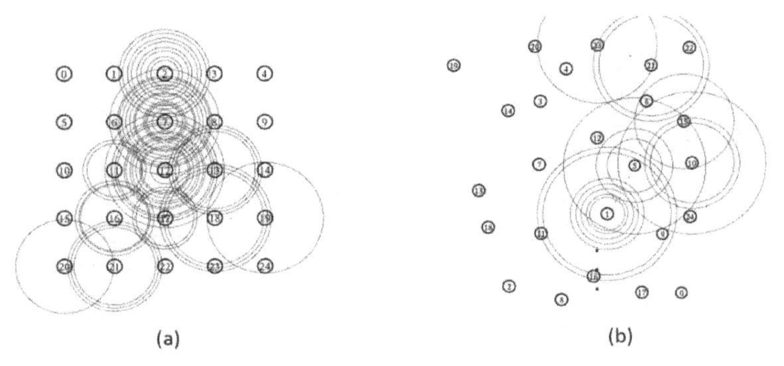

Figure 4. Grid (a) and random (b) topology of static WMN created in NS-2 simulator

3.2. Simulation results

In this section results of experiments are presented. The purpose of simulation was to determine the optimal number of radio interfaces for different WMN topologies, different number of data flows and different number of nodes to achieve the network capacity increasing expressed in enhancement of QoS parameters.

We chose four QoS parameters for simulation evaluation:

- *Average End-to-end Delay*: The average time taken for a packet to reach the destination. It includes all possible delays in the source node and in each intermediate host, caused by queuing at the interface queue, transmission at the MAC layer, routing discovery, etc. Only successfully delivered packets are counted.
- *Average Throughput*: The sum of data packets delivered to all nodes in the network in a given time unit (second).
- *Packet Loss*: Occurs when one or more packets being transmitted across the network fail to arrive at the destination.
- *Average Jitter*: The delay variations between all received data packets.

3.2.1. *Different network topologies*

In this simulation we created two different network topologies of WMN (grid topology and random topology). Ten random topologies were created and average values of chosen QoS parameters were computed.

Figure 5 shows the average values of end-to-end delay for various numbers of radio interfaces and two different network topologies. From results it is obvious that the highest value of end-to-end delay (0.92 sec) was reached in the grid WMN with one radio interface. The lowest value of delay (0.0097 sec) was achieved in grid WMN with seven radio interfaces. In WMN with random topology, the lowest value of delay (0.049 sec) was achieved in WMN with six radio interfaces. The best values of average delay were achieved in WMN with random topology, but differences between values of random and grid topologies were small

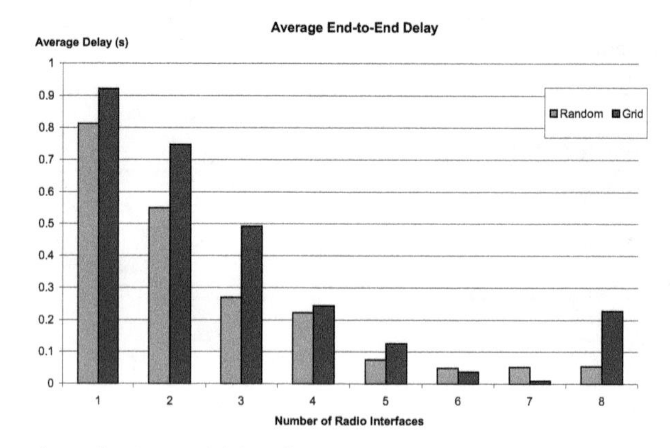

Figure 5. Average values of end-to-end delays for various radio interfaces and different network topologies

for higher number of radio interfaces. From results it may be concluded that optimal number of radio interfaces which guarantee the maximum allowable average delay 150 ms (ITU-T, 2003) for both network topologies is five, because more than five interfaces improved value of end-to-end delay only slightly, but the complexity of node is increased considerably.

Figure 6 shows the average values of network throughput for various numbers of radio interfaces and two different network topologies. The lowest value of average throughput was achieved in grid WMN, where nodes have used for transmission only one radio interface. In this case, the value of average throughput was 504.28 kbps. In the case where WMN with random topology and one radio interface was used, the lowest value of average throughput (739.3 kbps) was achieved. The highest value of throughput (2019.9 kbps) reaches the grid WMN with seven radio interfaces. The best value of average throughput in random WMN topology (1964.2 kbps) was achieved by WMN with seven radio interfaces. Again, the optimal number of interfaces for both network topologies was chosen as five.

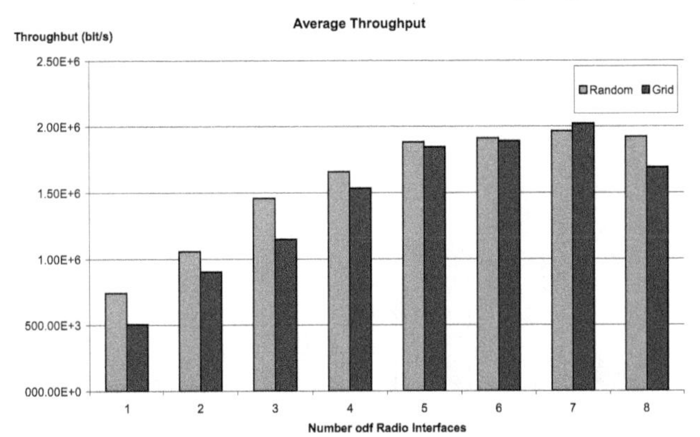

Figure 6. Average values of throughput for various radio interfaces and different network topologies

As we can see from Fig.7, the highest value of packet loss (75.1%) was reached in grid WMN with one radio interface. The lowest value of packet loss was achieved in WMN with seven radio interfaces. This value was 3.5% for the random topology and 2.5% for grid topology. As in the previous case, we can conclude the optimal number of radio interfaces as five, where grid topology achieved 9.8% of packet loss and 7.6% for random topology.

Figure 8 shows the average values of time jitter for different types of topologies and various number of radio interfaces. From results it is obvious that the highest value of average jitter was reached in the network with one radio interface. For the random topology this value was 0.7 sec and for grid topology it was 0.8 sec. On the other hand the lowest values of average jitter were achieved in grid WMN with seven interfaces (0.3 sec) and in random WMN with six interfaces (0.05 sec). As an optimal number of radio interfaces, the number of six was selected with average jitter value 0.11 sec for random topology and 0.14 sec for grid topology.

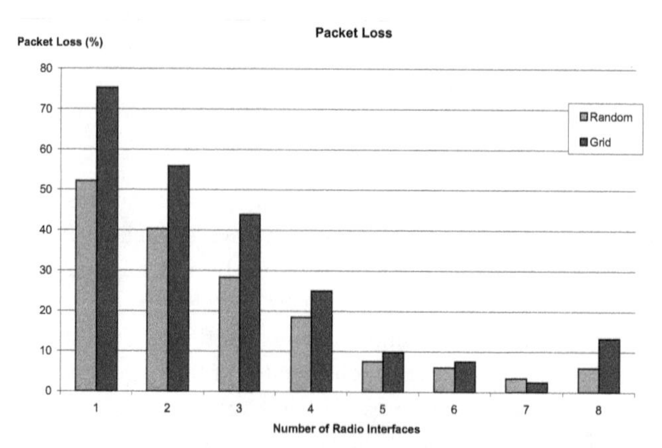

Figure 7. Values of packet loss for various radio interfaces and different network topologies

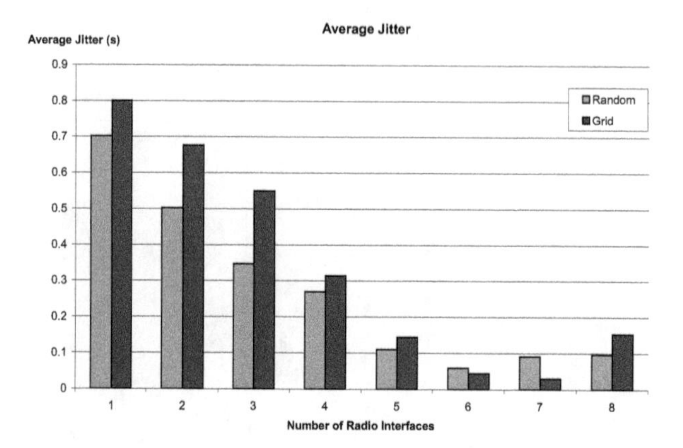

Figure 8. Average values of jitter for various radio interfaces and different network topologies

3.2.2. *Different number of data flows*

Simulation model consisted of 25 static wireless mesh nodes placed in grid in area 1000x1000 m (Fig.4a). Transmission range for each node was set to 200 m. As traffic transmission, the 5, 10, 15 and 20 CBR flows were simulated and packet size of 512 bytes was used. Data flows were created between random chosen node pairs.

Figure 9 shows the average values of end-to-end delay for different number of data flows. From results it is obvious that the best performance was achieved in multi-interface WMN with six interfaces, when the number of flows changed. The highest value of average end-to-end delay (for all data flows) was reached by WMN with one radio interface. For small number of data flows (5), WMN with 5 interfaces reached the best performance, whilst for 10 data flows the best performance was reached by 6 interfaces. For more data flows (15 and 20) the system performance is unsatisfactory regardless of number of interfaces.

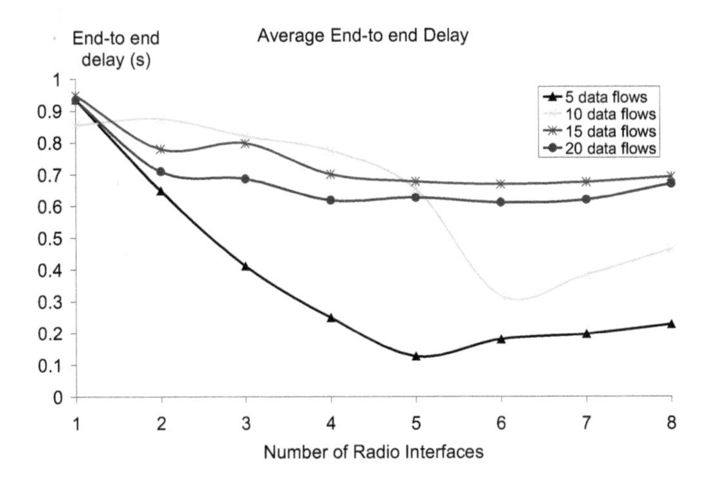

Figure 9. Average values of end-to-end delay for various radio interfaces and different number of data flows

Figure 10 shows the simulation results of average values of network throughput for the 5, 10, 15 and 20 data flows. The lowest value of average throughput was achieved in grid WMN with only one radio interface. From results it is obvious that the highest value of average throughput was reached in the multi-interface WMN with six radio interfaces. In the WMN with more than six interfaces the network performance is decreasing.

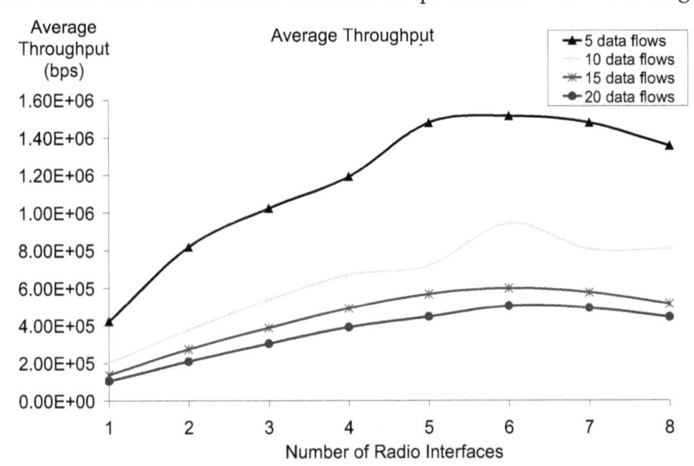

Figure 10. Average values of throughput for various radio interfaces and different number of data flows

As we can see from Figure 11, the best value of packet loss was reached in multi-interface WMN with six radio interfaces. The highest value of packet loss was reached in WMN, where nodes used for transmission one radio interface.

Figure 12 shows the average values of jitter for the different number of data flows. The highest values were achieved in WMN, where nodes have used for transmission only one radio interface. The best value of average jitter for all data flows was achieved in WMN with five or six radio interfaces.

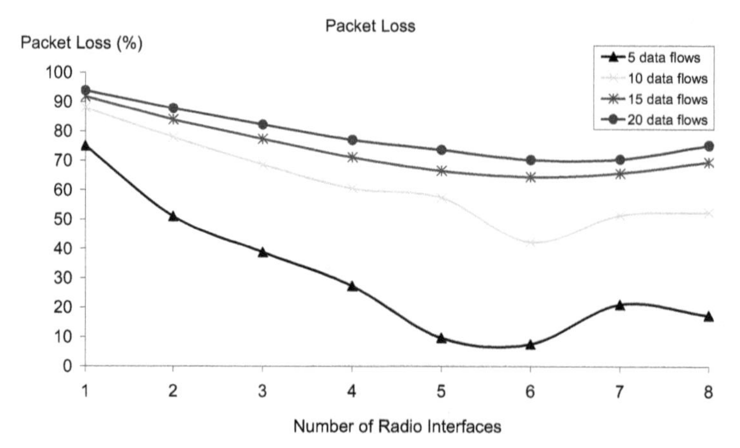

Figure 11. Values of packet loss for various radio interfaces and different number of data flows

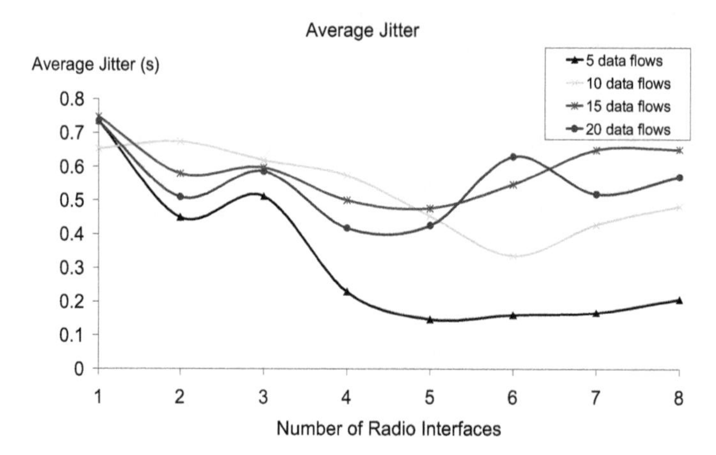

Figure 12. Average values of jitter for various radio interfaces and different number of data flows

3.2.3. Different number of nodes

In this simulation the static grid WMN was used (Fig. 4a), but with changing number of nodes. Six different *NxN* grid networks were created, where *N* was changed from five to ten. Transmission range for each node was set to 200 meters. For traffic transmission, 15 CBR flows were used and the packet size 512 bytes was set. Data flows were created between random chosen node pairs.

Results from previous sections (3.1.1 and 3.1.2) shows that the best values for almost all QoS parameters were achieved in WMN with six radio interfaces. For this reason the simulation model for different number of nodes only for WMN with six radio interfaces was created.

Figure 13 shows the average values of end-to-end delay for six radio interfaces and six different network topologies. The best value of average end-to-end delay was reached in multi-interface WMN with 25 nodes (5x5). The highest value of average end-to-end delay was achieved by WMN with 100 nodes (10x10). Results show that increasing number of nodes increase value of end to end delay.

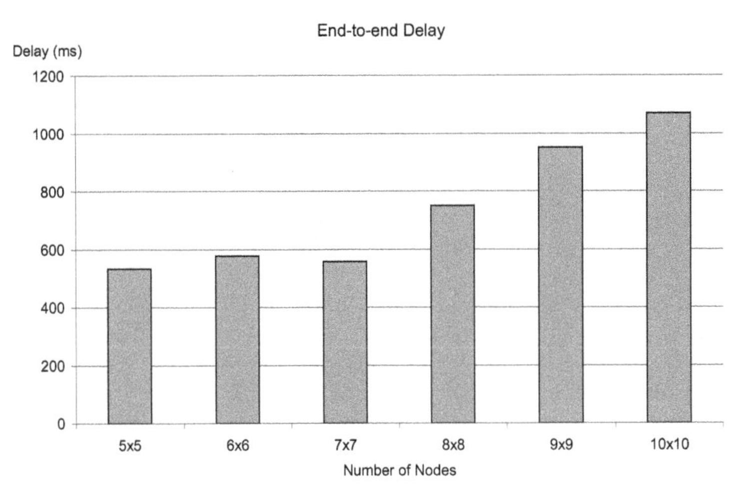

Figure 13. Average values of end-to-end delay for different number of nodes

The lowest value of average throughput (Fig. 14) was achieved in WMN with 100 static nodes. The best values of throughput were reached in configuration 6x6 and 7x7 nodes.

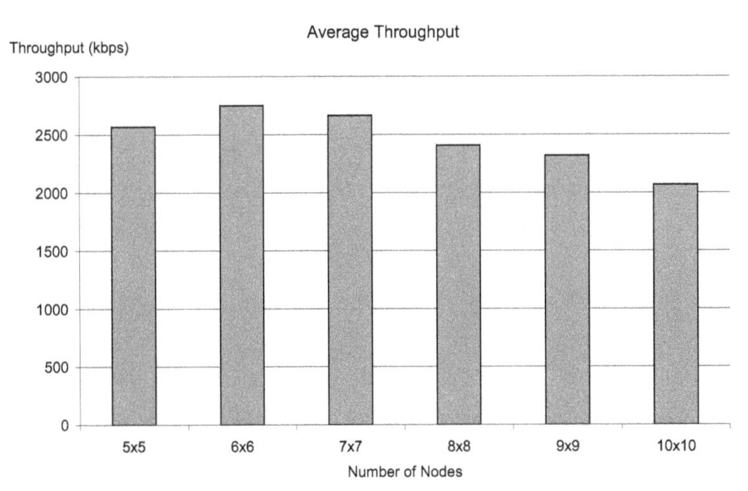

Figure 14. Average values of network throughput for different number of nodes

The highest values of packet loss (Fig. 15) were achieved in WMN with 10x10 nodes. The lowest value of packet loss was achieved in the WMN with 6x6 nodes.

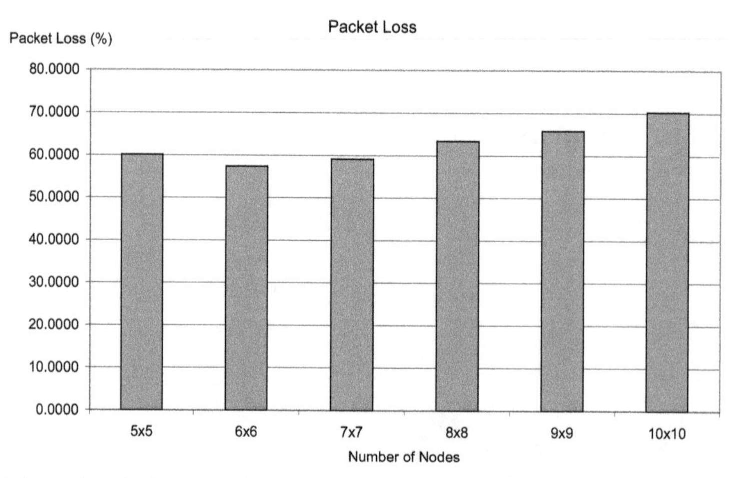

Figure 15. Values of packet loss for different number of nodes

As we can see from figure 16 the best value of average jitter was achieved WMN with 25 nodes and the highest value was reached in 9x9 grid network.

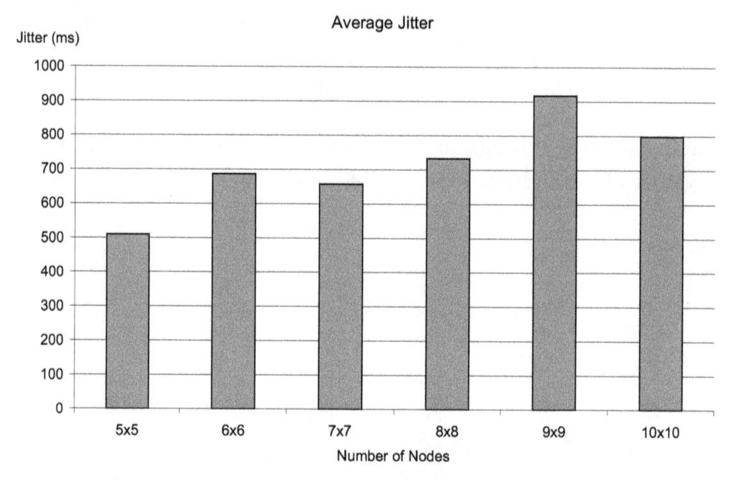

Figure 16. Average values of jitter for different number of nodes

These simulations showed unacceptable values for almost all simulated QoS parameters. Average delay combined with average jitter achieved in all networks (from 25 to 100 nodes) doesn't allow using several CBR services running simultaneously. This conclusion is certified by enormous packet loss in networks (over 55 % in the best solution).

3.3. Results summary

The results show the benefits of using multiple radio interfaces per node. This solution can improve the capacity of WMN. Simulation results show that by increasing the number of interfaces it is possible to increase network capacity by enhancing of QoS parameters. For all simulations of WMN with common channel assignment method, the number of six radio interfaces appears as an optimum solution, because further increasing of the number of interfaces improved the capacity of WMN only slightly and using more than seven radio interfaces decreased the network performance. These results can be used as a base to another research channel assignment methods, where using of suitable CA algorithm can additionally improve network performance.

4. Theoretical background

Optimal channel assignment in WMNs is an NP-hard problem (similar to the graph coloring problem). For this reason, before we present the channel assignment problem in WMNs, let us first provide some mathematical background about graph coloring problem.

4.1. Graph coloring

The graph coloring theory is used as a base for the theoretical modeling of channel assignment problem. At the beginning we must define two related terms: *communication range* and *interference range*. Communication range is the range in which a reliable communication between two nodes is possible. The interference range is the range in which transmission from one node can affect the transmission from other nodes on the same or partially overlapping channels. The interference range is always larger than the communication range (Fig. 17) (Prodan & Mirchandani, 2009).

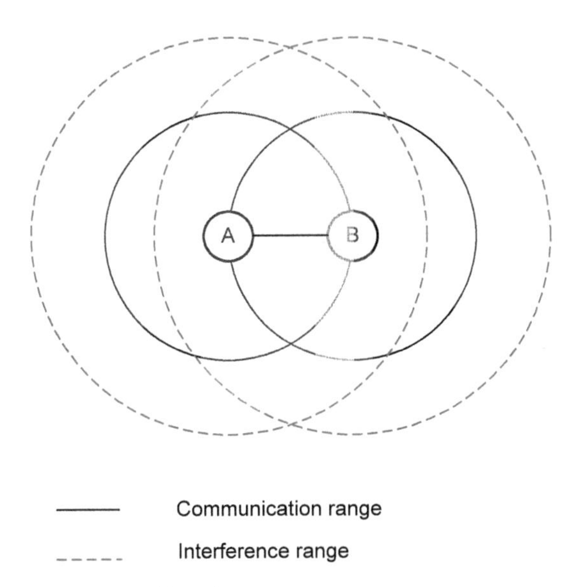

Figure 17. Communication range and interference range

Consider an undirected graph G(V, E) that models the communication network. A graph G, is defined as a set of vertices V and a set of edges E. Each vertex in graph represents a mesh router and each edge between two vertices represents a wireless link between two mesh routers. The color of each vertex represents a non-overlapping channel and the goal of the channel assignment is to cover all vertices with the minimum number of colors such that no two adjacent vertices use the same channel (Husnain Mansoor Ali et al., 2009).

4.2. Connectivity graph

The vertices set V consists of the network nodes, which may have multiple radio interfaces (not necessarily the same), while the edges/links set E includes all the communication links in the network. A link e between a pair of nodes (v_i, v_j); where $v_i, v_j \in V$ exists if they are within the communication range of each other and are using the same channel. The graph G described above is called the *Connectivity graph* (Fig. 5). The links presented in the network topology are referred to as the *logical links* (Husnain Mansoor Ali et al., 2009).

4.3. Interfering edges

To include the interference in network model, we introduce the concept of *Interfering edges.* Interfering edges for an edge e (IE(e)) are defined as the set of all edges which are using the same channel as edge e but cannot use it simultaneously in active state together with edge e. All edges are competing for the same channel hence the goal of channel assignment algorithm is to minimize the number of all edges e thereby increasing capacity (Husnain Mansoor Ali et al., 2009).

4.4. Conflict graph

In this subsection the concept of conflict graph is introduced. A conflict graph $G_c(V_c, E_c)$ consist of the set of edges E_c and the set of vertices V_c. The vortices V_c have a one relation with the set of edges E_c of the connectivity graph (i.e. for each edge $e \in E_c$, there exists a $v_c \in V_c$). As for the set E_c of the conflict graph, there exists an edge between two conflict graph vertices v_{ci} and v_{cj} if and only if the corresponding edges e_i and e_j of the connectivity graph, are in IE(e) set of each other. Hence, if two edges interfere in the connectivity graph, then there is an edge between them in the conflict graph. The conflict graph can now be used to represent any interference model. For instance, we can say that two edges interfere if they use the same wireless channel and they are within interference range. If we want use any other interference model based on signal power, then that can also be easily created by just defining the conditions of interference. Total interference can now be described as the number of links in the conflict graph (i.e. the cardinality of E_c).

The above mentioned concepts of connectivity graph, interfering edges and conflict graph are illustrated in Fig.18. For a graph $G(V, E)$, we find the IE for all the links and then create the conflict graph $G_c(V_c, E_c)$ (Husnain Mansoor Ali et al., 2009).

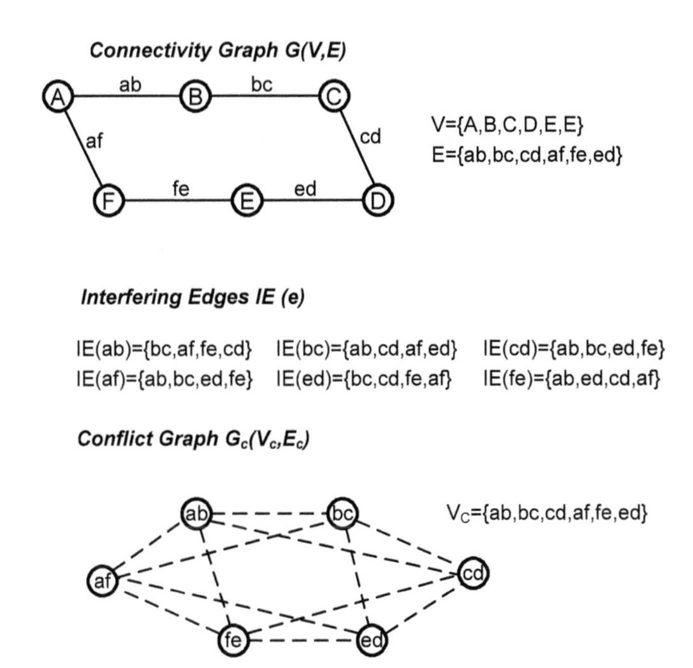

Figure 18. Connectivity graph, interfering edges and conflict graph

5. Channel assignment algorithms for WMN

As has been already mentioned, CA in a multi-interface WMN consists of assigning channels to the radio interfaces in order to achieve efficient channel utilization for minimizing interference and to guarantee an adequate level of connectivity. Nowadays, there exist many approaches to solve the channel assignment problem. These approaches can be divided into three main categories (Conti et al., 2007):

1. *Fixed (static) channel assignment approaches* – channels are statically assigned to different radio interfaces. The main concern includes the enhancement of efficiency and guaranteeing of the network connectivity.
2. *Dynamic channel assignment approaches* – a radio interfaces are allowed to operate on multiple channels, implying that a radio interfaces can be switched from one channel to another one. This switching depends on channel conditions, such as the value of interference. The basic issues are the switching delay and the switching synchronization.
3. *Hybrid channel assignment approaches* – in this approach the radio interfaces are divided into two groups, the first is fixed for certain channels and the second is switchable dynamically while deploying the channels.

In this section several channel assignment approaches are compared by QoS parameters mentioned in the previous section.

5.1. Common channel assignment

The Common channel assignment (CCA) is a simplest fixed channel assignment approach (Adya et al., 2004). In this CA approach all radio interfaces of each node were tuned to the same set of channels. For example, if every node has two radio interfaces then each node uses the same two channels (Fig. 19). The main benefit of this approach is the network connectivity. The connectivity is the same as that of a single interface approach, while the using of multiple radio interfaces can improve network throughput. However, if the number of non-overlapping channel is much higher than the number of radio interfaces, the gain of the CCA may be limited. CCA scheme presents a simplest channel assignment approach but it fails to account for the various factors affecting CA in a WMN. This solution will decrease the utilization of network resources (Yulong Chen et al., 2010).

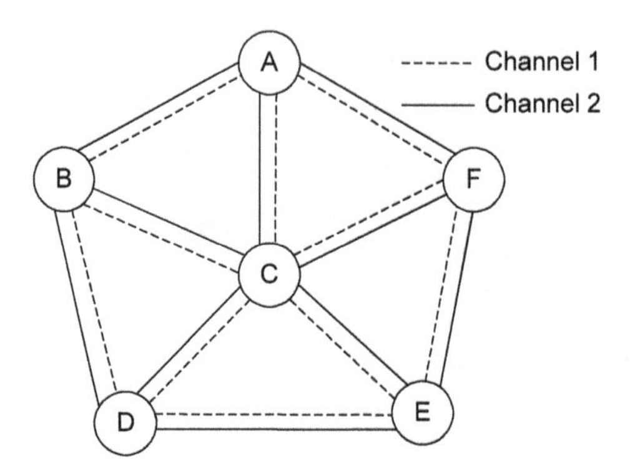

Figure 19. Example of common channel assignment approach

5.2. Load aware channel assignment

Load aware channel assignment (LACA) represents a dynamic centralized channel assignment and routing algorithm, where traffic is mainly directed toward gateway nodes (Raniwala et al., 2004), assuming that the offered traffic load on each virtual link is known. Algorithm assigns channels by such a way to ensure the network connectivity while takes into account the bandwidth limitation of each link. At the beginning, LACA estimates the total expected load on each virtual link based on the load imposed by each traffic flow. In the next step CA algorithm visits each virtual link in decreasing order of expected traffic load and greedily assigns it a channel. The algorithm starts with an initial estimation of the expected traffic load and iterates over channel assignment and routing until the bandwidth allocated on each virtual link matches its expected load. While this CA approach presents a method for CA that incorporates connectivity and flow patterns, the CA scheme on links may cause a *"ripple effect"*, whereby already assigned links have to be revisited, thus increasing the time complexity of the scheme.

An example of node revisiting is illustrated in Fig. 20. In this example each node has two radio interfaces. The channel list of node A is [1, 6] and channel list of node B is [2, 7]. Because nodes A and B have no common channel, a channel re-assignment is required. Link between nodes A and B needs to be assigned one of the channels from [1, 2, 6, 7]. Based on the channel expected loads, link between nodes A and B is assigned channel 6, and channel 7 assigned already to link between nodes B and D is reassigned to channel 6 (Raniwala et al., 2004, Yulong Chen et al., 2010).

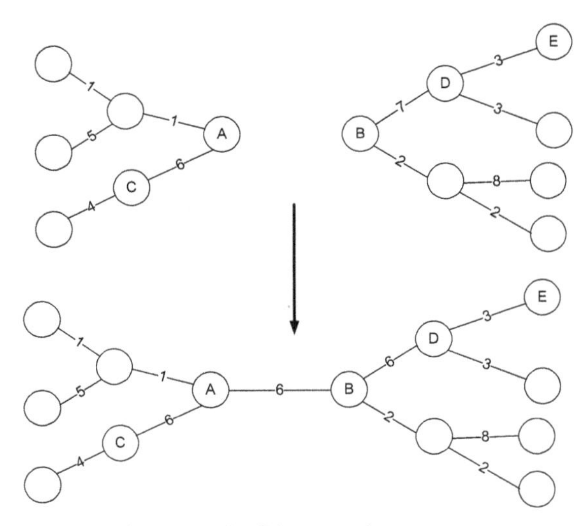

Figure 20. An example of channel revisit in LACA approach

5.3. First random channel assignment

The *First Random Channel Assignment algorithm* (FRCA) is a dynamic and centralized load aware channel assignment and routing algorithm for multi-interface multi-channel WMN (Pollak, Wieser, 2012). This approach takes into account the network traffic profile. FRCA algorithm assigns radio channels to links considering their expected loads and interference effect of other links, which are in interference range and which are tuned to the same radio channel.

FRCA algorithm consists of two basic phases:

1. Initial phase
2. Optimization phase

In the first phase, algorithm estimates initial loads on all links based on the initial routes created by routing algorithm. After load estimation, FRCA randomly assigns channels to all nodes for each radio interface.

In the second phase, FRCA algorithm uses similar steps as in the first phase, but channel assignment and routing iterations are based on results from the first phase. If some of the link load is higher than link capacity, the algorithm goes back and tries to find better solution. Algorithm's iterations end when no further improvement is possible. In optimization phase, FRCA uses greedy load-aware channel assignment algorithm similar to the one used in LACA algorithm (Raniwala et al., 2004). In this algorithm virtual links are visited in decreasing order of the link expected load. To find routes between nodes, FRCA uses shortest path routing based on minimum hop count metric (Kaabi et al., 2010).

5.3.1. Link load estimation

This approach is based on the concept of load criticality. The method assumes perfect load balancing across all acceptable paths between each communicating pair of nodes. Let $P(s, d)$ denote the number of acceptable paths between pair of nodes (s, d), $P_l\,(s, d)$ is the number of acceptable paths between (s, d) which pass a link l. And finally, let $B(s, d)$ be the estimated load between node pair (s, d). Then the expected traffic load Φ_l on link l is calculated as (Raniwala et al., 2004):

$$\Phi_l = \sum_{s,d} \frac{P_l(s,d)}{P(s,d)} \cdot B(s,d) \tag{1}$$

This equation implies that the initial expected traffic on a link is the sum of the loads from all acceptable paths, across all possible node pairs, which pass through the link. Because of the assumption of uniform multi-path routing, the load that an acceptable path between a pair of nodes is expected to carry is equal to the expected load of the pair of nodes divided by the total number of acceptable paths between them. Let us consider the logical topology as shown in Fig. 21 and assume that we have three data flows reported in table 2.

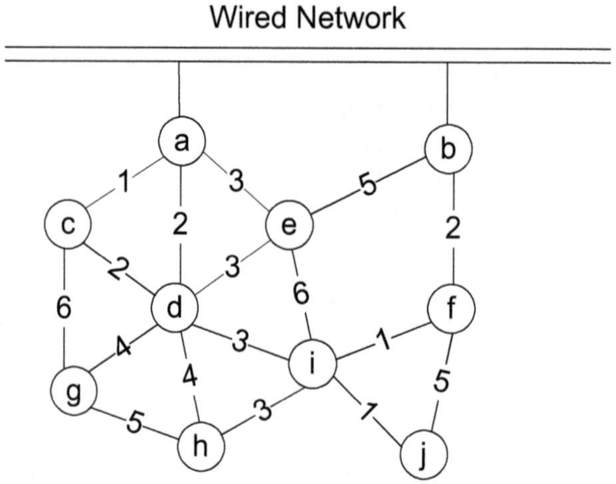

Figure 21. Multi-interface and multi-channel WMN

Because we have three different communications node pairs, we have

$$\Phi_l = \frac{P_l(a,g)}{P(a,g)} \cdot \gamma^{(a,g)} + \frac{P_l(i,a)}{P(i,a)} \cdot \gamma^{(i,a)} + \frac{P_l(b,j)}{P(b,j)} \cdot \gamma^{(b,j)} \tag{2}$$

Source (s)	Destination (d)	$\gamma^{(s, d)}$ (Mbps)
a	g	0.9
i	a	1.2
b	j	0.5

Table 2. Traffic profile with three data flows

(source, destination)	(a, g)	(i, a)	(b, j)
Possible paths	a-c-g	i-e-a	b-f-j
	a-c-d-g	i-e-d-a	b-f-i-j
	a-d-g	i-d-a	b-e-i-j
	a-d-c-g	i-d-c-a	b-e-i-f-j
	a-d-h-g	i-d-e-a	b-e-d-i-j
	a-d-i-h-g	i-d-g-c-a	
	a-e-d-g	i-h-d-a	
	a-e-i-h-g	i-h-g-c-a	
P (source, destination)	P(a, g) = 8	P(i, a) = 8	P(b, j) = 5

Table 3. Possible data flows between communicating nodes

In the next step we calculate $P(s, d)$ for each flow. We need to determine all the possible paths between source and destination. Table 3 shows all possible paths between communication node pairs for the WMN topology in Fig. 21. Values $P(s, d)$ and the corresponding link traffic load (Φ_l) is calculated using equation (2). Results are shown in table 4. Based on these calculations, we can estimate the load between each neighboring nodes. The result of calculation Φ_l is the expected traffic load of link l (i.e. the amount of traffic expected to be carried over a specific link) (Badia et al., 2009, Conti et al., 2007, Raniwala et al., 2004).

l	$P_l(a, g)$	$P_l(i, a)$	$P_l(b, j)$	Φl (Mbps)
a-c	2	3	0	0.675
c-g	2	2	0	0.525
c-d	2	1	0	0.375
d-g	2	1	0	0.375
a-d	4	3	0	0.9
g-h	0	1	0	0.15
d-h	1	1	0	0.2625
a-e	2	2	0	0.525
d-e	1	2	1	0.5125
d-i	1	3	1	0.6625
h-i	2	2	0	0.525
e-i	1	2	2	0.6125
b-e	0	0	3	0.3
b-f	0	0	2	0.2
f-i	0	0	2	0.2
i-j	0	0	2	0.2
f-j	0	0	2	0.2

Table 4. The results of calculation Φ_l on specific link l

5.3.2. Link capacity estimation

The link capacity (channel bandwidth available to a virtual link) is determined by the number of all virtual links in its interference range that are also assigned to the same radio channel. So when estimating the usable capacity of the virtual link, we should consider all traffic loads in its interference range. According to the channel assignment rules, the higher load a link is expected to carry, the more bandwidth it should get. On the other side, the higher loads its interfering links are expected to carry, the less bandwidth it could obtain. Thus, the link capacity should be proportional to its traffic load, and be inversely proportional to all other interfering loads. Thus, the capacity $bw_{(i)}$ assigned to link i can be obtained using the following equation:

$$bw_{(i)} = \frac{\Phi_i}{\sum_{j \in Intf(i)} \Phi_j} * C_{ch} \tag{3}$$

where Φ_i is the expected load on link i, Intf(i) is the set of all virtual links in the interference range of link i (i.e. links i and j operates on the same channel). C_{ch} is the sustained radio channel capacity (Badia et al., 2009, Conti et al., 2007, Raniwala et al., 2004).

5.4. Simulation results

In this section, the performance of proposed FRCA concept is evaluated and compared with CCA (Adya et al., 2004), LACA (Raniwala et al., 2004) and a single interface architecture by using NS-2 simulator (ns-2, 2008). Simulation model consisted of 25 static wireless mesh nodes placed in an area of 1000 x 1000 m (Fig.4a). The distance between nodes was set to 200 m. The capacity of all data links was fixed at 11Mbps. All nodes have the same transmission power and the same omni-directional antenna. The transmission range was set to 200 m and interference range was set to 400 m. For traffic generation, 25 CBR (Constant Bit Rate) flows with packet size 1000 bytes were used. Flows were created between randomly chosen node pairs. For simulation evaluation, the same metrics like in section 3.1 was used.

5.4.1. Different number of radio interfaces

From previous sections the conclusion about optimal number of six radio interfaces was gained. This conclusion was based on simple common channel assignment scheme CCA, which was used in simulations. With using more sophisticated channel assignment scheme it is possible to expect that the same results in QoS parameters may be reached with less number of interfaces. So the performance evaluation of chosen CA schemes was based on changing number of radio interfaces (between 2 to 8 radio interfaces for each node).

Figure 22 shows the average values of end-to-end delay for various number of radio interfaces. From results it is obvious that the highest value of delay (792.64 ms) was reached in WMN with CCA scheme. Lowest value (101.42 ms) reached WMN with FRCA algorithm for 4 radio interfaces. For CCA scheme the optimal number of radio interfaces was 6, but FRCA and LACA reached the best performance with only 4 radio interfaces. Results show that further increasing of number of radio interfaces didn't increase the network performance, so the optimal number of radio interfaces for LACA and FRCA algorithm is 4.

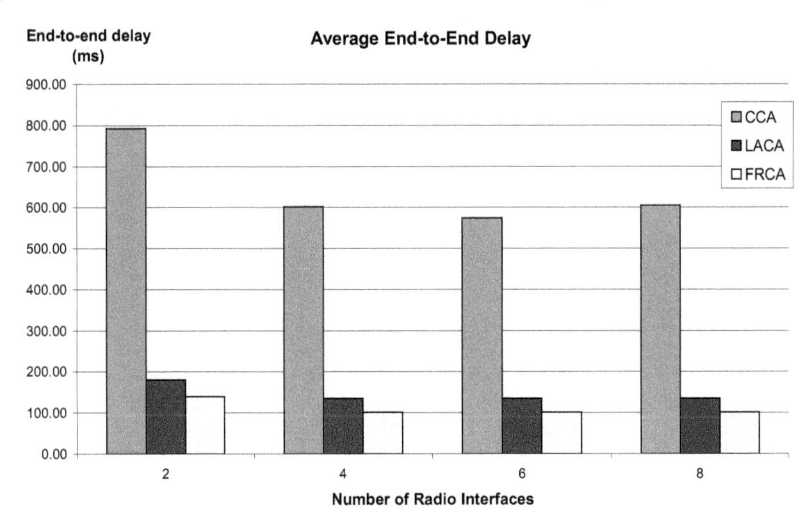

Figure 22. Average values of end-to-end delay for various radio interfaces and different CA schemes

Figure 23 shows the average values of network throughput. The lowest value of average throughput for all radio interfaces was achieved in WMN with CCA scheme. This approach reached the best results for 6 radio interfaces. Others CA algorithms (FRCA and LACA) achieved the best performance with only 4 radio interfaces, with FRCA slightly outperformed LACA algorithm.

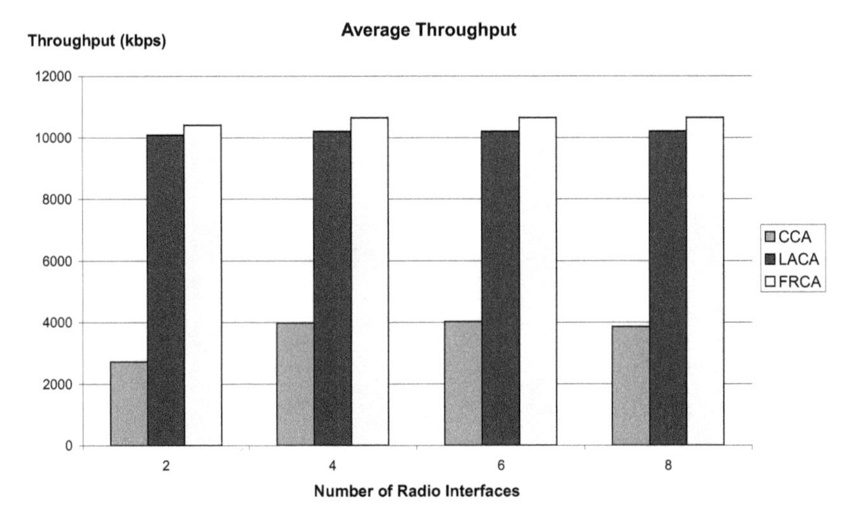

Figure 23. Average values of network throughput for various radio interfaces and different CA schemes

As we can see from figure 24 the highest value of packet loss for all number of interfaces was reached in WMN with CCA approach, with the best value reached for 6 radio interfaces (63.56 %). The best result (5.86 %) reached FRCA algorithm for 4 radio interfaces, whereas algorithm LACA with the same number of radio interfaces reached value 9.47%.

Figure 25 shows average values of average jitter. The best values of average jitter were again reached with FRCA algorithm for 4 radio interfaces (124.8 ms). CCA algorithm reached the best value for 6 radio interfaces (601.25 ms) and LACA approach for 4 radio interfaces (167. 27 ms).

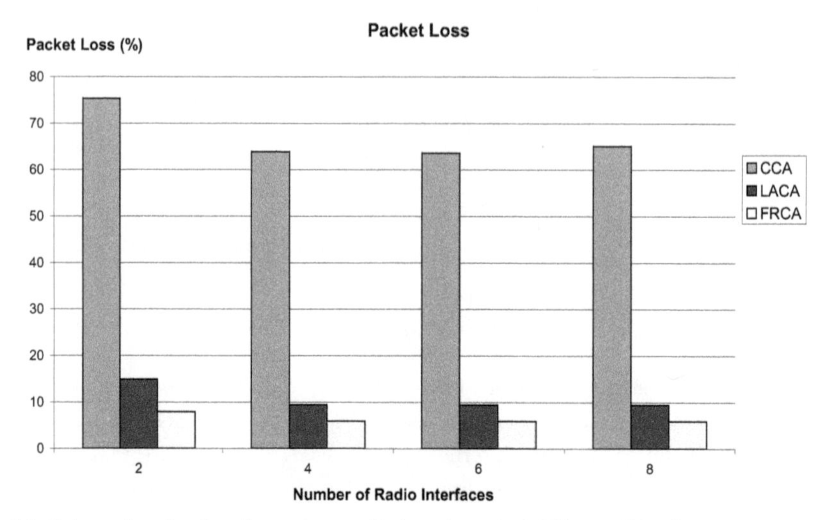

Figure 24. Values of packet loss for various radio interfaces and different CA schemes

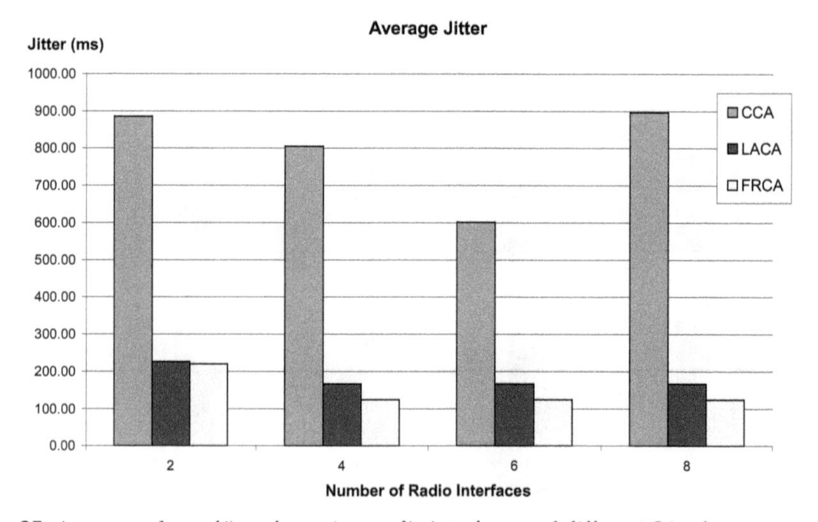

Figure 25. Average values of jitter for various radio interfaces and different CA schemes

6. Conclusion

In this chapter, the study of optimal number of radio interfaces and new channel assignment approach was presented (FRCA). The study of optimal number of radio interfaces was created for two different topologies (grid and random), different number of data flows and different number of nodes. The study was based on increasing number of radio interfaces (1 to 8) for each mesh nodes. The results show that by increasing the number of interfaces it is possible to increase network capacity by enhancing of QoS parameters. For all simulations of WMN with common channel assignment method CCA, the number of six radio interfaces appears as an optimum solution, because the further increasing of the number of interfaces improved the capacity of WMN only slightly and using more than seven radio interfaces decreased the network performance.

For further increasing of network performances more sophisticated channel assignment algorithms were used. The new channel assignment approach called First random channel assignment (FRCA) was compared with existing channel assignment algorithms (CCA, LACA). The results show that by using the suitable CA algorithm it is possible to further increase the network capacity. From all results it can be concluded that the multi interface approach with suitable CA algorithm can dramatically increase the whole network performance. In that case, if it is used the simplest CA approach (CCA), we need to assign for each node up to 6 radio interfaces to maximize network performance, but by using suitable dynamic CA algorithm (e.g. FRCA or LACA), the network performance may be maximized with only 4 radio interfaces.

Author details

Stefan Pollak and Vladimir Wieser
Department of Telecommunications and Multimedia, University of Zilina, Slovakia

Acknowledgement

This work was supported by the Slovak Scientific Grant Agency VEGA in the project No. 1/0704/12.

7. References

Adya, A.; Bahl, P.; Padhye, J.; Wolman, A. & Lidong Zhou (2004). A multi-radio unification protocol for IEEE 802.11 wireless networks, *Broadband Networks*, 2004. BroadNets 2004, pp. 344- 354, 25-29 Oct. 2004

Badia, L.; Conti, M.; Das, S. K.; Lenzini, L. & Skalli, H. (2009). Routing, Interface Assignment and Related Cross-layer Issues in Multiradio Wireless Mesh Networks, in *Handbook of Wireless Mesh Networks*, Springer Publishers, ISBN 978-1-84800-908-0, London 2009

Calvo R. A. & Campo J. P. (2007). Adding Multiple Interface Support in NS-2, Jan. 2007.

Chi Moon Oh; Hwa Jong Kim; Goo Yeon Lee & Choong Kyo Jeong (2008). A Study on the Optimal Number of Interfaces in Wireless Mesh Network, In *International Journal of Future Generation Communication and Networking*, IJFGCN Vol. 1, No. 1, pp. 59 – 66, Dec. 2008

Conti, M.; Das, S. K.; Lenzini, L. & Skalli H. (2007). Channel Assignment Strategies for Wireless Mesh Networks, In *Wireless Mesh Networks – Archi-tectures and Protocols*, Springer, ISBN 978-0-387-68839-8, New York

Draves, R.; Padhye, J. & Zill B. (2004). Routing in Multi-radio, Multi-hop Wireless Mesh Networks, Proc. *ACM MobiCom*, pp. 114-128, 2004

Gong, M. X. & Midkiff, S. F. (2005). Distributed Channel Assignment Protocols: A Cross-Layer Approach, *in IEEE WCNC*, 2005

Gupta, P. & Kumar, P. (2000). The Capacity of Wireless Networks, In *IEEE Transactions on Information Theory*, volume 46, pp. 388–404, March 2000

Husnain Mansoor Ali; Anthony Busson & Véronique Vèque (2009). Channel assignment algorithms: a comparison of graph based heuristics, In: *Proceedings of the 4th ACM workshop on Performance monitoring and measurement of heterogeneous wireless and wired networks*, p.120-127, October 26-26, 2009, Tenerife, Canary Islands, Spain

ITU-T (2003), ITU-T Recommendation G.114, 2003.

Kaabi, F.; Ghannay, S. & Filali F. (2010). Channel Allocation and Routing in Wireless Mesh Networks: A survey and qualitative comparison between schemes, In: *International Journal of Wireless & Mobile Networks (IJWMN)*, Vol. 2, No. 1, pp. 132-150, 2010

Marina, M. K. & Das, S. R. (2005). A topology control approach for utilizing multiple channels in multi-radio wireless mesh networks, *Broadband Networks Vol.1*, BroadNets 2005, pp. 381- 390, 3-7 Oct. 2005

ns-2 (2008), The Network Simulator ns-2, http://www.isi.edu/nsnam/ns/

Pollak, S. & Wieser, V. (2012). Interference Reduction Channel Assignment Algorithm for Multi-Interface Wireless Mesh Networks, In: *Proceedings of 22nd International Conference "Radioelektronika 2012"*, Brno, Czech republic (paper accepted)

Prodan, A. & Mirchandani, V. (2009). Channel Assignment Techniques for 802.11 based Multi-Radio Wireless Mesh Networks, in *Handbook of Wireless Mesh Networks*, Springer Publishers, ISBN 978-1-84800-908-0, London 2009

Ramachandran, K. N.; Belding, E. M.; Almeroth, K. C. & Buddhikot, M. M. (2006). Interference-Aware Channel Assignment in Multi-Radio Wireless Mesh Networks, INFOCOM 2006. *25th IEEE International Conference on Computer Communications. Proceedings*, pp.1-12, April 2006

Raniwala, A.; Gopalan, K. & Chiueh T. (2004). Centralized Channel Assignment and Routing Algorithms for Multi-Channel Wireless Mesh Networks, In Acm *Sigmobile MC2R*, vol. 8, no. 2, pp. 50-65, April 2004

Skalli, H.; Das, S. K.; Lenzini L. & Conti M. (2006). Traffic and interference aware channel assignment for multi-radio Wireless Mesh Networks, Technical report

So, J. & Vaidya, N. H. (2004). Multi-channel MAC for Ad Hoc Networks: Handling Multi-channel Hidden Terminals using a Single Transceiver, In *ACM Mobihoc*, 2004

Wei Yahuan; Taoshen Li & Zhihui Ge (2011). A Channel Assignment Algorithm for Wireless Mesh Networks Using the Maximum Flow Approach. In: *Journal of networks*, Vol. 6, No. 6, June 2011

Yulong Chen; Ning Xie; Gongbin Qian & Hui Wang (2010). Channel assignment schemes in Wireless Mesh Networks, In: *Mobile Congress (GMC)*, 2010 Global , vol., no., pp.1-5, 18-19 Oct. 2010

Section 2

Network Planning Aspects in WMNs

Chapter 6

Achieving Fault-Tolerant Network Topology in Wireless Mesh Networks

Svilen Ivanov and Edgar Nett

Additional information is available at the end of the chapter

1. Introduction

Wireless Mesh Network (WMN) is an ad-hoc network with a fixed network infrastructure (see an example in figure 1). The physical structure of a WMN includes base stations, a backbone and mobile stations. The *base stations* (also known as mesh routers or mesh points) are static wireless nodes, forming the network infrastructure and providing wireless network access to the mobile stations. The *backbone* is a wireless ad-hoc network among the base stations. The fixed network infrastructure provides wireless network access to the mobile stations in a service area. *Service area* is a finite three-dimensional space. The *mobile stations* are wireless

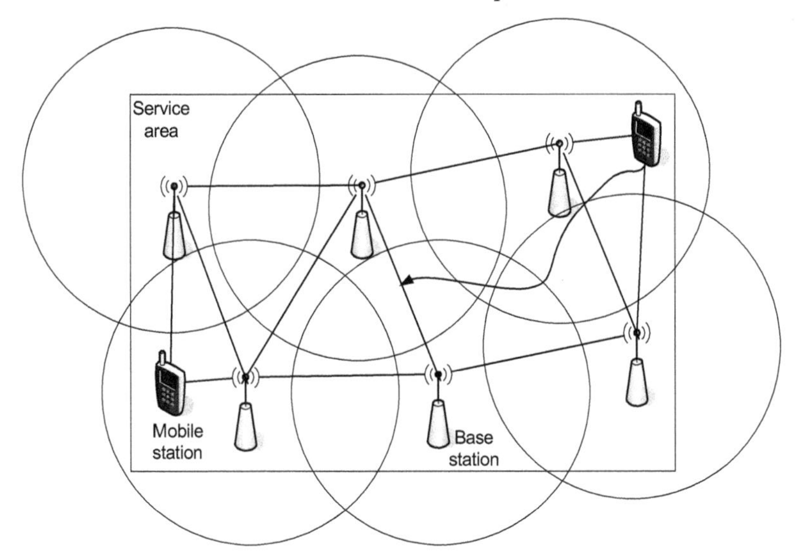

Figure 1. Wireless mesh networks and radio coverage

nodes which move within the service area and communicate to other stations via the WMN. The stations in a WMN use a *multi-hop routing protocol* for communication. This protocol automatically discovers the network topology and delivers the messages to the destination; if needed over multiple hops. We can think of a WMN as an infrastructure wireless network in which the backbone is replaced by a wireless one and the communication is done in a (multi-hop) ad-hoc way.

We consider a wireless mesh network which supports a business process and is under the administration of an organization. This is not a MANET (Mobile Ad-hoc Network) consisting of self-dependent mobile nodes, like it is often in the literature. The organization has control over the network infrastructure and aims at providing radio coverage and connectivity in a clearly defined service area. The *management appliance* is a central instance for basic configuration and diagnosis of the WMN, including topology monitoring, protocol settings, traffic management, etc.

Radio coverage and *connectivity* are basic *services* of a wireless mesh network which are required for communication. Radio coverage ensures that the mobile stations can access the network infrastructure (backbone) while they are located or moving in the service area. Connectivity ensures that the topology of the backbone is connected.

1.1. Radio coverage

The service *radio coverage* is *correct*, if the service area is *covered* by the base stations. The service area is covered, if the unification of radio cells of all base stations contains the whole service area. The radio cell of a base station is a part of the space around it, in which a mobile station observes the base station with a radio signal strength sufficient for communication. The sufficient radio signal strength in the service area is a basic requirement for the mobile stations to be able to access the WMN. The radio coverage service ensures this sufficient signal strength in the service area. Service location is a point of the service area, specified by its coordinates. A service location is covered, if the unification of radio cells of all base stations contains the service location.

1.2. Connectivity

The service *connectivity* is correct, if the backbone graph is connected. The *backbone graph* is a graph with the base stations as vertices and the routing layer links among them as edges. A *link* exists if two wireless devices can communicate through the wireless medium obeying some qualitative parameters (see section 4.3 for more information). The backbone graph represents the network topology at the routing layer. This graph is connected, if a *path* (a sequence of edges) exists between every two vertices. A connected backbone graph means a connected routing layer topology which is a basic requirement for communication through the WMN. The connectivity service ensures that the backbone graph is connected.

At the example WMN in figure 1 the radio coverage and the connectivity are correct. The unification of radio cells contains the service area and the backbone graph is connected.

1.3. Problem exposition and contributions

In this chapter, we address the problem of guaranteeing radio coverage of Wireless Mesh Networks, which are exposed to environmental dynamics.

The *environmental dynamics* are unpredictable changes of the radio propagation and radio attenuation properties of the environment (e.g. new obstacles, movement of obstacles, increased humidity). They occur due to reconfiguration of the plant layout. Environmental dynamics occur, for instance, in Reconfigurable Manufacturing Systems (RMS) [11, 26]. An RMS is a production system with an adjustable structure, that is able to meet the market requirements with respect to capacity, functionality and cost. This adjustable structure at the system level includes changes in the plant layout, for instance *"adding, removing or modifying machine modules, machines, cells, material handling units and/or complete lines"* [11]. In RMS the environmental dynamics are unpredictable at design time, because the system layout adjustments are made on the fly to meet the actual production demand. The environmental dynamics negatively affect the radio coverage (radio signal strength between mobile stations and base stations) and the backbone network connectivity among base stations of an WMN.

The first contribution of this chapter is a fault-tolerance method for guaranteeing radio coverage of wireless mesh networks in dynamic propagation environments. The basic idea of this approach is to automatically detect an *error state*, which is lack of redundancy in radio coverage and connectivity, and correct this error by reconfiguring the base stations before the radio coverage fails. The error detection is based on a radio propagation model: if an error is detected in the model, this is an indicator that an error in the real radio coverage exists. In order to be able to make this conclusion, this model is automatically calibrated to the real enviromnent by using radio signal strength measurements.

The second contribution of this chapter is an automatic base station planning algorithm for the reconfiguration phase of the fault-tolerance approach. In this phase base stations are added to the network in order to correct errors in the radio coverage and connectivity. The question is: what is the minimum number of base stations to be added and at which positions in order to restore the original state of radio coverage and connectivity. Our approach is an optimization algorithm, which uses knowledge from the calibrated radio propagation model and answers this question in a sufficient time frame.

1.4. Structure of the chapter

In section 2 we will discuss related work. In section 3 we will present our fault-tolerance approach for ensuring the availability of radio coverage and connectivity of wireless mesh networks. In section 4 we will present our approach for automatic base station planning in wireless mesh networks, which is used in the reconfiguration phase on the fault-tolerance approach. Section 6 provides a conclusion and a summary of future work.

2. Related work

Firstly, we will present related work aiming at availability of the radio coverage and connectivity. Then we will discuss related work to the automatic base station planning algorithm.

2.1. Availability of the radio coverage

The availability of the service *radio coverage* is a necessary condition for reliable communication in wireless networks. The issue of reliable communication via wireless medium has been extensively investigated during the design of every wireless communication system. Since the wireless medium is unshielded, the effect of the environment on the wireless communication is specific to the environment. Different methods have been developed for increasing the reliability of the communication through the wireless medium. Most of them are at the physical layer. For instance the robust modulation methods (e.g. MIMO), frequency hopping, spread spectrum transmission, redundancy in the antennas and redundancy of the transmitters. At the data link layer, error correction codes and retransmissions are typical measures. These methods mostly address the time-variability of the wireless channel caused by multi-path propagation. However, all these methods require some minimum radio signal strength at the receiver which is a basic requirement for decoding the frames successfully. Providing this minimum radio signal strength is a matter of network deployment and configuration in the particular environment.

The state-of-the-art method for ensuring radio coverage has a static nature (e.g. [8, 10, 35]). Figure 2 shows the general procedure of this method. The method ensures radio coverage

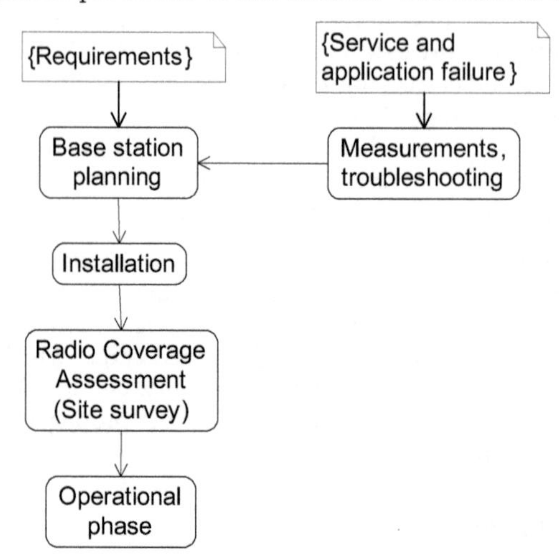

Figure 2. Static deployment method for radio coverage

during the network deployment before the network starts operation. Usually, an *expert* plans the base stations properties so that the requirements for the radio coverage are fulfilled. The expert makes this planning based on knowledge about the environment and the requirements. For this purpose, measurements in the particular environment are typically needed. Then, the base stations are installed. After the installation, a manual site survey is conducted with the purpose of proving that the requirements are satisfied. The site survey includes manual measurements of the radio signal strength on selected service locations in the whole area. If the requirements are not satisfied adjustments should be made. The adjustments are site-specific and may include removing obstacles, changing frequencies, or adding new equipment [10]. When the requirements are fulfilled, the wireless network enters the operational phase. In the operational phase, *there is no automatic function for monitoring and maintaining the radio coverage*. The only way to do this is by making a manual site survey which is expensive in terms of time and effort. The loss of radio coverage can only be detected by the mobile stations and the applications. The network connection is lost and no communication is possible. The repair of radio coverage is started when the applications report a problem of this kind. During the radio coverage repair the presence of a expert is required for troubleshooting and base station planning.

For compensating the dynamics of the environment, the static method uses static radio signal strength redundancy (called fade margin). In communication systems design the term *fade margin* (or margin) is the amount of signal strength reserve. This is the power, added to the needed minimum level for reception of the frames at the receiver. The fade margin is configured during the planning phase via adequate selection of transmitters and antennas [8, 37]. The fade margin is used for compensating temporal variations in the environment. When the environment changes, the radio coverage eventually degrades. But if the redundancy is sufficient, the radio coverage is still correct and the applications are not affected. However, the radio coverage could have entered a critical state; meaning that further changes in the environment may lead to service failure. Since there are no automatic monitoring functions for the radio coverage, this state of lost redundancy is not detected, and remains in the system. In this state, the next change in the environment can lead to service failures.

In the context of this chapter, we have high availability requirements. We have an environment which can change in unpredictable way during the network's life-cycle which is typically larger than 10-20 years. For this reason, it is hardly possible to plan sufficient static redundancy for all possible changes of the environment. They are not known at the deployment phase. Even if this would be possible, it would be extremely inefficient. Consequently, a new method is needed for guaranteeing radio coverage. When the factory-layout changes for adapting to a new market, the method should enable an easy adaption of the WMN and should guarantee high availability of the radio coverage and the connectivity.

2.2. Connectivity and base station planning

In this section we focus on the deployment and operation of the base stations which is an essential function for connectivity. For the routing protocol and the topology discovery we base on the research within our working group (e.g. [15, 29, 32]).

Industrial automation networks have usually been isolated, single-cell networks or classic infrastructure networks with multiple cells. This means that base station planning is required only for the 'last mile', i.e. the connection between a base station and a mobile station, e.g. [8]. In the case of multi-hop wireless mesh networks, the planning of the backbone network is a new research aspect that needs to be considered. Research on radio network planning consider network throughput as a main planning goal, e.g. [7]. However, the most common requirement of industrial networks is availability. With the introduction of technologies for multi-hop communication in industrial environments (e.g. Zigbee, Wireless HART), the base station planning problem gains importance. Paper [37], for instance, presents the challenges for developing a planning tool for industrial wireless sensor networks. However, to the best of our knowledge, no systematic approach exists for planning multi-hop wireless networks with respect to fault-tolerance requirements of industrial automation networks.

The existing algorithms for the base station planning in wireless mesh networks [2, 39] have a different goal. It is to design a mesh network with a minimum number of base stations such that the end-to-end throughput requirements of application flows are fulfilled. These requirements are typical for Internet access in areas with no alternative high-speed wired connection. The approach is to transform the planning problem into a linear optimization problem which is a combination of a set covering problem and a network flow problem. As a result, the backbone is a connected graph, but with no fault-tolerance. Another disadvantage is the intractability of the proposed approaches. For some inputs, the algorithm takes too much time for the result to be useful. This is because the underlying linear optimization problem is a binary integer problem which is well known for its NP-completeness. Paper [39] addresses this issue by a decomposition method, but the algorithm still runs about 22 hours for a network with 58 nodes. This is acceptable for the mentioned scenarios, but for network reconfiguration in automation scenarios a faster algorithm is required. Extending these algorithms to fault-tolerance would mean an additional increase in the complexity. Paper [41] considers the problem of coverage control in wireless sensor networks, including various aspects like activating/deactivating of the nodes, finding the coverage characteristics of a given network, and sensor node deployment. However, all considerations include only the aspect of last mile coverage, i.e. the sensing function of the nodes. They do not consider the problem of the backbone connectivity for communicating the sensed data to a central instance.

Our approach is to extend the existing methods from infrastructure network planning to planning multi-hop wireless mesh networks with fault-tolerance aspects. Other papers about fault-tolerance in wireless multi-hop networks can benefit from our approach for generating a fault-tolerant topology. Papers considering fault-tolerant routing, for instance [4, 19, 27], have a prerequisite of biconnected backbone network, but do not address the base station planning problem. The base station planning problem has been little addressed so far because in most mobile ad-hoc and sensor network scenarios the number and position of the nodes are considered uncontrolled or hardly controlled. However, in automation scenarios the networks are typically planned to provide service in some predefined geographical area (e.g. production hall). This requires careful base station planning for ensuring high availability of the radio coverage.

The topology control problem is to configure a given an instance of a multi-hop network such that it is connected and a quality of service property is fulfilled. Depending on the configured parameter, these methods adjust the transmission power [6] or the time of activity and sleeping periods of the nodes [5]. Paper [6] presents an algorithm for distributed adjustment of the transmission powers of the nodes with the purpose of minimizing the interference and keeping the network topology connected with a high probability. Paper [5] presents a distributed protocol for topology management which determines the active and sleeping periods for the nodes in such a way that the network is connected, the energy consumption is minimized, and the data is delivered with real-time guarantees. Paper [40] considers the issue of data forwarding in industrial wireless sensor networks and the integration in a wired backbone. It proposes a chain-based communication protocol for real-time communication over multiple hops. It is common for all topology control protocols that they operate on some existing instance of a multi-hop network. For achieving the required quality of service property, these protocols require some topological properties of the network (like connectivity or k-connectivity). The difference is that our base station planning algorithm plans a given network to be deployed with the desired topological properties. In this way, our algorithm can be used in the first phase of planning the topological properties of the network. In a second phase a topology control algorithm can be used to additionally adjust the transmission powers or active/sleep times of the nodes for achieving the required QoS property.

3. Fault-tolerant radio coverage and connectivity

This section presents our approach for fault-tolerant radio coverage and connectivity of wireless mesh networks in dynamic propagation environments. A premature version of this approach, considering only radio coverage, has been published in [22].

3.1. Fault-tolerance approach

We consider the goal of this chapter at a general abstraction level. It is to guarantee availability of the services (radio coverage and connectivity) of a system (wireless mesh network) which is exposed to dynamic external behavior (the dynamic propagation environment). The environmental dynamics is an external factor to the wireless network. It results from the changing surroundings of the wireless network.

For this general type of problem, a well-known method exists in the field of dependable computing. This is the *fault-tolerance* approach [3]. Fault-tolerance avoids service failures in the presence of faults. *Service failure*, or *failure*, is the inability of a system to perform a service according to the service specification. *Error* is a part of the system state which may lead to a subsequent service failure. A *fault* is the cause for an error. The fault-tolerant system design includes *fault model definition*, *error detection* and *system recovery*. The fault model definition identifies a set of faults, for which service failures do not occur. The error detection identifies errors in the system, caused by the faults. The system recovery transforms a system with errors to a system without errors. The idea is to detect errors and perform system recovery *before* the errors lead to failures. In this way, the fault-tolerance approach avoids failures if faults from the fault model occur. In this chapter we apply the fault-tolerance approach for guaranteeing availability of radio coverage and connectivity of wireless mesh networks in dynamic propagation environments.

Fault model definition

A fault in our system is the *environmental dynamics*. Environmental dynamics are changes of the radio attenuation properties of the environment (e.g. new obstacles, movement of obstacles, increased humidity). The *attenuation* describes the ability of the radio propagation environment to absorb and weaken the radio waves. An increased attenuation has a negative effect on radio coverage and connectivity. Regarding radio coverage, it reduces the radio signal strength at the service locations. This can lead to the fact that some service locations are not covered. The effect on connectivity is that some backbone links can be lost. This can disconnect the backbone network. If no measures are taken, the fault *environmental dynamics* can lead to service failures. A fault is the event of environmental dynamics which decreases the $ARSS$ (Average Radio Signal Strength) up to a user-specified amount $\Delta ARSS$.

Fault-tolerant system design

Our system design uses redundancy for tolerating the faults. Figure 3 shows the state machine of our fault-tolerant system. The figure shows the system states, their attributes and their entry actions. The initial state is the normal state. In addition to the *correct service*, the normal system state contains *redundancy* for compensating the faults at run-time. In this normal state the system performs *concurrent error detection*, meaning that the error detection takes place during the normal service delivery. In the *error* state the redundancy is lost due to a fault, but the service is correct because the initial redundancy has compensated the negative effects of the fault. In this state, the system performs system recovery. The system recovery restores the initial redundancy. In the following sections we will specify how we applied this concept to the services *radio coverage* and to *connectivity*. For each service we will define the correct service specification, the redundancy and the error. A failure for both services occurs when the service consumer (a mobile station) tries to use the service and the service is not correct. Our fault-tolerant system design avoids the failures.

3.1.1. Radio coverage

Correct service

Radio coverage is correct if every service is covered by at least one base station with a radio signal strength of at least $ARSS_{Min}$.

Redundancy

In order to ensure correct radio coverage in case of faults, the normal system state uses radio signal strength redundancy. This means that every service location is covered by at least one base station with a radio signal strength of at least $ARSS_{RED}$. $ARSS_{RED}$ is the value of the redundant radio signal strength needed for compensating the environmental dynamics during the error detection and system recovery ($ARSS_{RED} = ARSS_{Min} + \Delta ARSS$).

Error

In the error state, the radio coverage is not as good as the radio coverage in the normal state, but the radio coverage is still correct. An error exists, if at some service location the $ARSS$ is less than the redundancy value, but it exceeds the minimum threshold for correct coverage: $ARSS_{RED} > ARSS \geq ARSS_{Min}$.

3.1.2. Connectivity

Correct service

Connectivity is correct if the backbone graph is connected.

Redundancy

In order to ensure correct connectivity in case of faults, the backbone graph is *biconnected* (2-connected). A graph is biconnected if any two vertices can be joined by two independent paths [9]. This backbone redundancy compensates for the loss of a backbone link as a result of a fault.

Error

In the error state, the backbone graph is not biconnected, but it is connected. The loss of biconnectivity can be caused by environmental dynamics leading to link loss. The loss of a link is not necessarily a connectivity error. It is an error only if it leads to loss of the biconnectivity.

3.2. Error detection

When faults occur and lead to errors, the errors have to be automatically detected by the system. Since we are considering two services, radio coverage and connectivity, we need methods for detecting radio coverage errors and connectivity errors. Figure 4 shows our methods for error detection and their integration in our fault-tolerant system design.

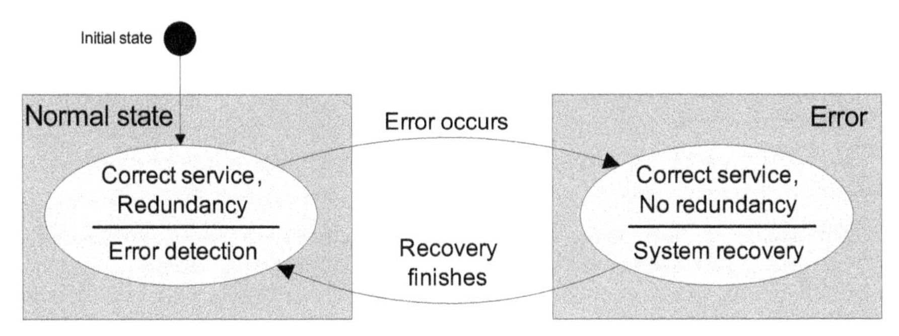

Figure 3. The states of our fault-tolerant system

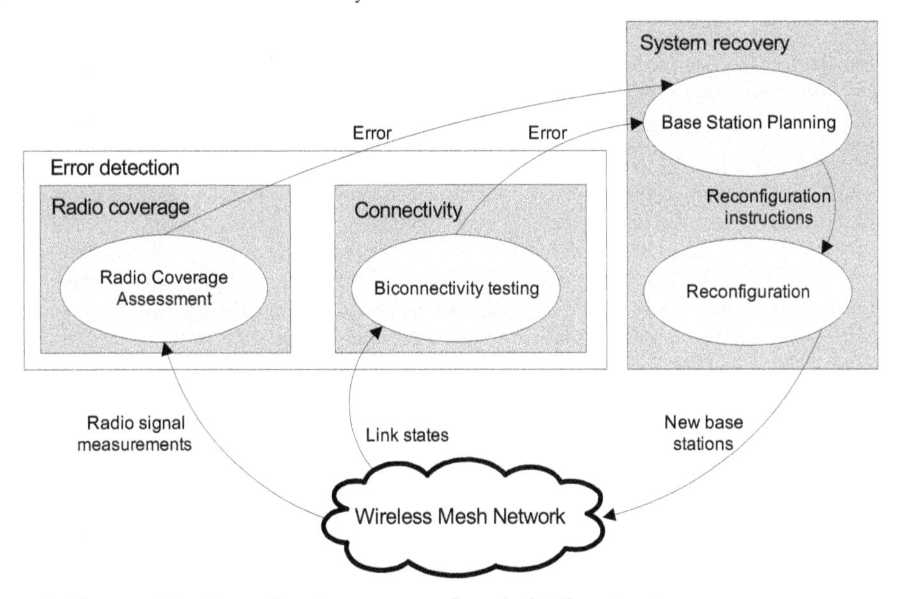

Figure 4. The error detection and system recovery of our fault-tolerant system

3.2.1. Connectivity error detection

For detecting connectivity errors we use a monitoring at the routing layer and a classic biconnectivity testing algorithm from graph theory [9]. This algorithm uses information about the backbone graph and determines whether it is biconnected or not. If the graph is not biconnected, then there is an error. The required information for biconnectivity testing are the edges (links) among the vertices (base stations) of the graph. In our scenario, this information is globally available at the management appliance. As a part of the routing protocol, the base stations monitor the backbone link states by exchanging control messages with other base stations [17]. The state of every link is determined by *two communication endpoints* (base stations). One of them sends control messages and the other one determines the link state based on a statistic on the received messages. The link state information is periodically updated and communicated, so the management appliance has an actual global view of the backbone network. Based on this global view, the management appliance performs biconnectivity testing. The fact that *every link state is determined by two communication endpoints* enables us to detect connectivity errors by *monitoring* at the routing layer. If the backbone link state information is not available globally, distributed biconnectivity testing algorithms can be used (e.g. [34]).

3.2.2. Radio coverage error detection

The information required for radio coverage error detection is the radio signal strength *at every service location*. However, a communication endpoint at every service location does not exist. Therefore, radio coverage errors can not be detected by monitoring, as with the connectivity errors. Nevertheless, a method for detecting these errors is needed because the environmental dynamics affect the radio coverage. The radio coverage should be guaranteed for every service location *before* a mobile station moves to those locations.

Our approach is to use a model-based assessment for detecting radio coverage errors at the physical layer. We use a radio propagation model for assessing the radio signal strength at every service location. This model has a tight relation to the propagation environment. We use measurements from the wireless network for calibrating the model to the reality.

In the state-of-the art assessment approaches the radio propagation models are static; meaning that they do not reflect the dynamics of the environment. The innovation of our approach is that the radio propagation model *automatically calibrates* to the real environment. *Radio model calibration* is the process of adjusting the model-parameters in such a way that the model reflects better a set of measurements from the actual propagation environment. *Radio coverage assessment* is the model-based estimation of the radio signal strength for the purpose of error detection. The radio model calibration method is out of scope of this chapter, but the reader can find a detailed description in [20, 23, 24].

3.3. System recovery

The system recovery transforms a system with errors to a system without errors. In our approach we use the same mechanism for recovery from radio coverage errors and for recovery from connectivity errors. This mechanism adds new base stations to the network. The new base stations improve the radio coverage by increasing the radio signal strength at the service locations. The new base stations also improve the connectivity by adding new links to the backbone network. Given a wireless mesh network with radio coverage and/or connectivity errors we have to decide how many base stations there is to install and and where to install them in order to correct the errors. For this purpose, we have developed an automatic base station planning algorithm (see section 4.

The error recovery includes automatic base station planning and manual reconfiguration (see figure 4). The management appliance runs the base station planning algorithm and gives instructions to the operating staff for the reconfiguration. The operating staff performs the reconfiguration which restores the redundancy of the services.

4. Automatic base station planning

This section describes our algorithm for automatic base station planning. It starts with a problem definition for the base station planning, followed by an overview of our approach in section 4.2. The following sections define the details of the algorithm, namely the used link state model, the optimization approach and the graph consolidation approach. This algorithm is published in [25]; in addition this section describes the integration with the presented fault-tolerance framework.

4.1. Problem definition

The problem of the base station planning algorithm is to find a minimum number of base stations to be installed which transform a wireless mesh network with radio coverage errors and/or connectivity errors to a system without errors. The existing algorithms for this type of problem in wireless mesh networks are computationally intractable, or do not provide the required fault-tolerance (see section 2.2 for a discussion). The following input information is given to the base station planning algorithm:

- Service location information. This is information about the service locations which have to be covered.
- Candidate sites information. This is information about possible locations of the base stations. The candidate sites and the service locations are specified by the deployment staff.
- Radio coverage information. This information is obtained from the radio propagation model. This is for every service location, the candidate sites which cover this service location, if base stations were installed at all candidate sites.
- Connectivity information: for every candidate site, the candidate sites which have a link in the backbone network, if base stations were installed at all candidate sites. For this purpose, we use our calibrated radio propagation model and a link state model (section 4.3).
- The currently installed base stations and their positions

The base station planning algorithm has to determine the number and positions of base stations to be installed such that:

- The radio coverage and the connectivity enter the normal state. The normal state includes redundancy in the services which has been defined in section 3.
- The algorithm should provide an acceptable relation between base stations minimality and running time. The running time of the algorithm should be appropriate for error detection and system recovery in a dynamic propagation environment.

The challenge of the defined problem is the connectivity requirement. The coverage requirement can be formally defined as a local property which depends only on the considered entities (e.g. a base station covers a service location). For the connectivity, the requirement is global. It includes all network paths among all pair of base stations. The existence of a path between two base stations depends not only on the considered base stations, but on the number and positions of all other base stations in the network. The fault-tolerance (biconnectivity) requirement increases the complexity of the problem. It has been shown that finding a minimum number of base stations for this type of problematic is an NP-complete problem. For this reason, we are looking for an approach, having a good balance between minimality and running time.

4.2. Overview of the algorithm

Our idea is to perform an optimization, satisfying a simple local network property which significantly affects the fulfillment of the global property (biconnectivity). This local property is the *minimum degree*. For the backbone (multi-hop) network, the *degree* of a base station is the number of links to other base stations. The minimum degree of the network is the least degree among all base stations. In graph theory, the minimum degree is a *necessary but not sufficient condition* for k-connectivity [9]. This means that a k-connected graph has a minimum degree of *k*, but a graph with minimum degree of *k* is not necessary k-connected. Formally, this rule applies to the backbone of wireless mesh networks. We consider both radio coverage and connectivity. The service locations are spread in some area (e.g. production hall). Hence, the probability that the necessary condition is also sufficient in mesh networks is significantly higher than the probability in graph theory. Therefore, our algorithm fulfills the local necessary condition and checks whether the global sufficient condition is also fulfilled. If not, the algorithm performs an incremental correction. The advantage of this approach is that it fulfills the connectivity requirement without increasing the complexity of the underlying optimization problem.

The algorithm operates in three steps: optimization, connectivity testing, and graph consolidation (figure 5). The optimization step finds an optimal solution for the optimization criteria. The optimization criteria are the radio coverage requirement and the *necessary condition for the connectivity* (the local property min. degree). The optimization uses the radio propagation model and the link state model. The connectivity testing step tests the resulted graph for biconnectivity (the sufficient condition). If the sufficient condition is true, the algorithm finishes. Otherwise the algorithm performs a graph consolidation step. The consolidation step maps biconnected parts of the to a single vertex. After the consolidation, the algorithm continues with the optimization step which is done based on the consolidated graph. After a few (expected 1-3) iterations, the algorithm produces a solution that satisfies the coverage requirements.

Example

The optimization step has produced a graph with minimum degree 2 (figure 6A) according to the necessary condition. This graph does not satisfy the biconnectivity requirements (one edge and two vertices exist whose removal disconnect the graph). The consolidation step identifies two sub-graphs which are biconnected, and maps them to vertices (figure 6B). Note that after the consolidation, the minimum degree of the graph is 1. Then the optimization step places a new base station, such that the consolidated graph plus the new vertex result in a graph with minimum degree of 2 (figure 6C). Finally, the deconsolidated graph satisfies the biconnectivity requirements.

4.3. Link state model

This section defines the used link model which models the link state based on the radio signal strength. The used link model in this chapter considers the operation of an ad-hoc routing protocol. We have shown in [17] that the communication in a mesh network is possible only if the links have some quality level.

The routing protocols determine the state of a link by analyzing the periodically received Hello packets from the neighbors. Depending on the mobility and the required stability of a link, different approaches for determining the link state at the routing layer exist [28, 31, 42]. What

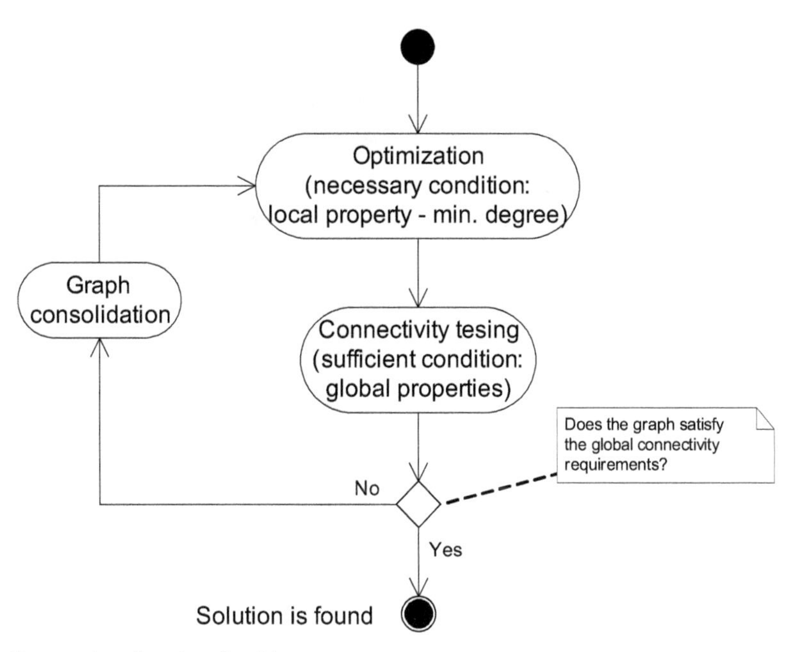

Figure 5. Base station planning algorithm

is common for all of them is the analysis of received Hello packets at the routing layer. The AWDS (Ad-hoc Wireless Distribution System) [1][17] routing software, for instance, identifies a link as existing if 10 consequent Hello packets in both directions are received correctly. A link is identified as non existing if 3 consequent Hello packets in either direction are not received.

The radio signal strength is one of the main factors which determine the reception of the packets at the receiver [14, 35]. This means that if the RSS is too low, then the wireless adapter can not decode the frame correctly. Therefore, to model the existence of a link, we use a threshold model based on $ARSS$. If the average radio signal strength exceeds the threshold ($ARSS \geq ARSS_{Min}$), then a link exists, otherwise a link does not exist. Remember that our fault-tolerance approach ensures that $ARSS \geq ARSS_{Min} + \Delta ARSS$.

There are other factors, influencing the packet loss and the link state (e.g. collision, radio interference). But the factor RSS is a *necessary condition* for successful frame decoding. In wireless mesh networks, it is one of the most influencing factors for the link state. This has been shown in our research in wireless mesh network routing [16–18], wireless network simulation and emulation [21]. Other researchers in our group are working on improving the link state model. They apply a data mining based approach for predicting the link state from various network monitoring information [28].

4.4. Optimization

4.4.1. Minimization approach

Our algorithm uses a minimization approach based on binary search for finding the minimum number of base stations (BS_{min}) which satisfies the optimization criteria. It searches iteratively

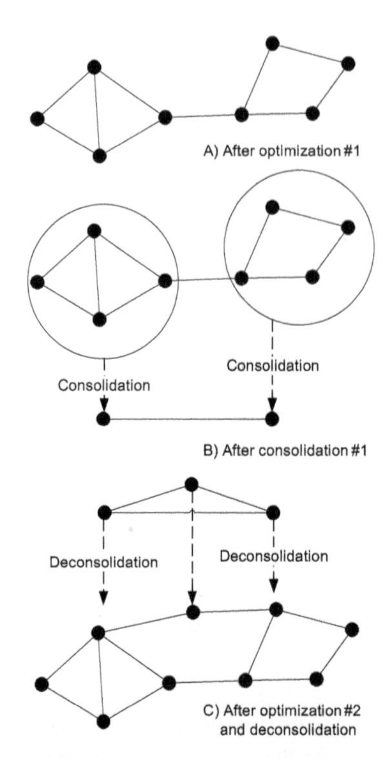

Figure 6. Example operation of the base station planning algorithm

the interval between a lower bound BS_{low} and an upper bound BS_{up}. At each iteration, the algorithm chooses the middle of the interval as a current value for BS and determines whether a solution is possible by solving an optimization problem. If the solution satisfies the optimization criteria, then the algorithm decreases BS by searching the lower half of the interval, otherwise it increases BS by searching the upper half of the interval. Finally, the algorithm finds a minimum value for BS which satisfies the optimization criteria.

4.4.2. Optimization problem formulation

The optimization performed at each iteration can be defined by the following:

- Variables

  The optimization variables are the positions of the base stations $(X, Y, Z)_{BS}$. We consider a typical multi-hop network, operating in a single frequency. Therefore, the frequency assignment is a constant for all base stations.

- Bounds

  The variables have lower and upper bounds according to the candidate sites information, provided by the user. For instance, if the base stations are to be installed on the ceiling of a production hall with dimensions 200x300x6m, then the bounds are: $0 \leq X \leq 200$, $0 \leq Y \leq 300, Z = 6$. For the currently installed base stations, the lower and upper bounds are equal to the base stations coordinates. In this way, they are considered in the solution, but are not relocated by the algorithm.

Algorithm 1 Objective function of the optimization step

```
function Objective (X, Y, Z)
{
PenaltyCoverage = 50;
PenaltyConnectivity = 100;
Coverage = Model.RadioCoverage(X, Y, Z);
Connectivity = Model.BSDegree(X, Y, Z);
ShortfallCoverage = sum(Nlm -
      Coverage(find(Coverage < Nlm)));
ShortfallConnectivity = sum(Nbb -
      Connectivity(find(Connectivity < Nbb)));
Objective = mean(Coverage)
            + mean(Connectivity)
            - PenaltyCoverage*ShortfallCoverage
            - PenaltyConnectivity*ShortfallConnectivity;
}
```

- Service locations
  The service locations, defined by their coordinates, are stored in the set SL.
- Radio coverage model
  From the values of the variables $(X,Y,Z)_{BS}$ the radio coverage model provides the radio coverage by the function *Model.RadioCoverage*$((X,Y,Z)_{BS})$. The result is a vector. For every service location in the set SL, it contains the number of base stations that cover this service location. The calculation is based on the calibrated radio propagation model.
- Connectivity model: *Model.BSDegree*$((X,Y,Z)_{BS})$. The result is a vector. For every base station, it contains the number of links to other base stations. The calculation is based on the calibrated radio propagation model and the link state model.
- Objective function
  The objective function (Matlab pseudo code in algorithm 1) influences the solution in a direction which satisfies the optimization criteria (the coverage requirements and the necessary condition for connectivity). In addition, the objective function maximizes the mean radio coverage degree and the mean backbone degree. The radio coverage degree is the number of base stations covering a service location. From the input coordinates, the radio coverage model and the link state model, the function calculates the radio coverage degree and the backbone degree. For base stations which have less than $N_{bb} = 2$ links to other base stations, the function calculates the backbone shortfall. This is the sum of the differences between the required and the current degree over all base stations. The shortfall is weighted by a backbone penalty factor and subtracted by the objective function. The penalty factor is a relatively large number, compared to the mean values which influences the solution to a direction of a zero shortfall. The processing for the radio coverage links is similar. The objective function should be maximized.

4.4.3. Optimization problem solving

In order to solve this optimization problem, we apply an optimization method. Specially for this problem is that the objective function can not be differentiated. This is because the objective function, can not be represented as an algebraic function of only the optimization parameters $(X, Y, Z)_{BS}$. This is because the objective function contains the radio coverage model which includes the geometry of the model. Several algorithms exist for solving this type of problem (pattern search, genetic algorithm, simulated annealing). We have selected pattern search, because it has a proven convergence and supports any type of constraints [33].

4.5. Connectivity testing

For k-connectivity testing in a graph with n vertices, we use existing algorithms from the graph theory [9]. The complexity of this algorithm is $O(k * n^3)$, under the condition that $k < \sqrt{n}$ which is true in our case.

4.6. Graph consolidation

In this step, the algorithm finds sub-graphs satisfying the connectivity requirements and transforms each subgraph into a single vertex. The formal specification of the graph consolidation step is described by pseudo code in algorithm 2 which is explained in the following list. Figure 7 shows an example of the operation of the graph consolidation step.

1. Given a graph G, identify all biconnected components G_c containing at least 3 vertices and store them in a set BC. For finding biconnected components, existing graph theory algorithms are used.
2. Identify the *special articulation points* which are articulation points shared between the biconnected components in the set BC. An articulation point is a vertex whose removal disconnects a graph. On figure 7B) vertices 1, 2 and 3 are articulation points. Vertex 1 is a special articulation point, since it is shared between two biconnected components of size of at least 3. For identifying biconnected components and articulation points existing graph algorithms are used [9].
3. Every vertex which is either a special articulation point or other vertex, not belonging to a biconnected component in BC, is directly transformed into a vertex in the consolidated graph. The consolidated vertex inherits all edges of the original vertex.
4. For every biconnected component in the set BC:
    (a) If it contains special articulation points, then they are removed from the component.
    (b) All vertices from the component are transformed into a single vertex in the consolidated graph.
    (c) The consolidated vertex inherits all edges of the original vertices to other vertices in the graph. Other vertices are vertices not belonging to the same biconnected component.
    (d) Duplicated edges in the consolidated graph are removed.

5. Evaluation approach and implementation

We will present an evaluation of the base station planning algorithm according to the following evaluation criteria:

- Fault-tolerance: this shows the algorithm's ability to generate a network configuration that satisfies the fault-tolerance coverage requirements.

Algorithm 2 Pseudo code of the graph consolidation step

1. $BC = find.biconnected.components(G, |G_c| \geq 3)$
2. $V_{sap} = find.articulation.points(G, shared.among(G_c \in BC))$
3. $foreach\ \ v \in V_{sap} \cup (V(G) - V(BC))$:
   (a) $v \rightarrow v'$
   (b) $E(v') = E(v)$
4. $foreach\ \ G_c \in BC$:
   (a) $G_c = G_c - V_{sap}$
   (b) $G_c \rightarrow v'$
   (c) $E(v') = ExternalEdges(G_c)$
   (d) $remove.duplicate.edges(v')$

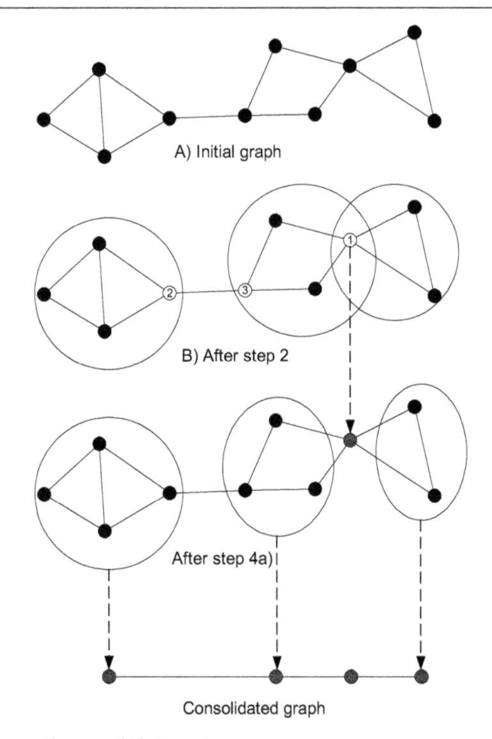

Figure 7. Example of the graph consolidation step

- Termination: this shows the number of iterations the algorithm needs to complete and the running time.
- Minimality: this shows the ability of the algorithm to use minimum number of base stations.

Parameter	Values
Transmit power P_{tx} [dBm]	20
Required receive power P_{min} [dBm]	-78
Path loss exponent	3
Area size (X/Y) [meters]	(50/50),(100/100),...,(300/300)
Shadowing deviation σ [dB]	5,6,7,8,9,10

Table 1. Evaluation parameters

We performed a model-based evaluation of the algorithm. We generated different inputs to the algorithm, then executed the algorithm and observed the evaluation criteria. As an input of the algorithm, we used a service area with various sizes; typical for a production environment (see table 1 for the parameter values). The service locations comprise of the entire floor. The candidate sites comprise of the entire ceiling. We also varied the attenuation of the propagation environment. For the radio connectivity model, we used the log-normal shadowing propagation model [36] which is used for radio coverage assessment. The path loss exponent has been fixed in these experiments. The shadowing factor X_σ models the inhomogeneity of the propagation environment and it has been varied in these experiments. The other parameters of the propagation model are fixed. To determine the connectivity, we used our threshold-based link state model. The base station planning algorithm has been implemented in Matlab (about 600 lines of code). The algorithm has been tested on all the combinations of input parameters (area size and shadowing deviation) which make a total of 36 executions. At the end of each algorithm execution, we performed a requirements test. We tested whether the radio coverage and the connectivity were in normal (redundant) state.

5.1. Results for fault-tolerance

With all the inputs, the algorithm has generated a network topology in which the radio coverage and the connectivity were in the normal (redundant) state, as defined in section 3. An example graph of the network topology, generated by the algorithm for area size 200/200m and shadowing deviation 8 is shown on figure 8. The related work algorithms [2, 39] generated topologies which are not fault-tolerant. Their topologies optimized the network throughput, but the backbone network war not biconnected (see figure 3 in [2], and figure 4 in [39]). Figure 8 clearly shows the effect of the shadowing (inhomogeneous environment) on the base station planning. Because of the shadowing, some links are shorter than others and in some areas, more base stations are needed to provide coverage.

5.2. Results for termination, minimality and running time

Figure 9 shows the measured termination property of the algorithm within the performed evaluation. The figure shows the cumulative termination, i.e. the percentage of the algorithm executions that have terminated *up to* some number of iterations. 30% of the algorithm executions generated a correct fault-tolerant solution directly after the first iteration. This means that in these cases, the graph consolidation step was not performed at all. These were the cases when the area sizes were smaller (50/50m and 100/100m). 80% of the algorithm executions generated a correct fault-tolerant topology after the second iteration. This means that only two optimizations and one graph consolidation were needed. The algorithm needed a maximum four iterations to complete all the inputs.

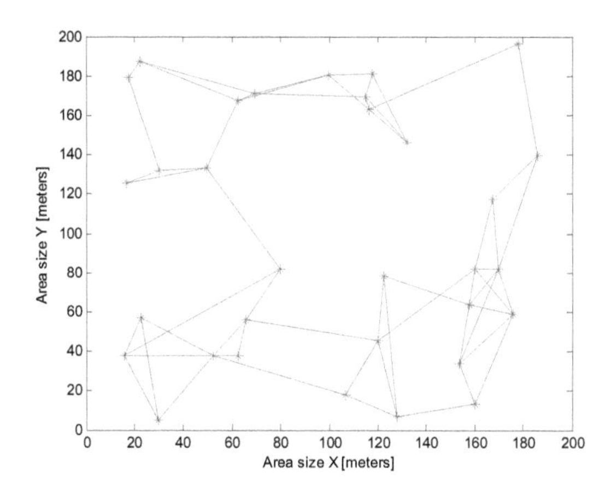

Figure 8. Example fault-tolerant (biconnected) topology produced by the algorithm

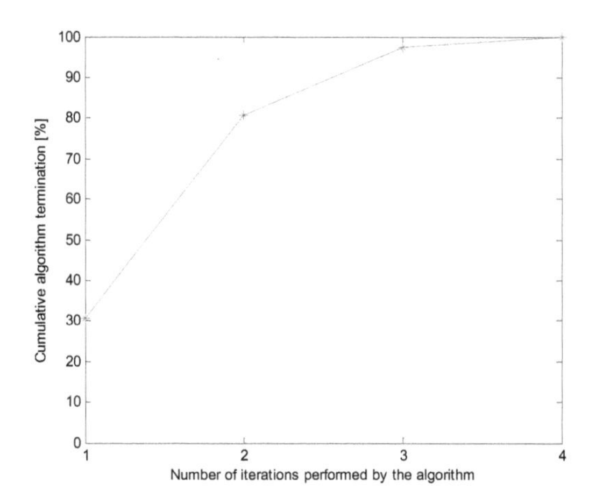

Figure 9. Algorithm termination: 80% of all algorithm executions terminated after 2 iterations. The algorithm needed a maximum of 4 iterations to complete.

90% of the base stations were selected at the first algorithm iteration. This means that 90% were selected according to the global optimization function and were optimally placed. The remaining 10% of the base stations were selected during the subsequent algorithm iterations in order to ensure the biconnectivity of the backbone. Figure 10 shows the result after the first iteration for area size 150/150m and shadowing deviation 7. In the middle of the graph (around coordinates 65/44), a base station exists, whose removal would disconnect the network. In the next iteration the algorithm corrected this by inserting one base station in proximity of the first one (see figure 11).

For the total 36 executions, the algorithm needed about 25 minutes to complete on a laptop with a dual core 2.5GHz processor and 3GB operating memory. This means that the average running time was 42 seconds. As a comparison, a related work algorithm in [39] needed 22

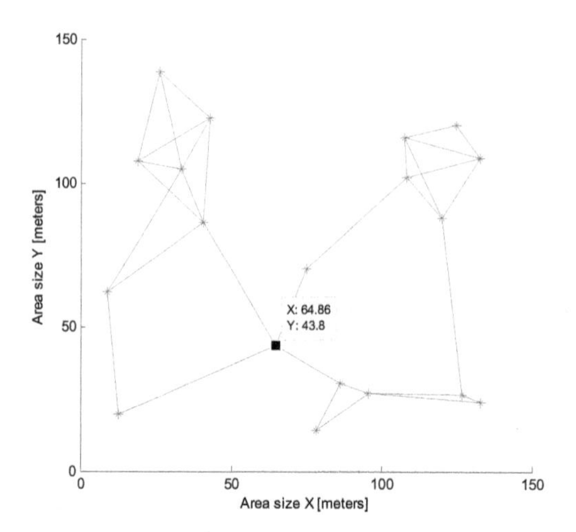

Figure 10. Example network topology after the first algorithm iteration

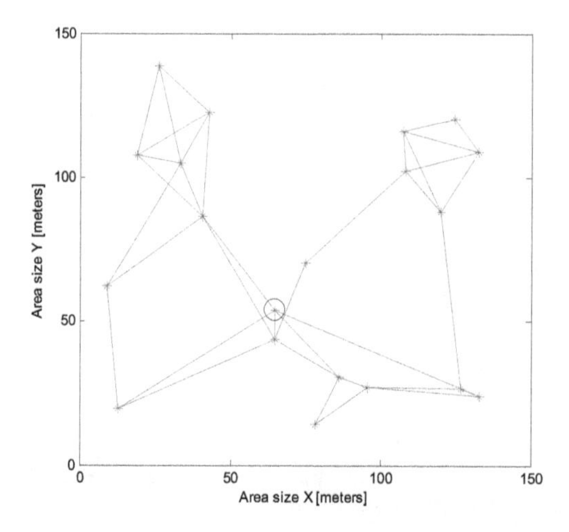

Figure 11. Example network topology after the second algorithm iteration. Only one additional base station results in a biconnected topology.

hours for a 58-node scenario because of the intractability of the approach. This means that for the purpose of the system recovery, our algorithm has an acceptable running time.

6. Conclusion

In this chapter, we developed a new approach for guaranteeing the availability of the services radio coverage and connectivity of Wireless Mesh Networks in dynamic propagation environments. Our approach is to apply fault-tolerance for avoiding service failures in the presence of environmental dynamics. Differing from the existing methods, we use reconfigurable redundancy of the services. As the radio propagation environment changes, our method changes the redundancy of services.

When the environmental dynamics is detected, the system recovery adds base stations to the network for restoring the redundancy of the services. But firstly, it has to be decided what the minimum number of base stations would be (and respectively their positions) which will restore the redundancy. For this purpose, we developed a new base station planning algorithm which takes the required decision and proposes reconfiguration instructions. Since the underlying optimization problem is NP complete, our algorithm is a trade-off between minimum base stations and minimum running time. The operating staff performs the network reconfiguration which restores the redundancy of the services.

In future work the presented concept will be integrated in a system for dependable end-to-end communication in wireless mesh networks. This system will incorporate other ongoing research works within our working group [30, 31] developing concepts for end-to-end quality of service guarantees (throughput, packet loss, latency) in Wireless Mesh Networks. Another aspect of our future work is to integrate the developed concepts in components for industrial wireless communication in cooperation with german product manufacturers.

Acknowledgement

This work has been partially funded by the European Commission within the EU-project flexWARE, grant number 224359.

The project flexWARE (Flexible Wireless Automation in Real-Time Environments) develops a communication system for factory-wide wireless real-time control [12, 13, 38]. The methods, presented in this chapter, are used in flexWARE to achieve availability of the wireless medium by radio coverage monitoring and prediction and network engineering.

Author details

Svilen Ivanov
rt-solutions.de GmbH, Oberländer Ufer 190a, D-50968 Cologne, Germany

Edgar Nett
Institut of Distributed Systems, Otto von Guericke University of Magdeburg, Universitätsplatz 2, 39106 Magdeburg, Germany

7. References

[1] *Ad-Hoc Wireless Distribution System* [n.d.]. http://awds.berlios.de/.

[2] Amaldi, E., Capone, A., Cesana, M., Filippini, I. & Malucelli, F. [2008]. Optimization models and methods for planning wireless mesh networks, *Elsevier Journal on Computer Networks* 52(11): 2159–2171.

[3] Avizienis, A., Laprie, J.-C., Randell, B. & Landwehr, C. [2004]. Basic Concepts and Taxonomy of Dependable and Secure Computing, *IEEE TRANSACTIONS ON DEPENDABLE AND SECURE COMPUTING* 1: 11–33.

[4] Avresky, D. & Natchev, N. [2005]. Dynamic reconfiguration in computer clusters with irregular topologies in the presence of multiple node and link failures, *IEEE Transactions on Computers* 54(5): 603–615.

[5] Bello, L. L. & Toscano, E. [2009]. An Adaptive Approach to Topology Management in Large and Dense Real-Time Wireless Sensor Networks, *IEEE TRANSACTIONS ON INDUSTRIAL INFORMATICS* 5(3): 314–324.

[6] Blough, D. M., Leoncini, M., Resta, G. & Santi, P. [2006]. The k-neighbors approach to interference bounded and symmetric topology control in ad hoc networks, *IEEE Transactions on Mobile Computing* 5(9): 1267–1282.

[7] Bosio, S., Capone, A. & Cesana, M. [2007]. Radio Planning of Wireless Local Area Networks, *IEEE/ACM Transactions on Networking* 15(6): 1414–1427.

[8] Chan, T. M., Man, K. F., Tang, K. S. & Kwong, S. [2007]. A Jumping-Genes Paradigm for Optimizing Factory WLAN Network, *IEEE TRANSACTIONS ON INDUSTRIAL INFORMATICS* 3: 33–43.

[9] Diestel, R. [2005]. *Graph Theory*, Springer-Verlag, Heidelberg.

[10] *Ekahau Site Survey – Wi-Fi Planning and Site Survey Tool* [n.d.]. http://www.ekahau.com/products/ekahau-site-survey/overview.html, accessed 02.05.2012.

[11] ElMaraghy, H. A. [2009]. *Changeable and Reconfigurable Manufacturing Systems*, Springer Series in Advanced Manufacturing.
URL: *http://www.springer.com/engineering/production+eng/book/978-1-84882-066-1*

[12] *flexWARE - Flexible Wireless Automation in Real-Time Environments. Research project funded under the EU-FP7 programme.* [n.d.]. http://www.flexware.at/.

[13] Gaderer, G. & Trsek, H. [2010]. flexWARE - Drahtlose Echtzeitkommunikation für die Fertigungsautomatisierung, *Kommunikation in der Automation, KommA*.

[14] Gungor, V., Lu, B. & Hancke, G. [2010]. Opportunities and challenges of wireless sensor networks in smart grid, *Industrial Electronics, IEEE Transactions on* 57(10): 3557 –3564.

[15] Herms, A. [2009]. *Dienstg"ute in Wireless Mesh Networks*, PhD thesis, Otto-von-Guericke-Universit"at.

[16] Herms, A., Ivanov, S. & Lukas, G. [2007]. Precise admission control for bandwidth reservation in wireless mesh networks, *Proceedings of 4th Intl. Conference on Mobile Ad-Hoc and Sensor Networks MASS'07*, Pisa, Italy.

[17] Herms, A., Lukas, G. & Ivanov, S. [2006]. Realism in design and evaluation of wireless routing protocols, *in* O. Spaniol (ed.), *Proceedings of First international Workshop on Mobile Services and Personalized Environments (MSPE'06)*, Vol. P-102, Lecture Notes in Informatics (LNI), Aachen, Germany, pp. 57–70.
URL: *http://ivs.cs.uni-magdeburg.de/EuK/forschung/publikationen/pdf/2006/awds-mspe.pdf*

[18] Herms, A., Lukas, G. & Ivanov, S. [2007]. Measurement-based detection of interfering neighbors for QoS in wireless mesh networks, *16th IST Mobile and Wireless Communications Summit 2007, Proceedings of.* URL: *http://ivs.cs.uni-magdeburg.de/EuK/forschung/publikationen/pdf/2007/interf-mobilesummit2007.pdf*

[19] Hsu, C.-Y., Wu, J.-L. C., Wang, S.-T. & Hong, C.-Y. [2008]. Survivable and delay-guaranteed backbone wireless mesh network design, *Journal of Parallel and Distributed Computing* 68(3): 306 – 320. Wireless Mesh Networks: Behavior, Artifacts and Solutions.
URL: *http://www.sciencedirect.com/science/article/B6WKJ-4NRT39Y-2/2/65080a92e72ed991ef6964c8430ddab1*

[20] Ivanov, S. [2011]. *Fault-tolerant Radio Coverage and Connectivity in Wireless Mesh Networks*, PhD thesis, University of Magdeburg.
URL: *http://edoc2.bibliothek.uni-halle.de/hs/content/titleinfo/2030*

[21] Ivanov, S., Herms, A. & Lukas, G. [2007]. Experimental Validation of the NS-2 Wireless Model using Simulation, Emulation, and Real Network, *Proceedings of the 4th Workshop on*

Mobile Ad-Hoc Networks (WMAN'07), in conjunction with the 15th ITG/GI - Fachtagung Kommunikation in Verteilten Systemen (KiVS'07), VDE Verlag, pp. 433 – 444. URL: *http://ivs.cs.uni-magdeburg.de/EuK/forschung/publikationen/pdf/2007/wman07magdeburg.pdf*

[22] Ivanov, S. & Nett, E. [2008]. Fault-tolerant Coverage Planning in Wireless Networks, *27th IEEE International Symposium on Reliable Distributed Systems (SRDS 2008), 6-8 October, Napoli, Italy.*

[23] Ivanov, S. & Nett, E. [to appear 2012]. Localization-Based Radio Model Calibration for Fault-tolerant Wireless Mesh Networks, *IEEE Transactions on Industrial Informatics* .

[24] Ivanov, S., Nett, E. & Schemmer, S. [2007]. Planning Available WLAN in Dynamic Production Environments, *7th IFAC International Conference on Fieldbuses and Networks in Industrial and Embedded Systems.*

[25] Ivanov, S., Nett, E. & Schumann, R. [2010]. Fault-tolerant Base Station Planning of Wireless Mesh Networks in Dynamic Industrial Environments, *15th IEEE International Conference on Emerging Technologies and Factory Automation (ETFA), Bilbao, Spain.*

[26] Koren, Y. [2006]. *Reconfigurable Manufacturing Systems and Transformable Factories*, Springer, chapter General RMS Characteristics. Comparison with Dedicated and Flexible Systems, pp. 27–45.

[27] Lin, B. & Ho, P.-H. [2009]. Dimensioning and location planning of broadband wireless networks under multi-level cooperative relaying, *IEEE Transactions on Wireless Communications* 8(11): 5682–5691.

[28] Lindhorst, T. [2009]. Schicht"ubergreifende fr"uherkennung von verbindungsausf"allen in drahtlosen mesh-netzwerken, *in* P. H. Wolfgang A. Halang (ed.), *Software-intensive verteilte Echtzeitsysteme - Echtzeit 2009*, Informatik aktuell, Fachtagung des GI/GMA - Fachausschusses Echtzeitsysteme, Springer, p. 67 ff.

[29] Lindhorst, T., Lukas, G. & Nett, E. [2010]. Modeling fast link failure detection for dependable wireless mesh networks, *IEEE International Symposium on Network Computing and Applications*, IEEE.

[30] Lindhorst, T., Lukas, G., Nett, E. & Mock, M. [2010]. Data-mining-based link failure detection for wireless mesh networks, *IEEE International Symposium on Reliable Distributed Systems*, IEEE.

[31] Lukas, G., Lindhorst, T. & Nett, E. [2011]. Modeling medium utilization for admission control in industrial wireless mesh networks, *IEEE International Symposium on Reliable Distributed Systems*, IEEE.

[32] Mahrenholz, D. [2006]. *Providing QoS for Publish/Subscribe Communication in Dynamic Ad-hoc Networks*, Doctoral dissertation, University of Magdeburg. online `http://diglib.uni-magdeburg.de/Dissertationen/2006/danmahrenholz.pdf`.

[33] MathWorks [2010]. Matlab Global Optimization Toolbox – User's Guide.

[34] Milic, B. [2009]. *Distributed Biconnectivity Testing in Wireless Multi-hop Networks*, PhD thesis, Humboldt University – Berlin.

[35] Pathak, P. & Dutta, R. [2011]. A survey of network design problems and joint design approaches in wireless mesh networks, *Communications Surveys Tutorials, IEEE* 13(3): 396 –428.

[36] Rappaport, T. S. [2002]. *Wireless Communications - Principles and Practice*, Prentice Hall PTR.

[37] Ray, A. [2009]. Planning and analysis tool for large scale deployment of wireless sensor network, *International Journal of Next-Generation Networks (IJNGN)* 1(1): 29–36.

[38] Sauter, T., Jasperneite, J. & Bello, L. L. [2009]. Towards New Hybrid Networks for Industrial Automation, *ETFA'09: Proceedings of the 14th IEEE international conference*

on Emerging technologies & factory automation, IEEE Press, Piscataway, NJ, USA, pp. 1141–1148.
[39] So, A. & Liang, B. [2009]. Optimal placement and channel assignment of relay stations in heterogeneous wireless mesh networks by modified bender's decomposition, *Elsevier Journal on Ad Hoc Networks* 7(1): 118–135.
[40] Toscano, E. & Bello, L. L. [2010]. A novel approach for data forwarding in industrial wireless sensor networks, *15th IEEE International Conference on Emerging Technologies and Factory Automation (ETFA), Bilbao, Spain.*
[41] Wang, B. [2010]. *Coverage Control in Sensor Networks*, Springer London.
[42] Yackoski, J. & Shen, C.-C. [2006]. Cross-layer inference-based fast link error recovery for manets, *Wireless Communications and Networking Conference*, Vol. 2, IEEE, pp. 715 – 722.

Chapter 7

A Correctness Proof of a Mesh Security Architecture

Doug Kuhlman, Ryan Moriarty, Tony Braskich, Steve Emeott and Mahesh Tripunitara

Additional information is available at the end of the chapter

1. Introduction

We discuss our proof of security properties of a standards-track protocol suite for authentication and key establishment using a formal verification technique. Our technique is Protocol Composition Logic (PCL) [15] (see Section 2.1). Our setting is the IEEE 802.11 Mesh Networking task group, known as 802.11s, which was formed to define extensions to IEEE 802.11 [1] to support wireless mesh networking [25]. A goal of the task group is to secure a mesh by utilizing existing IEEE 802.11 security mechanisms and extensions.

The Mesh Security Architecture (MSA) proposal [4–7] to 802.11s consists of a definition of a key hierarchy and a suite of protocols to enable security in a wireless mesh network. The proposal includes detailed information to implement MSA within the framework defined by 802.11s, including key derivation, protocol execution, and message formatting. The suite of protocols encompasses all the necessary components to create and maintain a mesh of nodes.

We describe the following three major contributions in this chapter:

- We conduct a comprehensive assessment of all 10 protocols (averaging 4 messages and 8 components) of the MSA proposal from a security standpoint and proven its correctness. We present an overview of the protocol suite and the main insights from the proof. The full details are generally unenlightening; a companion technical report [28] complements this chapter.

  As this is one of few instances of the proof of correctness of a substantial, standards-track protocol suite of which we are aware, we feel that this is an important contribution.
- PCL has been used to prove the correctness of the IEEE 802.11i protocol suite [26]. However, 802.11s presents new challenges that have necessitated extensions to PCL for us to be able to carry out our correctness proof. We present these extensions and details from the MSA proposal that illustrate their necessity (see Section 3). We believe that the extensions are general enough to be useful in other work in protocol verification.

- In the course of carrying out our proof, we discovered two security issues with protocols in the proposal. We discuss these issues and our suggestions for changes to address them. Our suggestions have since been incorporated into the proposal. As we point out in Section 5, our proof would not have been possible without these changes.

The remainder of this chapter is organized as follows. In Section 2, we provide a background on PCL, 802.11s and the MSA proposal. In Section 3 we present our additions to PCL; for each addition we illustrate its need via components from the protocol suite we have analyzed. We provide an overview of the proof in Section 4. In Section 5, we discuss our recommendations for changes to the original design of the protocol suite in the MSA proposal based on our proof efforts. We conclude with future work and general conclusions in Section 6.

2. Preliminaries

In this section, we provide some background and motivations for our work.

2.1. Overview of proof method

We use Protocol Composition Logic (PCL) to prove the correctness and security of the Mesh Security Architecture. We provide a brief overview of PCL in this section. PCL has been used for a security analysis of 802.11i [26], Kerberos [32], and the Group Domain of Interpretation (GDOI) protocol [29].

2.1.1. Terminology

Protocols in PCL are modeled using a particular syntax. A *role* specifies a sequence of actions performed by an honest party. A matching set of roles (two, in this chapter) define a *protocol*. A particular instance of a role run by a specific principal is a *thread*. Possible actions inside a thread include nonce generation, signature creation, encryption, hash calculation, network communication, and pattern matching (which includes decryption and signature verification). Each thread consists of a number of *basic sequences*, each of which has pre- and post-conditions. A basic sequence is a series of actions, which may not include a blocking action (like receive) except as the first action. Pre- and post-conditions are assertions expressed as logic predicates that must be true before and after a protocol run, respectively. Each basic sequence may have pre- and post-conditions as well, allowing for additional reasoning about certain actions.

2.1.2. Notation

We use the following notation in this chapter. Our notation is consistent with previous work on PCL except for the extensions that we propose in this chapter (see Section 3 for a discussion of the extensions).

$X, Y, Z, \ldots$ are used to denote threads.

$\hat{X}, \hat{Y}, \hat{Z}, \ldots$ denote the principals associated with the corresponding threads.

send, receive, new, ... are *actions*. Actions are things that principals do in a thread.

MKHSH, TLS:CLNT, 4WAY, ... denote protocols. We use the convention of protocol:role to note both the protocol and the associated role that a principal plays in an instance of the protocol; for example, in **TLS:CLNT**, **CLNT** denotes that it is the client's portion of the **TLS** protocol.

$pmk_{X,Z}$, gtk_X, ... denote cryptographic keys. We use subscripts to indicate the principal(s) with whom a key is associated.

θ, Φ, Γ, ... are used to denote logic formulae that express pre- or post-conditions, or *invariants*.

Has(), KOHonest(), SafeMsg(), ... are logic predicates that are used in assertions (pre- and post-conditions, and invariants).

Many of the predicates follow a $Pred(actor, action)$ format. Thus, Has(X, m) means that thread X has information m. Similar predicate formats follow for Send, Receive, New, and Computes. Other predicates can be more complicated. Honest$(\hat{X})$ means that the principal $(\hat{X})$ running the thread is honest. KOHonest$(s, \mathcal{K})$ essentially means that all principals with access to any key $k \in \mathcal{K}$ or to the value s are honest. Contains(m, t) is equivalent to $t \subseteq m$ and means that information t is a subterm of m.

2.1.3. *Proof methodology*

The proof methodology of PCL is described by Durgin et al. [21, 22] and Datta et al. [12–18, 26, 32]. We use the standard syntax of $\theta[P]_X\Phi$. This means that with preconditions θ before the run of actions P by thread X, the result (postcondition) Φ is proven to hold. θ is always used to denote a precondition, Φ a postcondition, and Γ an invariant.

The proof system is built on three fundamental building blocks. The first is a series of first-order logical axioms [15]. A first-order logical axiom is a natural logical assumption (e.g., creation of a value implies possession of that value). The second is a series of cryptographic/security axioms [15, 22, 26]. Cryptographic axioms provide formal logic equivalents of standard cryptography (e.g., possession of a key and a value provides possession of the encryption of the value with that key). These assume idealized cryptographic functionality which most cryptographic primitives do not achieve in practice. For example, the hash of two different values is assumed to never be the same.

The third building block is the fundamental principle of *honesty*. Honesty imposes certain restrictions on roles – that they follow protocol descriptions correctly and do not send out particular information assigned to that role. Honesty is a special type of rule that allows an instance of a thread to reason about the actions of another, corresponding thread that participates in the same protocol. The actions of an attacker are not assumed to be honest. We do, however, assume that the attacker does not violate an assumption, condition or invariant (e.g., the possession of a private key) that is necessary for a protocol to run to completion. This notion of an attacker model is the same as that considered in previous work that uses approaches based on mathematical logic to verify protocols (c.f. [26]).

All but one of the axioms on which we depend have been proposed previously [12, 15, 16, 26]; space constraints preclude the presentation of a comprehensive list of all PCL axioms in this chapter. We provide a few frequently used axioms in Figure 1. We do, however, point out that we need an additional axiom: a node which generates a signature over some (previously-defined) information has that information and the key with which the signature is generated. The existence of information m outside of the computation is important to eliminate concerns about existential signature forgery.

SIG1: $m \wedge \text{Computes}(X, SIG_k(m)) \supset \text{Has}(X, m) \wedge \text{Has}(X, k)$.

(Computes() and Has() are predicates, $\supset$ can be read as "implies," $\wedge$ is logical conjunction and $a < b$ indicates a occurred temporally before b.)

AA1	$\phi[a]_X \mathsf{a}$
AA4	$\phi[a_1; a_2; \ldots; a_k]_X \mathsf{a}_1 < \mathsf{a}_2 \wedge \ldots \wedge \mathsf{a}_{k-1} < \mathsf{a}_k$
AN2	$\phi[\text{New } x]_X \text{Has}(Y, x) \supset Y = X$
AN3	$\phi[\text{New } x]_X \text{Fresh}(X, x)$
ARP	$\text{Receive}(X, p(x))[\text{match } q(x)/q(t)]_X$
	$\text{Receive}(X, p(t))$
FS1	$\text{Fresh}(X, t)[\text{send } t']_X$
	$\text{FirstSend}(X, t, t') \forall t \subseteq t'$
FS2	$\text{FirstSend}(X, t, t') \wedge \mathsf{a}(Y, t'') \supset \text{Send}(X, t') < \mathsf{a}(Y, t'')$, where $X \neq Y \wedge t \subseteq t''$
HASH1	$\text{Computes}(X, HASH_K(x)) \supset \text{Has}(X, x) \wedge \text{Has}(X, K)$

Figure 1. Some PCL Axioms Used in MSA Proofs

The methodology of PCL has proven very successful in dealing with large-scale architectures. A recent paper by Cremers looked at the soundness of the various axioms of PCL [11]. For the problem of preceding actions, we have consistently used implicit pre- and post-conditions at the basic sequence level, leading to a tighter joining of actions. Another issue arises with the **HASH3** axiom. We propose a straight-forward generalization of the **HASH3** axiom, following earlier work on signatures. We define a new axiom, which is sound.

HASH3$'$: $\text{Receive}(X, HASH_K(x)) \supset \exists Y.\text{Computes}(Y, HASH_K(x)) \wedge \text{Send}(Y, m) \wedge \text{Contains}(m, HASH_K(x))$.

2.1.4. Composing proofs

An important feature of PCL is that with it, we can reason about how protocols interact. As this chapter covers an entire architecture, it is imperative that the large number of individual protocols be proven secure not only independently, but also working together in conjunction as a complete system. To this end, we extensively use the methodology of protocol composition developed by Datta et al. [15]. We discuss this in Section 4.3. Alternate composition methods are available, in certain circumstances [10].

2.2. Overview of the MSA proposal for 802.11s

The 802.11s task group is working to develop a mesh networking protocol that sets up auto-configuring, multi-hop paths between wireless stations to support the exchange of data packets. A goal of the task group is to utilize existing IEEE 802.11 security mechanisms [1], with extensions, to secure a mesh in which all the stations are controlled by a single logical administrative entity from the standpoint of security [25]. The 802.11s task group continues to refine its draft specification through the resolution of comments received during a review of the specification that began in late 2006 [4–7].

A mesh network is a collection of network nodes, each of which can communicate with the others. Several kinds of nodes are specified in the MSA proposal. One is a *Mesh Point* (MP), a member of the mesh that can communicate with other nodes. Each mesh has at least one *Mesh Key Distributor* (MKD) which is an MP that is responsible for much of the key management within its *domain* (a MKD's domain is the set of nodes with which it has a secure connection). The MKD also provides a secure link to an external authentication server (e.g., a RADIUS [30] server). A *Mesh Authenticator* (MA) is an MP which has been authorized by the MKD to participate in key distribution within the mesh. A *Candidate MP* is an entity that wishes to join the mesh but is not yet an MP.

Differences from 802.11i Part of the MSA proposal is very similar to the 802.11i protocol suite [1]. In 802.11i, connections are established between authenticators and supplicants in a server-client topology. An authenticator is connected to a backbone infrastructure, and each supplicant may use an Extensible Authentication Protocol (EAP) [3] method such as EAP-TLS [34] to authenticate with the infrastructure. Each supplicant then uses a four-message handshake to secure a session with an authenticator, allowing subsequent use of its resources. The authenticator also maintains a broadcast key that is given to each of its successful supplicants. These protocols were examined in [26] and proven to be secure.

In addition to the 802.11i functionality, the MSA proposal allows the mesh to be a peer-to-peer network. Nodes in an MSA mesh may play different roles at different times. Thus, the proof of security of the 802.11i 4-way handshake [26], which assumes limitations on the messages a node can send, does not hold. The peer-to-peer nature also poses some difficulties with timing. The 802.11i proofs adopt *matching conversations* [2] as the authenticity property. As we discuss in Section 3.1.1, the notion of matching conversations imposes a rather strict ordering of messages in a protocol run, and is too rigid for our purposes. In MSA, we must provide for the case that both parties simultaneously start instances of a protocol and messages are not necessarily well-ordered. Thus, the proofs from [26] do not carry over directly.

The key hierarchy Each node in Figure 2 represents a key in principal X's key hierarchy. An edge from key k_1 to k_2 shows that k_2 is either derived from or delivered using k_1 (that is, k_1 *protects* k_2, as knowledge of k_1 is required to obtain k_2). The edge's label is the protocol that is used to derive or deliver the key. The subscript in a key (for example, the subscript X in $pmkmkd_X \{X, T\}$) is used to denote the principal(s) associated with the key. Principals listed in curly brackets are the honest entities that may possess the key. The subscripts are ordered (i.e., $pmk_{X,Y}$ is different from $pmk_{Y,X}$).

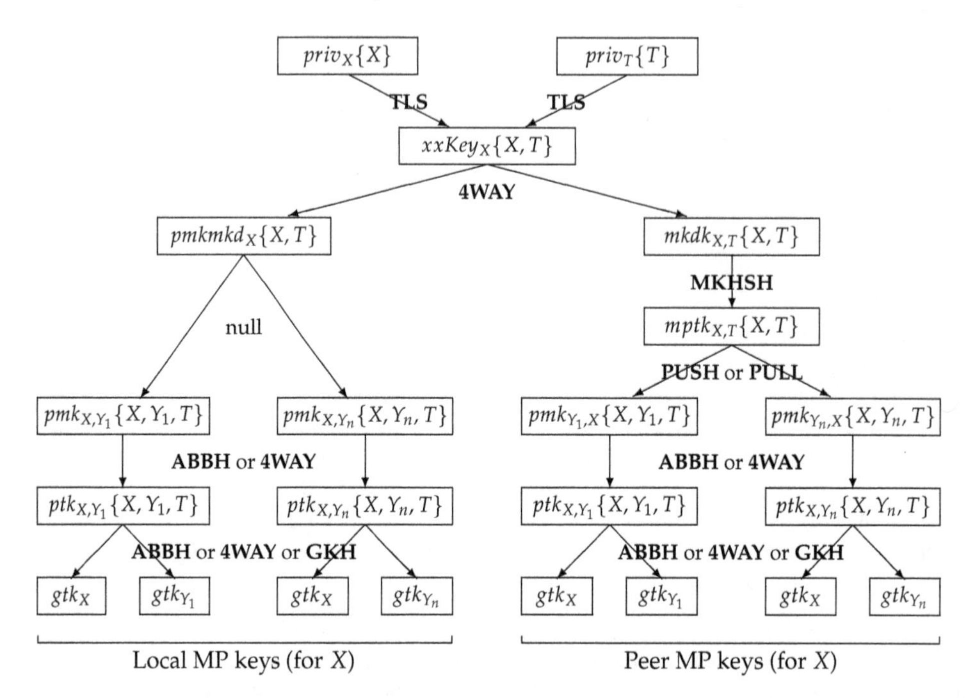

Figure 2. The key hierarchy of the MSA proposal

Key derivation (one-way) functions are utilized rather than key generation for efficiency. The MSA suite's use of key derivation also provides the potential for certain protocols to complete successfully when neither principal has connectivity to the rest of the mesh. This results in a key hierarchy, with each node being associated with several keys. The key hierarchy is an excellent avenue for understanding and summarizing the various protocols in the MSA proposal, and for demonstrating which keys protect other keys [6]. The complete descriptions of the protocols in PCL and prose are in the companion technical report [28].

We start our progression through the key hierarchy and the protocols at the top of Figure 2. Let X be a Candidate MP and let T be the MKD. The *MSA Authentication Protocol* allows X to join the mesh and become an MP, and consists of three stages: *Peer Link Establishment* (**PLE**), **TLS** [19], and a *Four-Way Handshake* (**4WAY**). X either has a shared $xxKey_X$ with T or it shares public key credentials with T. If it shares public credentials with T, then T and X run **TLS** to derive the $xxKey_X$; otherwise, **TLS** is omitted. To derive the $pmkmkd_X$, X needs a nonce; it is delivered to X from T when X runs **4WAY** with an MA (which we denote Y, noting that Y may be T). Subsequently, X can derive the $pmkmkd_X$, $pmk_{X,Z}$ for any Z and the $mkdk_{X,T}$. When X completes **4WAY** with Y, X will have derived $ptk_{X,Y}$, received the gtk_Y from Y, and delivered gtk_X to Y.

At this point, X is a Mesh Point (MP) but is not yet a Mesh Authenticator (MA). To become an MA, X needs to run the *Mesh Key Holder Security Handshake* (**MKHSH**) with T, and derive the $mptk_{X,T}$, which is a session key between X and T. This enables X to run the **PUSH** and

PULL protocols with T. **PUSH** is started by T to tell X to retrieve $pmk_{Z,X}$ for some Z. **PULL** is started by X to request $pmk_{Z,X}$ from T.

The 802.11s task group has expressed interest in developing an Abbreviated Handshake (**ABBH**) [9, 35]. **ABBH** is used by an MP or an MA X to derive $ptk_{X,Z}$, and exchange gtk_X and gtk_Z with another MA or MP Z. Without an **ABBH**, the method of exchanging these credentials is to have the MP or MA run the full MSA Authentication Protocol with the other MA or MP. In this chapter we discuss a candidate **ABBH**, which has been presented to 802.11s [8], and its proof of security and composability with the rest of the MSA architecture. The full **ABBH** is presented in the full paper [28] and comprises two variants. One denoted is **ABBH.INIT** and the other **ABBH.SIMO**, depending on the timing of the first messages. We

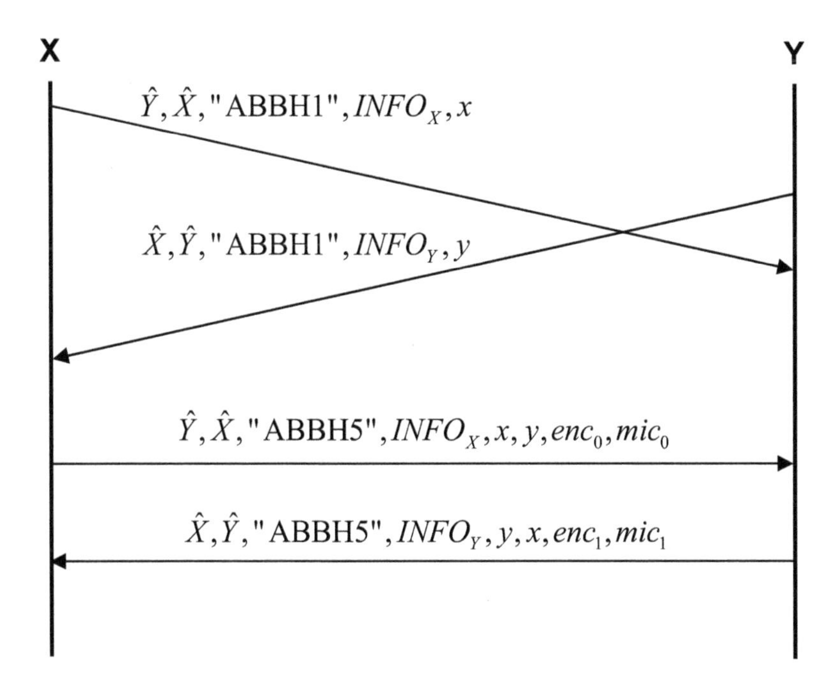

Figure 3. A SIMO Abbreviated Handshake

Additional MSA protocols include the *Group Key Handshake* (**GKH**) and the **DEL**ete protocol (for key management). **GKH** is used by X to update its group key (gtk_X) at Z. The protocol only works with nodes with which X maintains a security association (i.e., shares $ptk_{X,Z}$). **DEL** is started by T to tell X to delete a particular $pmk_{Z,X}$.

We note that each protocol message has a unique identifier. These identifiers must be unique amongst all protocols at a node, so that no other protocol at a node can use those unique identifiers.

Further work has been done on the security of the 802.11s suite of protocols. See [23] for additional details.

3. Additions to PCL and proof methodology

In modeling the MSA protocol suite in PCL, we found a number of situations for which the current language model had no support. We provide a motivating example from MSA and discuss our proposed additions to the language. None of the additions modify the existing language, so all previous proofs and work should not need re-examination. We also broadened the proof methodology slightly.

Many of the additions can be explained by looking at the abbreviated handshake protocol with a simultaneous open (**SIMO**). The purpose of this protocol is to establish a linkage between two nodes which are already part of a mesh. Therefore, the two nodes already have authenticated to the MKD and need only authenticate to each other and establish a fresh, unknown session key. One possible instantiation of this protocol is presented in Figure 3. Values x and y are nonces generated by X and Y respectively. $INFO_X$ contains additional information about node X's configuration. The *enc* values are broadcast keys encrypted with session keys derived from x, y and a shared key. The *mic* values provide integrity and authentication verification. We note that the messages labeled "ABBH5" do not have to occur in the listed order. Node X can receive message 5 before or after it sends its message 5. Note, too, that X might receive its message 5 after it sends its own message 5, even if node Y sends its message 5 before X does.

The thread for this protocol is symmetric (though it does not have to be) and is presented in Figure 4. Some additions to PCL were used in the thread description, which are fully described below. The precondition $\theta_{ABBH,1}$, invariant $\Gamma_{\mathbf{ABBH},1}$ and a sample security goal $\Phi_{\mathbf{SIMO},AUTH}$ are also presented. This protocol is useful in demonstrating the necessity of the additions, as well as providing a sample of how the addition is used in the proof of the MSA proposal.

3.1. Flexible temporal ordering

The temporal ordering of actions in the original PCL definition is too strict for our applications. In the **SIMO** protocol presented in Figure 4, the order of sending and receiving the message labeled "ABBH5" is nondeterministic. Once the initial messages have been exchanged, the final messages could be sent/received in either order. The change to PCL is realized as an addition to the language. The proposed modification does not change any other aspect of PCL; therefore all previous proofs are still valid.

We add an *action group* and redefine the notion of a strand. We define an action group as: (action group) $g ::= (a; \ldots ; a)$, where a is an action as defined in [15]. We also redefine a strand as: (strand) $s ::= [g(; \text{or} :) \ldots (; \text{or} :)g]$. Thus a strand is now composed of an arbitrary number of action groups separated by colons or semicolons. The idea behind the action group is that the actions in an action group must be done in the order they appear. However, the action groups within a strand separated by a colon (:) can be done in any order and action groups separated by a semicolon (;) must be done in the order they appear. Note that any strand defined previous to this addition to the language can still be defined exactly the same

ABBH.SIMO $= (X, \hat{Y}, INFO_X, gtk_X)$
$[$New x; send $\hat{Y}, \hat{X},$ "ABBH1", $INFO_X, x$;
receive $\hat{X}, \hat{Y},$ "ABBH1", $INFO_Y, y$;
match select$(INFO_X, INFO_Y)/CS, pmkN$;
match retrieve$(pmkN)/pmk$;
match $\text{HASH}_{pmk}(x, y)/ptk_{X,Y}$;
(match $\text{ENC}_{ptk_{X,Y}}(gtk_X)/enc_0$;
match $\text{HASH}_{ptk_{X,Y}}(\hat{Y}, \hat{X},$ "ABBH5", $INFO_X, x, y, enc_0, INFO_Y)/mic_0$;
send $\hat{Y}, \hat{X},$ "ABBH5", $INFO_X, x, y, enc_0, mic_0)$:
(receive $\hat{X}, \hat{Y},$ "ABBH5", $INFO_Y, y, x, enc_1, mic_1$;
match $enc_1/\text{ENC}_{ptk_{X,Y}}(gtk_Y)$; match $mic_1/\text{HASH}_{ptk_{X,Y}}(\hat{X}, \hat{Y},$ "ABBH5", $INFO_Y, y, x, enc_1, INFO_X))]_X$

$\theta_{\mathbf{ABBH,1}} := \text{Has}(X, pmk_{X,Y}) \wedge \text{Has}(Y, pmk_{Y,X}) \wedge (\text{Has}(X, mptk_{X,T}) \vee \text{Has}(Y, mptk_{X,T}))$

$\Gamma_{\mathbf{ABBH,1}} := \text{Honest}(\hat{X}) \wedge \text{Send}(X, m) \wedge$
$(\text{Contains}(m, \text{Hash}_{ptk}((\text{"ABBH2"}, \hat{Y}, \hat{Z}))) \vee \text{Contains}(m, \text{Hash}_{ptk}((\text{"ABBH3"}, \hat{Y}, \hat{Z}))) \vee$
$\text{Contains}(m, \text{Hash}_{ptk}((\text{"ABBH4"}, \hat{Y}, \hat{Z}))) \vee \text{Contains}(m, \text{Hash}_{ptk}((\text{"ABBH5"}, \hat{Y}, \hat{Z})))) \supset$
$\hat{Z} = \hat{X}$

$\Phi_{\mathbf{SIMO,AUTH}} :=$
$\text{KOHonest}(ptk_{X,Y}, \{pmk_{X,Y}, pmk_{Y,X}\}) \supset$
$(\text{Send}(X, \text{SIMO1X}) < \text{Receive}(Y, \text{SIMO1X})) \wedge (\text{Send}(Y, \text{SIMO1Y}) < \text{Receive}(X, \text{SIMO1Y})) \wedge$
$(\text{Send}(Y, \text{SIMO5Y}) < \text{Receive}(X, \text{SIMO5Y})) \wedge (\text{Send}(X, \text{SIMO1X}) < \text{Receive}(X, \text{SIMO1Y}) <$
$(\text{Send}(X, \text{SIMO5X}) \wedge \text{Receive}(X, \text{SIMO5Y})) \wedge (\text{Send}(Y, \text{SIMO1Y}) < \text{Receive}(Y, \text{SIMO1X}) <$
$\text{Send}(Y, \text{SIMO5Y}))$

Figure 4. Protocol Description of the Abbreviated Handshake Simultaneous Case

way by defining each action group to be one action and by setting all the separators inside a strand to ';'.

We update Axiom **AA4** [15] to reflect this addition to the language. The original version is **AA4**: $\top[a; \ldots; b]_X a < b$. We redefine it to be

AA4: $\top[a : b; \ldots; c : d]_X a \wedge b < c \wedge d$

where a, b, c and d are action groups. Thus nothing about the temporal order of a compared to b or c compared to d is indicated. We also include a new axiom

AA5: $\top[(a; \ldots; b)]_X a < b$

where a and b are actions, to deal with the temporal ordering of action groups. If each action group is exactly one action and only semicolons are used in the new strands our **AA4** becomes exactly the **AA4** previously defined and **AA5** is redundant.

A consequence of the above addition is that protocols whose definition includes ':' have an additional complication in the determination of basic sequences. Recall that a basic sequence is defined as any actions before a receive. With the : notation, two different sets of actions may occur before a receive, corresponding to the potential temporal ordering of the action groups. Thus we must ensure that invariants and preconditions hold over all possible basic sequence orderings and compositions.

3.1.1. Generalized matching conversations and generalized mutual authentication

The proofs of mutual authentication used in many previous work that use PCL for protocol verification have adopted the notion of matching conversations [2] for the authenticity property. This is natural as these protocols are "turn-by-turn" protocols in which one a participant receives a message and then responds to the message, which is received by the other party who responds to it, and so on. However, some of the peer-to-peer protocols analyzed in this chapter can never correspond to matching conversations as the order in which messages are sent and received is flexible, as a functional requirement. We generalize the properties of matching conversations and define two new notions which we call maximal conversations and generalized matching conversations. We feel these definitions are of general interest beyond this work. Recall the definitions of conversation and matching conversation from [2].

We define the *maximal conversation* for a participant A. We first determine the maximal possible temporal ordering. To do this we consider all legal orderings in an ideal world (one with no adversarial interference) from the view of a participant A in a protocol. Given this maximal temporal ordering, we note the existence of messages for which A can never confirm reception. We take the maximal temporal ordering and remove any send or receive for which A cannot confirm reception in the ideal world – the remaining actions represent the maximal conversation for participant A.

We now define *generalized matching conversations* for a participant A. We say A has generalized matching conversations, if in every run of the system, every action in the maximal conversation for participant A has a corresponding action at participant A (e.g., A does all its actions) and at the appropriate other participant in the system. For two-participant protocols (like all those in this chapter), this means that the maximal conversation for participant A has messages exactly matching the other participant's maximal conversation, with the strictest time ordering possible.

We now define *generalized mutual authentication*. In a world in which an adversary has access to every message and can act on them within the restraints of the proof system (symbolic or computational), generalized mutual authentication means that generalized matching conversations for every participant implies acceptance and acceptance implies generalized matching conversations for every participant. For the purpose of this chapter we wish to keep the definition of generalized mutual authentication general. We explore all these definitions in detail in separate work.

When our definition is applied to a "turn-by-turn" two-party protocol it becomes exactly the definition from [2]. In every other instance our definition requires that the ordering of actions be maximal with respect to what is possible in the ideal world. As this definition imposes maximal temporal ordering on a protocol, this definition is at least sufficient for mutual authentication. Most protocols in the MSA are turn-by-turn and thus the [2] definition suffices for those cases. The three exceptions are **SIMO** (which is a peer-to-peer protocol and has some timing flexibility), **PLE** (which is not a cryptographic protocol in itself and requires no temporal ordering), and **PUSH** (which is a composition of two protocols).

We note that the generalized matching conversations property encompasses the matching record of runs property [20]. Also, this property guarantees all desired properties from [27] and implies all the possible authentication definitions in [24].

3.1.2. Generalized matching conversations For ABBH.SIMO

We apply the generalized matching conversations definition to **SIMO**, which is a protocol for establishing a secure connection between two nodes already in a mesh.

Let X be the principal from whose view we seek to establish the proof of generalized matching conversation and Y be the other principal. SIMO1X and SIMO5X represent X's messages, and similarly for Y's messages. We need to determine the maximal timing relations in the ideal world (no adversaries) when only **SIMO** is run. X cannot confirm whether Y has received SIMO5X even in the ideal world, because it may be the last message sent. Therefore, SIMO5X is not part of X's maximal conversation. Note that every message must be sent by the correct party before it is received by the other party in an ideal world. Now we simply list what actions must happen after other actions and omit receives after sends that are irrelevant (e.g., $\text{Send}(X, \text{SIMO1X}) < \text{Receive}(Y, \text{SIMO1X})$).

$$\text{Send}(X, \text{SIMO1X}) < \text{Receive}(X, \text{SIMO1Y}) < (\text{Send}(X, \text{SIMO5X}) \wedge \text{Receive}(X, \text{SIMO5Y}))$$
$$\text{Send}(Y, \text{SIMO1Y}) < \text{Receive}(Y, \text{SIMO1X}) < \text{Send}(Y, \text{SIMO5Y})$$

This temporal ordering is inherently maximal for X's view of an arbitrary run of **SIMO**, so it satisfies the definition of generalized matching conversations for X (Y's view is similar). The enforcement of the order of the send messages within a node can be accomplished by waiting for acknowledgements from the MAC layer before proceeding. If X has not sent its message 1, it initiates the corresponding thread for **ABBH.INIT**, not for **ABBH.SIMO**, so this ordering is maximal.

3.2. Modeling information exchange

In the full paper [28], we provide detailed PCL equivalents of the protocols presented in the MSA submissions. Such detailed examinations are necessary to prove protocol correctness. For example, the presence of $INFO_Y$ in mic_0 in **SIMO** (Figure 4) is not intuitively obvious but is essential to the security of the protocol. Modeling the protocol at a higher level of abstraction would have missed this subtle requirement.

Real protocol implementations such as MSA require more than simple key agreement. Additional information must be exchanged and agreed on before secure communication can happen. Examples of information of this type are basic network functions (e.g., bandwidth selections) and security information (e.g., cipher suite selection). The two principals in the protocol must agree in each case, and an attacker must not be able to influence the selection. That is, the agreed-upon value in all protocol runs must match the agreed-upon value in an ideal world with no adversary. The peer-to-peer nature of certain protocols such as **SIMO** do not allow pre-defined protocol roles of principals to always allow one principal to make this selection and dictate the choice to the other. The two principals must independently choose matching values from two lists.

A new construct, *INFO* is used to capture this. The information principal *X* contributes to a protocol is $INFO_X$. It contains ordered lists of acceptable selections for one or more fields. The contents of $INFO_X$ may vary for different protocols. In MSA, the **PLE**, **ABBH**, and **MKHSH** protocols require the use of $INFO$.

The Select() action We have added a new action, Select(), to PCL. Two principals *X* and *Y* must make identical but independent selections of link and protocol options from exchanged information $INFO_X$ and $INFO_Y$. The Select() action deterministically retrieves information from two lists, independent of the order of the lists (i.e., select($INFO_X, INFO_Y$) = select($INFO_Y, INFO_X$)). During Peer Link Establishment and Abbreviated Handshake protocols, Select() determines the key to be used based on information each principal sends about the keys it has cached and whether it is an MA capable of retrieving the key from the MKD. Thus, the function ensures that a key is either locally cached or may be retrieved from the MKD if the protocol is to continue. The Select() action is used in other contexts as well, such as to determine which principal initiates the 4-way handshake, or which pairwise cipher suite to use after completing a protocol.

This level of detail is necessary to provide protection against downgrade attacks (wherein the attacker chooses the protocol selection suite) and other attacks where public information can be subverted by an attacker to weaken the final strength of the protocol. Additionally, modeling the interactions at the lower level, demonstrated in the description of SIMO in Figure 4, allows us to provide additional guarantees against attack vectors which may be non-obvious to a lay implementer.

Without modeling at this detailed level, a nearly-equivalent SIMO protocol, which only creates a keyed hash across the information it sends, would appear viable and secure. The cryptographic components would be identical. However, without node $\hat{X}$ including $INFO_Y$ in its *mic* (and equivalent for $\hat{Y}$), attack vectors become possible. In particular, the strong requirement that the messages sent exactly match the messages received no longer holds. This loss directly leads to potential manipulation of the $INFO_X$ and $INFO_Y$ fields by an attacker. Not only can an attack user such manipulation to mount a straightforward denial-of-service attack, but the attacker could also compromise the selection of the shared cipher suite, a dangerous form of a downgrade attack.

3.3. Calling one protocol in another

For many of the protocols in MSA, the protocol may instantiate another protocol partway through its run. This second protocol must complete before the first protocol can continue. For example, in a key exchange protocol such as **SIMO**, if both parties that try to establish a session key do not have the other party's pairwise master key cached locally (e.g., *X* does not have the current $pmk_{Z,X}$), then one of the parties must pause its protocol run and run a key pull protocol. Furthermore, the second protocol could potentially be triggered in the middle of a basic sequence.

This is new ground for PCL and we have devised a system and proof (see section 4.3) that enables us to frame this complex action in PCL and develop sound proofs. Essentially, we define the inception of functions that may need to run a separate protocol to be basic sequences, as they may involve blocking actions (like receive). Then, before and after these actions we define basic sequence pre- and post-conditions that must be satisfied for a successful completion of the protocols. The idea of basic sequence pre- and post-conditions were give in [26] to enable staged composition and remain standard in the language [15], although they have not been previously used in this way to enable mid-protocol composition.

The retrieve action We have added a new action, retrieve, to PCL. The retrieve action provides the key to the strand, after key selection is complete. The retrieve action takes a key name (*pmkN*, corresponding to a specific *pmk*) as its input. If the principal that executes the function does not have the key locally cached on disk, but is an MA (and has a connection with the MKD), retrieve initiates the **PULL** protocol. If the key is not on disk and the principal is not an MA, retrieve fails and the protocol that called it aborts. As the retrieve action may or may not perform a key pull, we create a break in the basic sequence directly before and directly after the retrieve.

The retrieve action has inherent pre- and post-conditions as it is a series of one or more basic sequences (e.g., a protocol). As a precondition, retrieve must have the *pmk* cached locally or it must have the $mptk_{X,T}$. Thus the precondition is $\text{Has}(X, mptk_{X,T}) \vee \text{Has}(X, pmk)$ where *pmk* matches the input *pmkN*. The postcondition is simply $\text{Has}(X, pmk)$. The retrieve function itself has security requirements only if the principal must perform a key pull, when it inherits the requirements of the **PULL** protocol.

In Figure 4, retrieve is used to get the selected *pmk*. This provides two potential paths of execution through the protocol, one which runs a key pull mid-protocol and one which simply fetches some stored memory (equivalent to a match action).

4. Overview of the proof

In this section, we provide an overview of our proof efforts by highlighting three aspects of it. In Section 4.1 we discuss our approach to proving key secrecy in the MSA proposal. In Section 4.2 we present additional security goals and a theorem that culminates our proof efforts. Finally, in Section 4.3, we discuss our approach to protocol composition.

4.1. Key secrecy in MSA

Key secrecy is a critical security requirement. Some previous work [26] has proven key secrecy as a protocol postcondition. We show that proving key secrecy as a postcondition is insufficient by providing an example of a protocol which has key secrecy as a postcondition (i.e., upon protocol completion, key secrecy can be proven) but is insecure, because key secrecy can be lost. The Insecure Key Transfer Protocol in Figure 5 illustrates this point. If we assume protocol completion from the point of view of RESP we can prove that the secret key is

Inputs and Parties:
- Two parties: INIT and RESP.
- Shared input: confirmation key (ck).
- INIT private input: INIT public key (PK_{INIT}).
- RESP private input: secret key (sk).
- Goal: $\text{Has}(Z, sk) \supset Z = \text{INIT} \vee Z = \text{RESP}$

Insecure Key Transfer Protocol:

1. INIT sends PK_{INIT} to RESP.
2. RESP receives PK_{INIT}; encrypts sk under PK_{INIT}, computes the keyed hash of the encryption with key ck; and sends $(\{sk\}_{PK_{INIT}}, HASH_{ck}(\{sk\}_{PK_{INIT}}))$ to INIT.
3. INIT receives $(\{sk\}_{PK_{INIT}}, HASH_{ck}(\{sk\}_{PK_{INIT}}))$, verifies the keyed hash; decrypts sk; computes the keyed hash of sk and PK_{INIT} with the ck and sends $HASH_{ck}(sk, PK_{INIT})$ to RESP.
4. RESP verifies the signature.

Figure 5. Insecure Key Transfer

distributed correctly, as the validity of INIT's public key is established once RESP receives the third message. However RESP uses the public key in the second message before the validity of the public key can be established. Thus if the protocol aborts after the RESP sends the second message, it may be the case that the public key sent in message 1 is an adversary's public key. It is therefore possible for the adversary to intercept the secret key. While this protocol is contrived, in larger protocols with complex security goals, it may be the case that a subtle insecurity such as this goes unnoticed. Thus, for certain assertions related to secrecy, we advocate showing that they hold at every critical point in a protocol.

We prove the security of MSA's key hierarchy using the work of Roy et al. [31–33]. We present the key secrecy postconditions relevant to the MSA key hierarchy in Figure 6. We prove that these conditions hold at every point during any protocol execution of MSA, as long as the indicated principals are honest. We also claim the new axiom

SAF5: $\text{SafeMsg}(HASH_s(M), s, \mathcal{K})$

Informally, this states that a keyed hash of a message does not reveal the key.

It seems natural to prove key secrecy in an inductive manner, locally for each thread and role. But that's not sufficient in the proposed MSA system, because key information is not generally held purely locally and other nodes with the information may be abusing it. Key secrecy must be maintained locally, of course, but it requires a global proof.

The techniques of [32] would note the deficiencies of the protocol in Figure 5, because the first message sent by RESP could not be proven to be a SafeMsg, inductively. We utilize this notion and extend the SafeNet concept across the entire suite of MSA protocols, to show that no protocol violates the key secrecy of any other protocol. Doing this step before examining other desired protocol security goals (postconditions) provides for more elegance and correctness in the other proofs.

$\theta_{TLS,SI,1} :=$
$\text{KOHonest}(xxKey_X, \{priv_X, priv_T, xxKey_X\}) \supset \text{Has}(Z, xxKey_X) \supset \hat{Z} = \hat{X} \vee \hat{Z} = \hat{T}$

$\theta_{4WAY,SI,1} :=$
$\text{KOHonest}(pmkmkd_X, \{xxKey_X\}) \supset \text{Has}(Z, pmkmkd_X) \supset \hat{Z} = \hat{X} \vee \hat{Z} = \hat{T}$

$\theta_{4WAY,SI,2} :=$
$\text{KOHonest}(mkdk_{X,T}, \{xxKey_X\}) \supset \text{Has}(Z, mkdk_{X,T}) \supset \hat{Z} = \hat{X} \vee \hat{Z} = \hat{T}$

$\theta_{PPD,SI,1}, \theta_{MKHSH,SI,1} :=$
$\text{KOHonest}(mptk_{X,T}, \{mkdk_{X,T}\}) \supset \text{Has}(Z, mptk_{X,T}) \supset \hat{Z} = \hat{X} \vee \hat{Z} = \hat{T}$

$\theta_{PPD,SI,2} :=$
$\text{KOHonest}(pmk_{X,Y}, \{pmkmkd_X, mptk_{Y,T}\}) \supset \text{Has}(Z, pmk_{X,Y}) \supset \hat{Z} = \hat{X} \vee \hat{Z} = \hat{Y} \vee \hat{Z} = \hat{T}$

$\theta_{ABBH,SI,1}, \theta_{4WAY,SI,3}, \theta_{GKH,SI,1} :=$
$\text{KOHonest}(ptk_{X,Y}, \{pmk_{X,Y}, pmk_{Y,X}\}) \supset \text{Has}(Z, ptk_{X,Y}) \supset \hat{Z} = \hat{X} \vee \hat{Z} = \hat{Y} \vee \hat{Z} = \hat{T}$

$\theta_{ABBH,SI,2}, \theta_{4WAY,SI,4}, \theta_{GKH,SI,2} :=$
$\text{KOHonest}(gtk_X, \{ptk_{X,Y_1}, \ldots ptk_{X,Y_n}\}) \supset \text{Has}(Z, gtk_X) \supset \text{Has}(Z, ptk_{X,Y_i})$

Figure 6. MSA Key Secrecy Conditions

We introduce new notation $\vdash_U$. The meaning of $P \vdash_U \theta$ is that postcondition θ must hold at every intermediate point of the relevant protocols in program set P. That is, if the terms in θ are defined and bound at the end of a basic sequence in P, then θ holds.

Theorem 1. *Let MSA represent all the protocols in the Mesh Security Architecture and $\theta_{SI,ALL}$ represent all of the key secrecy conditions in Figure 6. Then $\theta_{SI,ALL}$ are satisfied by MSA. That is,*

$$MSA \vdash_U \theta_{SI,ALL}$$

Proof sketch: This theorem is proven in two steps. The first step is a massive induction over all the basic sequences of all the protocols that could be run by any participant in a mesh. This induction guarantees that all messages sent are "safe," in that critical information is protected by keys. In the key secrecy goals of Figure 6, the critical information protected is another key, lower in the hierarchy. From this, we argue the invariant nature of multiple SafeNet axioms over the entire MSA protocol suite, limiting various goals to the protocols where the terms are instantiated/defined. Then, we use the **POS** and **POSL** axioms [32] to state who can potentially have access to other keys. By proceeding in this way through the entire key hierarchy, we establish all the necessary key secrecy goals, at any point in a run where the keys may be defined. The full proof is generally unenlightening and we do not provide it. We stress that this proof does not depend on any of the analysis done in proceeding sections. It is simply induction over all basic sequences and application of secrecy axioms. This proves key secrecy is maintained by all protocols in MSA.

This theorem guarantees that the parties listed in Figure 6 are the only principals with those keys. This proves that an attacker could not learn any key in the entire hierarchy from the MSA protocols.

4.2. Goals and correctness result

We present important security postconditions (goals) below. For each goal, we point out the kinds of protocols to which it applies. Goals are customized for each protocol; formal instances of each kind of goal we discuss below are in Appendix 6. We keep our discussions in this section more informal for clarity.

AUTH: Authentication as realized by the generalized matching conversations property (see Section 3.1.1). In practice, this confirms peer liveness and peer possession of a particular key. This goal applies to all protocols in the MSA proposal. This goal is expressed as:

$$\Phi_{AUTH} := \exists Y.ActionsInOrder(\text{Send}(X,\hat{X},\hat{Y},msg_1), \text{Receive}(Y,\hat{X},\hat{Y},msg_1), \text{Send}(Y,\hat{Y},\hat{X},msg_2), \cdots, \text{Receive}(X,\hat{Y},\hat{X},msg_n))$$

KF: Key freshness as realized by a freshly-generated nonce from each party as a term in the agreed-upon key. This goal applies only to protocols which create a joint (session) key.

$$\Phi_{KF} := \text{KOHonest}(k,\mathcal{K}) \supset (\text{New }(\hat{X},x) \wedge x \subseteq k \wedge \text{New }(\hat{Y},y) \wedge y \subseteq k) \wedge \text{FirstSend}(X,x,\hat{X},x,m) \wedge \text{FirstSend}(Y,y,\hat{Y},y,m)$$

KA: Key agreement as realized by the Has predicate. This ensures that both parties have the session key. This goal applies to only those protocols that establish a session key.

$$\Phi_{KA} := \text{KOHonest}(k,\mathcal{K}) \supset \text{Has}(X,k) \wedge \text{Has}(Y,k)$$

KD: Transfer of secret information (key delivery) as realized by the key secrecy goals and the Has predicate. This applies only to those protocols which transmit keys (either a group transfer key (*gtk*) or a pairwise master key (*pmk*)).

$$\Phi_{KD} := \text{KOHonest}(k,\mathcal{K}) \supset \text{Has}(X,k) \wedge \text{Has}(Y,k)$$

INFO: Authentic exchange of non-secret information and authenticated selection of sub-elements as realized in detailed protocol description and validated return information. This applies only to protocols which must exchange non-security information and agree on parameters.

$$\Phi_{INFO} := \text{KOHonest}(k,\mathcal{K}) \supset \text{Select}(INFO_X, INFO_Y) = CS, pmkN \wedge \text{Has}(X,CS,pmkN) \wedge \text{Has}(Y,CS,pmkN)$$

Our goals are extensions and clarifications of the goals adopted by He et al. [26], which in turn are adapted from the list of desired security properties for 802.11i [1]. No security goals have been explicitly specified for the general 802.11s protocol suite; however, we anticipate that the security goals for 802.11i are meaningful for 802.11s as well, provided they are adapted appropriately. Furthermore, we feel that the goals we present above have intrinsic intuitive appeal. We recommend that these goals, in addition to the key secrecy goals discussed in Section 4.1, be formally adopted by the 802.11s task group.

In the following Theorem, we introduce some notation ($\rightsquigarrow$) for ease of exposition. $TLS \rightsquigarrow AUTH, KD$ means $\Gamma_{\mathbf{TLS},\{1,2,\}} \vdash \theta_{\mathbf{TLS}}[\mathbf{TLS}:\mathbf{CLNT}]_X \Phi_{\mathbf{TLS},\{AUTH,KD\},CLNT}$ and the corresponding goal for the other node, namely $\Gamma_{\mathbf{TLS},\{1,2,\}} \vdash \theta_{\mathbf{TLS}}[\mathbf{TLS}:\mathbf{SRVR}]_X \Phi_{\mathbf{TLS},\{AUTH,KD\},SRVR}$. These state that, with the proper invariants, the protocol from each perspective provably satisfies the security goals *AUTH* and *KD* , particular to **TLS**. Similar expansions have been made for each protocol and the details are in our full paper [28].

With the changes that we discuss in Section 5, we are able to prove the component-wise correctness of each of the protocols of the MSA proposal.

Theorem 2. *The following are true, with the notation described above.*

(i) $TLS \rightsquigarrow AUTH, KD$
(ii) $4WAY \rightsquigarrow AUTH, KF, KA, KD, INFO$
(iii) $MKHSH \rightsquigarrow AUTH, KF, KA, KD, INFO$
(iv) $GKH \rightsquigarrow AUTH, KD$
(v) $PUSH \rightsquigarrow AUTH, KD$
(vi) $PULL \rightsquigarrow AUTH, KD$
(vii) $DEL \rightsquigarrow AUTH$
(viii) $ABBH \rightsquigarrow AUTH, KF, KA, KD, INFO$

This theorem was one of the major driving forces behind the work. It asserts that the full protocol suite in the MSA proposal, with a rather complex key hierarchy, is secure. Each protocol achieves the maximal security goals for its type. Appendix 6 contains a proof of part of (viii) and provides a feel for the proof methodology. The proof of Theorem 2 depends on the PCL additions of Section 3.

4.3. Composition

The MSA architecture allows for significant variation in how protocols compose together [4]. Once an established state is reached, many protocols (which may have been run previously to reach the established state) may be chosen. Reaching an established state may take a variety of paths, depending on the authentication mechanism (TLS or pre-shared key) used. Error-handling strategies will cause protocols to restart, or, potentially, different protocols to be run. This introduces a complex state diagram and complexities of composition.

While *staged composition* proofs have been presented previously [26, 31], the presentation in each case has differed. Staged composition allows arbitrary back arrows and paths through possible protocol execution paths. This allows for protocol restarts, lost connections, and other real-world considerations about the order in which protocols are run. We provide a slightly different presentation of similar ideas in Section 4.3.1. Readers primarily interested in the proof of MSA may skip this section and proceed to Section 4.3.2 where the overall MSA security theorem is presented.

4.3.1. Consistent composition

The concept of branches within protocols or between protocols has not been explicitly mentioned in previous PCL composition theorems. We require this functionality, to denote how a particular staging can be accomplished within the MSA framework. One of our motivations is to allow such possibilities as are represented in Figure 7. After basic sequence A, either sequence B or sequence C may follow. Sequence D follows C and both B and D lead to E. The consistent composition theorem provides the requirements under which such branches will still compose. This also provides for all manner of if/then functionality within PCL, if it can be properly created in semantics and the various results of the if/then statement are properly modeled in terms of basic sequence breaks. We believe this fills a gap in the span of PCL.

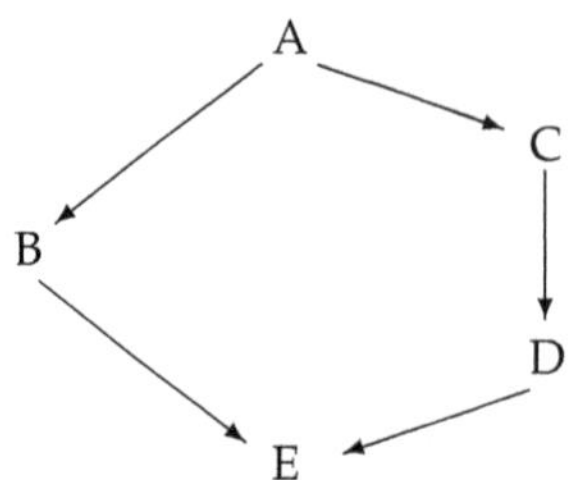

Figure 7. Branches in PCL

We utilize the definitions of role-prefix, staged role, and staged composition from [26], suitably augmented for the retrieve action. Informally, role-prefix defines which sets of basic sequences can lead to a particular basic sequence. A staged role is a particular, legitimate sequence of basic sequences leading to a particular execution point. And staged composition allows for sequential implementation with arbitrary returns to earlier execution points, with the branching of retrieve potentially following different paths on each iteration.

We use θ_{P_i} to indicate the precondition for basic sequence P_i. Additionally, to add simplicity to our exposition, we use Γ to denote the conjunction of all invariants within a staged composition of protocols. That is, Γ is the totality of all the invariants from each of the protocols Q_i that make up a composition of protocols Q. This allows us to state the following theorem succinctly.

Theorem 3. *Let* Q *be a staged composition of protocols* $Q_1, Q_2, \ldots, Q_n$ *and* $P; P_i \in SComp(\langle Q_1, Q_2, \ldots, Q_n \rangle)$ *and* $P_i \in Q_i$. *Then* $Q \vdash \theta_{P_0}[P; P_i]_X \theta_{P_{i+1}}$, *if for all* $RComp(\langle P_1, P_2, \ldots, P_n \rangle) \in Q$, *all of the following hold:*

(Invariants)
(i) $\forall i. \forall S \in BS(Q_i). \vdash \theta_{P_i} \wedge \Gamma[S]_X \Gamma$
(Preconditions)
(ii) $Q_1 \otimes Q_2 \otimes \cdots \otimes Q_n \vdash \forall i. \theta_{P_i}[P_i]_X \theta_{P_{i+1}}$
(iii)$\forall i. \forall S \in \bigcup_{j \geq i} BS(P_j). \theta_{P_i}[S]_X \theta_{P_i}$
(iv) $Q_1 \otimes Q_2 \otimes \cdots \otimes Q_n \vdash Start(X) \supset \theta_{P_1}$

Theorem 3 states the conditions under which a particular run through a set of actions reaches its ultimate goal. The "Invariants" condition requires that no basic sequence violate any invariant of any basic sequence, with its proper preconditions, and invariants holding before the basic sequence. The "Preconditions" conditions require that each basic sequence's postconditions imply the next basic sequence's preconditions, that no basic sequence ever violate any preceding basic sequence's preconditions, and that the start state is valid.

We point out that Theorem 3 is dependent on basic sequences as its fundamental building block. The protocols themselves, while useful distinctions in understanding and modeling the system, are not critical. In particular, the Q_i's could be single basic sequences and the entire theorem still holds. This allows us to model at the level of basic sequences. This level of granularity has been suggested before [26], but we make it explicit.

This allows, for example, the behavior of the retrieve action that we discuss Section 3.3. Retrieve allows two different paths through a larger staged composition. In one path, a locally stored value is returned. In the other path, an entire protocol is run. As protocols compose consistently at the granularity of basic sequences in the initial protocol, retrieve fundamentally denotes alternate methods of staging the composition. In all protocols that use retrieve, the invariants and various preconditions in the protocol are proven against all possible stagings of the retrieve action.

4.3.2. Composition in MSA

We wish to apply Theorem 3 to the protocols of the MSA proposal. We view the protocols of staged composition as the protocols given previously. As mentioned, we consider arbitrary breaks at the basic sequence level, for mid-protocol composition as well as overall composition. We need to prove that all protocols within MSA (comprising **PLE**, **TLS**, **4WAY**, **MKHSH**, **GKH**, **PULL**, **PUSH**, **DEL**, and **ABBH**, (both **ABBH.INIT** and **ABBH.SIMO**) satisfy the necessary conditions for composition.

Theorem 4. *Let Q be a specific composition of protocols from MSA and* $RComp(\langle P_1, P_2, \ldots, P_n \rangle) \in Q$ *and* $\Gamma = \Gamma_{TLS,\{1,2\}} \wedge \Gamma_{4WAY,1} \wedge \Gamma_{MKHSH,1} \wedge \Gamma_{GKH,\{1,2\}} \wedge \Gamma_{PPD,\{1,2\}} \wedge \Gamma_{ABBH,1}$. *Then:*

(i) $\forall i.\forall S \in BS(Q_i). \vdash \theta_{P_i} \wedge \Gamma[S]_X \Gamma$
(ii) $\Phi_{4WAY} \vdash \theta_{MKHSH} \wedge \theta_{GKH}$
$\Phi_{MKHSH} \vdash \theta_{PUSH} \wedge \theta_{PULL} \wedge \theta_{DEL}$
$\Phi_{MKHSH} \vdash \theta_{ABBH}$
$\Phi_{ABBH} \vdash \theta_{GKH}$
(iii) $\forall i.\forall S \in \bigcup_{j \geq i} BS(P_j).\theta_{P_i}[S]_X \theta_{P_i}$
(iv) θ_{P_1}

Proving that all the protocols securely compose is a lengthy induction process, which we omit owing to space constraints. We briefly discuss the meaning of the various subpoints. No portion of any protocol in MSA violates the invariants (i) or changes the preconditions (iii) of any MSA protocol. All nodes in a mesh start with the correct information, by assumption (iv). Point (ii) gives the protocols which guarantee certain subsequent protocols can be completed with other legitimate nodes, via pre and post condition matching.

This theorem states that, given any MSA protocol, if the MKD and the players in the protocol are honest (that is, they conform to the protocol specification), then the security of that protocol is ensured, regardless of what other protocols may be running in the system. By extension, a mesh of honest nodes guarantees our security goals; the Mesh Security Architecture is sound.

5. Modifications to MSA

Our analysis of the protocols and key hierarchy of the MSA proposal indicate that it was largely well-designed. We have two recommendations that have been incorporated into the

802.11s draft (as of March 2008) and are necessary for Theorem 2 to hold; otherwise, the protocols are insecure.

5.1. Include mesh nonce in 4WAY

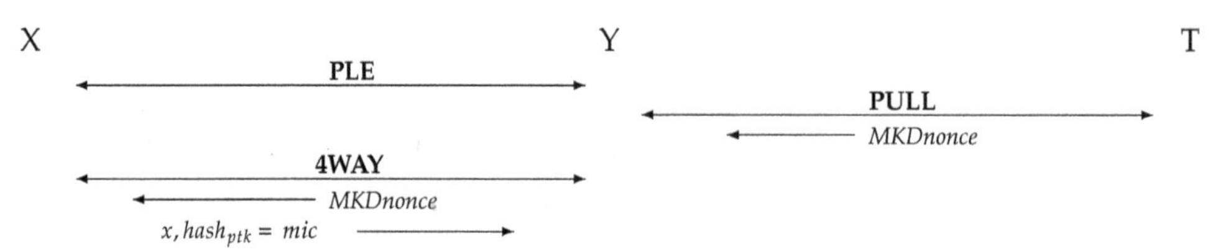

Figure 8. MSA Authentication. The text above a double-headed arrow (e.g., **4WAY**) is a protocol, and text below (e.g., *MKDnonce*) is some data that is sent as part of the protocol.

The draft specification of **4WAY** during MSA authentication does not properly provide key freshness. The proposal has the key generation nonce (*MKDnonce*) provided by the MKD used both to derive the *pmkmkd* (see Figure 2) and as the nonce to derive the session key (*ptk*). This is shown in Figure 8.

This enables an attack that proceeds as follows. At some point, a legitimate node (X) disconnects from the mesh. The attacker then starts MSA authentication with the same MA with which X connected before. The rogue node does **PLE** (claiming to be X) and then continues to the **4WAY** protocol, where the *MKDnonce* is the same. The rogue node re-uses the nonce X used and now the same *ptk* is derived. The attacker may then utilize some information recorded from the legitimate conversation or otherwise abuse the mesh. If **TLS** is used and not a pre-shared key, then this particular attack no longer works.

The solution adopted by 802.11s is to modify the derivation of *pmkmkd* so that it does not require an *MKDnonce*, so that 4WAY is responsible only for transporting nonces used to derive the *ptk*. The *MKDnonce* was removed as it did not provide significant benefit to the architecture and was not required for our key freshness goal. At this point, key freshness (for the *ptk*) can be proven and the attack outlined above is thwarted.

5.2. Include MAC address in the Group Key Handshake protocol (GKH)

In the original proposal, **GKH** does not provide authentication. Recall from section 4.2 that the *AUTH* goal requires matching conversations between two different nodes. In the proof of this property, it became apparent to us that the proposal did not protect against a reflection attack.

MAC addresses were contained in the **GKH** message headers (to facilitate transport of the messages) but were not incorporated in the calculation of the message integrity code (mic) included in each **GKH** message. **GKH** messages are protected using the *ptk*, a pairwise key known only to two parties, but either party may initiate the **GKH**. Owing to this symmetry, the first **GKH** message could be reflected back to the sender, and would be accepted as valid because of the presence of a valid mic. This reflection attack could change the security state at the MP that sends the first message of **GKH**, such as by installing a stale *gtk* or installing its own *gtk* as if it were its neighbor's *gtk*.

The proposed modification includes the explicit identification of sender and receiver in the protected portion of the message, and updates the processing of **GKH** messages to verify this information upon reception. This prevents the replay attack because the sender and receiver MAC addresses would not match if a reflection attack is attempted.

6. Conclusions and future work

We have proven the security of the MSA, under standard assumptions. We provided and justified a few recommendations that were incorporated (as of March 2008) into the 802.11s draft standard, which is still being developed. We also hope that providing a security proof during the design and review process will lead to additional efforts in that regard. We feel that protocol design is important and an analysis of a system should be done before implementation, not after. In the process of this analysis, we made a number of contributions to PCL.

The most important contribution, from our perspective, is the ability to handle simultaneity, with the introduction of action groups and associated axioms and proof techniques. The definition of generalized authentication using generalized matching conversations is also required for simultaneous peer-to-peer protocols. The select and retrieve actions were also designed to extend naturally to examinations of other architectures.

This chapter also takes a deeper dive into the details of the protocols than is often undertaken. While examining only the security components (nonces, keys, etc.) simplifies analysis, it also leaves a gap. Our experience leads us to believe that gaps in analysis are often dangerous, as they lead to assumptions about security, implementation difficulties, and unforeseen attack vectors. Some level of abstraction is necessary, but adding a model for authenticated information exchange is critical for many applications.

This chapter opens opportunities for applying PCL to other peer-to-peer protocols, where ordering may not be as strict as in server-client models. Other protocol systems, particularly those on standard-track, would be natural candidates for additional analysis.

Finally, we provide a new, more general composition theorem, which explicitly allows for mid-protocol composition and branching. As it is not unusual for protocols to intermix, explicitly allowing multiple potential paths through basic sequences is important, and should naturally extend to other situations.

Acknowledgements

The authors would like to thank Anupam Datta and Arnab Roy for their helpful comments and suggestions.

Author details

Doug Kuhlman
Motorola Mobility 600 US Hwy 45 E1-40Y Libertyville, IL 60048

Ryan Moriarty[1]
Computer Science Department
University of California at Los Angeles

Tony Braskich, Steve Emeott and Mahesh Tripunitara
Motorola, Schaumburg, IL 60196

Appendix A: SIMO Security

A.1. Security goals for SIMO

Here we detail the five PCL security goals for the SIMO abbreviated handshake protocol (the *AUTH* goal was presented in Figure 4 and is repeated here). These directly correspond to the security goals detailed in Section 4.2. Unlike the generic goals presented there, these are the specific instances for the SIMO protocol.

Goals SIMO:

$\Phi_{SIMO,AUTH} :=$
$\text{KOHonest}(ptk_{X,Y}, \{pmk_{X,Y}, pmk_{Y,X}\}) \supset$
$(\text{Send}(X, \text{SIMO1X}) < \text{Receive}(Y, \text{SIMO1X})) \wedge (\text{Send}(Y, \text{SIMO1Y}) < \text{Receive}(X, \text{SIMO1Y})) \wedge$
$(\text{Send}(Y, \text{SIMO5Y}) < \text{Receive}(X, \text{SIMO5Y})) \wedge (\text{Send}(X, \text{SIMO1X}) < \text{Receive}(X, \text{SIMO1Y}) <$
$(\text{Send}(X, \text{SIMO5X}) \wedge \text{Receive}(X, \text{SIMO5Y})) \wedge (\text{Send}(Y, \text{SIMO1Y}) < \text{Receive}(Y, \text{SIMO1X}) <$
$\text{Send}(Y, \text{SIMO5Y}))$

$\Phi_{SIMO,KF} :=$
$\text{KOHonest}(ptk_{X,Y}, \{pmk_{X,Y}, pmk_{Y,X}\}) \supset$
$(\text{New}\ (\hat{X}, x) \wedge x \subseteq ptk_{X,Y} \wedge \text{New}\ (\hat{Y}, y) \wedge y \subseteq ptk_{X,Y}) \wedge$
$\text{FirstSend}(X, x, \hat{X}, x, \text{SIMO1X}) \wedge \text{FirstSend}(Y, y, \hat{Y}, y, \text{SIMO1Y})$

$\Phi_{SIMO,KA} :=$
$\text{KOHonest}(ptk_{X,Y}, \{pmk_{X,Y}, pmk_{Y,X}\}) \supset$
$\text{Has}(X, ptk_{X,Y}) \wedge \text{Has}(Y, ptk_{X,Y})$

$\Phi_{SIMO,KD} :=$
$\text{KOHonest}(ptk_{X,Y}, \{pmk_{X,Y}, pmk_{Y,X}\}) \wedge$
$\text{Receive}(Y, SIMOX5) \supset \text{Has}(X, gtk_Y) \wedge \text{Has}(Y, gtk_X)$

$\Phi_{SIMO,INFO} :=$
$\text{KOHonest}(ptk_{X,Y}, \{pmk_{X,Y}, pmk_{Y,X}\}) \supset$
$SELECT(INFO_X, INFO_Y) = CS, pmkN \wedge \text{Has}(X, CS, pmkN) \wedge \text{Has}(Y, CS, pmkN)$

A.2. Proof security goals, SIMO

Proof sketch generalized authentication, SIMO

We only need to show the proof from a single point of view as the roles are symmetric. Let principal X be the principal from whose view we are establishing the proof from and let Y be the other principal. As the proof assumes X has completed the protocol successfully, we know that SIMO1X was sent before SIMO5X and SIMO1Y was received before SIMO5Y. Thus to complete the proof we must show that Y sent exactly SIMO1Y before SIMO5Y and received

[1] Funded by Motorola while working on this project

exactly SIMO1X before sending SIMO5Y. We can determine the MIC in SIMO5Y could have only been sent by Y if X, Y and T are honest. Since all the variables used in the protocol are contained in the MIC of SIMO5Y, we know that X and Y share identical variables. Now using the honesty of Y we are sure that Y sent SIMO1Y and received SIMO1X before sending SIMO5Y and that it was sent exactly as X received it. Again if Y is honest since X and Y share variables, then Y must have received SIMO1X exactly as X had sent it. This gives us generalized authentication.

Generalized Authentication:

$$\begin{aligned}&\mathbf{AA1}, \mathbf{ARP}, \mathbf{AA4}, \boldsymbol{\theta}_{\mathbf{ABBH,1}}\\&[\mathbf{ABBH}:\mathbf{SIMO}]_X\\&\text{Send}(X,\hat{Y},\hat{X},\text{“ABBH1”},INFO_X,x) < \text{Receive}(X,\hat{X},\hat{Y},\text{“ABBH1”},INFO_Y,y) <\\&(\text{Receive}(X,\hat{X},\hat{Y},\text{“ABBH5”},INFO_Y,y,x,enc_1,mic_1) \wedge \text{Send}(X,\hat{Y},\hat{X},\text{“ABBH5”},INFO_X,x,y,enc_0,mic_0))\end{aligned} \quad (1)$$

$$\begin{aligned}&\mathbf{ARP}, \mathbf{HASH3'}, \boldsymbol{\theta}_{\mathbf{ABBH,1}}\\&[\mathbf{ABBH}:\mathbf{SIMO}]_X\text{Receive}(X,\hat{X},\hat{Y},\text{“ABBH5”},INFO_Y,y,x,enc_1,mic_1) \supset\\&\exists Z.\text{Computes}(Z,\text{HASH}_{ptk_{X,Y}}(\hat{X},\hat{Y},\text{“ABBH5”},INFO_Y,y,x,enc_1,INFO_X))\wedge\\&\text{Sends}(Z,\text{HASH}_{ptk_{X,Y}}(\hat{X},\hat{Y},\text{“ABBH5”},INFO_Y,y,x,enc_1,INFO_X)) <\\&\text{Receive}(X,\hat{X},\hat{Y},\text{“ABBH5”},INFO_Y,y,x,enc_1,mic_1)\end{aligned} \quad (2)$$

$$\begin{aligned}&\boldsymbol{\theta}_{ABBH,SI,1}, HASH1\\&\text{KOHonest}(ptk_{X,Y},\{pmk_{X,Y},pmk_{Y,X}\}) \supset\\&\text{Computes}(Z,\text{HASH}_{ptk_{X,Y}}(\hat{X},\hat{Y},\text{“ABBH5”},INFO_Y,y,x,enc_1,INFO_X)) \supset\\&\text{Has}(Z,ptk_{X,Y}) \supset \hat{Z}=\hat{X} \vee \hat{Z}=\hat{Y} \vee \hat{Z}=\hat{T}\end{aligned} \quad (3)$$

$$\begin{aligned}&2,3,\mathbf{AA1}, \boldsymbol{\Gamma}_{\mathbf{ABBH,1}}, \boldsymbol{\theta}_{\mathbf{ABBH,1}}\\&[\mathbf{ABBH}:\mathbf{SIMO}]_X\\&\text{KOHonest}(ptk_{X,Y},\{pmk_{X,Y},pmk_{Y,X}\}) \supset\\&\text{Send}(Z,\text{HASH}_{ptk_{X,Y}}(\hat{X},\hat{Y},\text{“ABBH5”},INFO_Y,y,x,enc_1,INFO_X)) \supset \hat{Z}=\hat{Y}\end{aligned} \quad (4)$$

$$\begin{aligned}&2,4,\boldsymbol{\theta}_{\mathbf{ABBH,1}}\\&[\mathbf{ABBH}:\mathbf{SIMO}]_X\\&\text{KOHonest}(ptk_{X,Y},\{pmk_{X,Y},pmk_{Y,X}\}) \supset\\&\text{Computes}(Y,\text{HASH}_{ptk_{X,Y}}(\hat{X},\hat{Y},\text{“ABBH5”},INFO_Y,y,x,enc_1,INFO_X))\wedge\\&\text{Send}(Y,\text{HASH}_{ptk_{X,Y}}(\hat{X},\hat{Y},\text{“ABBH5”},INFO_Y,y,x,enc_1,INFO_X))\end{aligned} \quad (5)$$

$$\begin{aligned}&5,\mathbf{HASH1}, \boldsymbol{\theta}_{\mathbf{ABBH,1}}\\&[\mathbf{ABBH}:\mathbf{SIMO}]_X\\&\text{Has}(Y,ptk_{X,Y}) \wedge \text{Has}(Y,\hat{X},\hat{Y},\text{“ABBH5”},INFO_Y,y,x,enc_1,mic_1)\end{aligned} \quad (6)$$

$$\begin{aligned}&5,6,\boldsymbol{\phi}_{\mathbf{HONESTY}}, \boldsymbol{\theta}_{\mathbf{ABBH,1}}\\&[\mathbf{ABBH}:\mathbf{SIMO}]_X\\&\text{KOHonest}(ptk_{X,Y},\{pmk_{X,Y},pmk_{Y,X}\}) \supset\\&\text{Send}(Y,\hat{X},\hat{Y},\text{“ABBH1”},INFO_Y,y) < \text{Receive}(Y,\hat{Y},\hat{X},\text{“ABBH1”},INFO_X,x) <\\&\text{Send}(Y,\hat{X},\hat{Y},\text{“ABBH5”},INFO_Y,y,x,enc_1,mic_1)\end{aligned} \quad (7)$$

$2,7,\boldsymbol{\theta}_{\mathbf{ABBH,1}}$
$[\mathbf{ABBH : SIMO}]_X$
$\text{KOHonest}(ptk_{X,Y}, \{pmk_{X,Y}, pmk_{Y,X}\}) \supset$
$\text{Send}(Y, \hat{X}, \hat{Y}, \text{"ABBH5"}, INFO_Y, y, x, enc_1, mic_1) < \text{Receive}(X, \hat{X}, \hat{Y}, \text{"ABBH5"}, INFO_Y, y, x, enc_1, mic_1)$ (8)

$\mathbf{FS1}, \mathbf{AN3}, \theta_{\mathbf{ABBH,1}}$
$[\mathbf{ABBH : SIMO}]_X$
$\text{FirstSend}(X, x, \hat{Y}, \hat{X}, \text{"ABBH1"}, INFO_X, x)$ (9)

$9, \mathbf{FS2}, \boldsymbol{\theta}_{\mathbf{ABBH,1}}$
$[\mathbf{ABBH : SIMO}]_X$
$\text{Send}(X, \hat{Y}, \hat{X}, \text{"ABBH1"}, INFO_X, x) < \text{Receive}(Y, \hat{Y}, \hat{X}, \text{"ABBH1"}, INFO_X, x)$ (10)

$\mathbf{FS1}, \mathbf{AN3}, \boldsymbol{\theta}_{\mathbf{ABBH,1}}$
$[\mathbf{ABBH : SIMO}]_X$
$\text{Honest}(\hat{Y}) \supset \text{FirstSend}(Y, y, \hat{X}, \hat{Y}, \text{"ABBH1"}, INFO_Y, y)$ (11)

$7, 11, \mathbf{FS2}, \boldsymbol{\theta}_{\mathbf{ABBH,1}}$
$[\mathbf{ABBH : SIMO}]_X$
$\text{KOHonest}(ptk_{X,Y}, \{pmk_{X,Y}, pmk_{Y,X}\}) \supset$
$\text{Send}(Y, \hat{X}, \hat{Y}, \text{"ABBH1"}, INFO_Y, y) < \text{Receive}(Y, \hat{X}, \hat{Y}, \text{"ABBH1"}, INFO_Y, y)$ (12)

$1, 7, 8, 10, 12, \boldsymbol{\theta}_{\mathbf{ABBH,1}}$
$[\mathbf{ABBH : SIMO}]_X$
$\text{KOHonest}(ptk_{X,Y}, \{pmk_{X,Y}, pmk_{Y,X}\}) \supset$
$(\text{Send}(X, \hat{Y}, \hat{X}, \text{"ABBH1"}, INFO_X, x) < \text{Receive}(Y, \hat{Y}, \hat{X}, \text{"ABBH1"}, INFO_X, x)) \wedge$
$(\text{Send}(Y, \hat{X}, \hat{Y}, \text{"ABBH1"}, INFO_Y, y) < \text{Receive}(Y, \hat{X}, \hat{Y}, \text{"ABBH1"}, INFO_Y, y)) \wedge$
$(\text{Send}(Y, \hat{X}, \hat{Y}, \text{"ABBH5"}, INFO_Y, y, x, enc_1, mic_1) < \text{Receive}(X, \hat{X}, \hat{Y}, \text{"ABBH5"}, INFO_Y, y, x, enc_1, mic_1)) \wedge$
$(\text{Send}(X, \hat{Y}, \hat{X}, \text{"ABBH1"}, INFO_X, x) < \text{Receive}(X, \hat{X}, \hat{Y}, \text{"ABBH1"}, INFO_Y, y) <$
$(\text{Receive}(X, \hat{X}, \hat{Y}, \text{"ABBH5"}, INFO_Y, y, x, enc_1, mic_1) \wedge \text{Send}(X, \hat{Y}, \hat{X}, \text{"ABBH5"}, INFO_X, x, y, enc_0, mic_0))) \wedge$
$(\text{Send}(Y, \hat{X}, \hat{Y}, \text{"ABBH1"}, INFO_Y, y) < \text{Receive}(Y, \hat{Y}, \hat{X}, \text{"ABBH1"}, INFO_X, x) <$
$\text{Send}(Y, \hat{X}, \hat{Y}, \text{"ABBH5"}, INFO_Y, y, x, enc_1, mic_1))$ (13)

7. References

[1] 802.11-2007, I. S. [2007]. Local and metropolitan area networks – specific requirements – part 11: Wireless LAN medium access control and physical layer specifications.

[2] Bellare, M. & Rogaway, P. [1993]. Entity authentication and key distribution., *CRYPTO*, pp. 232–249.

[3] Blunk, L., Vollbrecht, J., Aboba, B., Carlson, J. & Levkowetz, H. [2004]. Extensible authentication protocol (EAP), http://tools.ietf.org/html/draft-ietf-eap-rfc2284bis-09.

[4] Braskich, T. & Emeott, S. [2007a]. Clarification and update of MSA overview and MKD functionality text, https://mentor.ieee.org/802.11/documents doc 11-07/2119r1.

[5] Braskich, T. & Emeott, S. [2007b]. Initial MSA comment resolution, https://mentor.ieee.org/802.11/documents doc 11-07/0564r2.

[6] Braskich, T. & Emeott, S. [2007c]. Key distribution for MSA comment resolution, https://mentor.ieee.org/802.11/documents doc 11-07/0618r0.

[7] Braskich, T. & Emeott, S. [2007d]. Mesh key holder protocol improvements, https://mentor.ieee.org/802.11/documents doc 11-07/1987r1.

[8] Braskich, T., Emeott, S., Barker, C. & Strutt, G. [2007]. An abbreviated handshake with sequential and simultaneous forms, https://mentor.ieee.org/802.11/documents doc 11-07/2535r0.

[9] Braskich, T., Emeott, S. & Kuhlman, D. [2007]. Security requirements for an abbreviated MSA handshake, https://mentor.ieee.org/802.11/documents doc 11-07/0770r0.

[10] Cortier, V. [2012]. Secure composition of protocols, *in* S. MÃűdersheim & C. Palamidessi (eds), *Theory of Security and Applications*, Vol. 6993 of *Lecture Notes in Computer Science*, Springer Berlin / Heidelberg, pp. 29–32.

[11] Cremers, C. [2008]. On the protocol composition logic PCL, *Proc. of the Third ACM Symposium on Information, Computer & Communication Security (ASIACCS '08)*, ACM Press, Tokyo. To appear.

[12] Datta, A., Derek, A., J.C.Mitchell & B.Warinschi [2006]. Computationally sound compositional logic for key exchange protocols, *Proceedings of 19th IEEE Computer Security Foundations Workshop*, pp. 321–334.

[13] Datta, A., Derek, A., Mitchell, J. C. & Pavlovic, D. [2003a]. Secure protocol composition., *FMSE*, pp. 11–23.

[14] Datta, A., Derek, A., Mitchell, J. C. & Pavlovic, D. [2005]. A derivation system and compositional logic for security protocols, *J. Comput. Secur.* 13(3): 423–482.

[15] Datta, A., Derek, A., Mitchell, J. C. & Roy, A. [2007]. Protocol composition logic (PCL)., *Electr. Notes Theor. Comput. Sci.* 172: 311–358.

[16] Datta, A., Derek, A., Mitchell, J. C. & Warinschi, B. [n.d.]. Key exchange protocols: Security definition, proof method and applications.
URL: *citeseer.ist.psu.edu/datta06key.html*

[17] Datta, A., Derek, A., Mitchell, J. & Pavlovic, D. [2003b]. A derivation system for security protocols and its logical formalization, *16th IEEE Computer Security Foundations Workshop (CWFW-16)*, pp. 109–125.
URL: *citeseer.ist.psu.edu/datta03derivation.html*

[18] Datta, A., Mitchell, J., Roy, A. & Stiller, S. [2011]. Protocol composition logic, *Formal Models and Techniques for Analyzing Security Protocols*, IOS Press.

[19] Dierks, T. & Rescorla, E. [April 2006]. The Transport Layer Security (TLS) Protocol, version 1.1 – RFC 4346, http://tools.ietf.org/html/rfc4346.

[20] Diffie, W., van Oorschot, P. C. & Wiener, M. J. [1992]. Authentication and authenticated key exchanges., *Des. Codes Cryptography* 2(2): 107–125.

[21] Durgin, N., Mitchell, J. & Pavlovic, D. [2001]. A compositional logic for proving security properties of protocols, *Proceedings of 14th IEEE Computer Security Foundations Workshop*, pp. 241–255.
URL: *citeseer.ist.psu.edu/article/durgin02compositional.html*

[22] Durgin, N., Mitchell, J. & Pavlovic, D. [2004]. A compositional logic for proving security properties of protocols, *J. Comput. Secur.* 11(4): 677–721.

[23] Fu, J., Jiang, X., Ping, L. & Fan, R. [2009]. A novel rekeying protocol for 802.11s key management, *Proceedings of the 2009 International Conference on Information Management and Engineering*, ICIME '09, IEEE Computer Society, Washington, DC, USA, pp. 295–299. URL: *http://dx.doi.org/10.1109/ICIME.2009.14*

[24] Gollmann, D. [1996]. What do we mean by entity authentication?, *SP '96: Proceedings of the 1996 IEEE Symposium on Security and Privacy*, IEEE Computer Society, Washington, DC, USA, p. 46.

[25] Haasz, J. & Hampton, S. [2006]. Amendment: Mesh networking, http://standards.ieee.org/board/nes/projects/802-11s.pdf.

[26] He, C., Sundararajan, M., Datta, A., Derek, A. & Mitchell, J. C. [2005]. A modular correctness proof of IEEE 802.11i and TLS., *ACM Conference on Computer and Communications Security*, pp. 2–15.

[27] Krawczyk, H. [2003]. SIGMA: The 'SIGn-and-MAc' approach to authenticated diffie-hellman and its use in the IKE-protocols., *CRYPTO*, pp. 400–425.

[28] Kuhlman, D., Moriarty, R., Braskich, T., Emeott, S. & Tripunitara, M. [2007]. A proof of security of a mesh security architecture, http://eprint.iacr.org/2007/364.pdf.

[29] Meadows, C. & Pavlovic, D. [2004]. Deriving, attacking and defending the GDOI protocol, *ESORICS*, pp. 53–72.

[30] Rigney, C., Willens, S., Rubens, A. & Simpson, W. [June 2000]. Remote Authentication Dial In User Service (RADIUS) – RFC 2865, http://tools.ietf.org/html/rfc2865.

[31] Roy, A., Datta, A., Derek, A. & Mitchell, J. C. [2007]. Inductive proof method for computational secrecy, *Proceedings of 12th European Symposium On Research In Computer Security*.

[32] Roy, A., Datta, A., Derek, A., Mitchell, J. C. & Seifert, J.-P. [2006]. Secrecy analysis in protocol composition logic., *Proceedings of 11th Annual Asian Computing Science Conference*.

[33] Roy, A., Datta, A., Derek, A., Mitchell, J. C. & Seifert, J.-P. [2008]. Secrecy analysis in protocol composition logic., *Formal Logical Methods for System Security and Correctness*, IOS Press.

[34] Simon, D., Aboba, B. & Hurst, R. [2007]. The EAP TLS authentication protocol, http://www.ietf.org/internet-drafts/draft-simon-emu-rfc2716bis-11.txt.

[35] Zhao, M., Walker, J. & Conner, W. S. [2007]. Overview of abbreviated handshake protocol, https://mentor.ieee.org/802.11/documents doc 11-07/1998r01.

Chapter 8

Autonomous Traffic Balancing Routing in Wireless Mesh Networks

Sangsu Jung

Additional information is available at the end of the chapter

1. Introduction

Wireless mesh networks (WMNs) are attractively deployed as a backhaul for public Internet access with the advantages of network performance and cost efficiency. With the cooperation of multiple mesh nodes, a packet is transmitted through multi-hops to reach a destination. Wireless medium experiences relatively an unstable environment due to interferences of wireless signals. As traffic increases, a communication environment becomes worse. Similarly, as more mesh nodes join a WMN, a network performance is also degraded due to increasing interferences. Therefore, challenges of a WMN are how to accommodate a dynamic nature of wireless medium and achieve the scalability.

A typical WMN serves hub-and-spoke type accesses, where a mesh gateway (hub) connects to the Internet for mesh clients as shown in Fig. 1. In other words, a mesh node is required to communicate with just one-of-many mesh gateways similar to anycast communications [1, 2]. For anycasting, conventional routing schemes [3–5] are developed with modifications of existing unicast routing protocols. That is, the schemes usually select the closest destination among multiple service gateways. Thus, they are inefficient in taking benefits of having multiple gateways. Even though some protocols [6–8] utilize multiple gateways for load balancing, they convey flooding overheads to collect traffic load information for re-routing and require associations among the gateways.

Classically, back-pressure routing [9] and geographic routing [10] are considered as alternatives for traditional hop-count-based routing. Back-pressure routing is well-known to achieve throughput-optimal by adaptively selecting paths depending on queuing-dynamics. However, it unnecessarily chooses long paths and degrades network performance by keeping old data packets. This problem manifests critically in lightly- or moderately-loaded cases [11]. On the other hand, conventional geographic routing is scalable with no flooding overhead, but it is vulnerable to avoiding congested hot spots due to its simple geographical routing metric. Even though some enhancements of geographic routing for congestion mitigation, it entails similar overheads such as perimeter routing or other face routing.

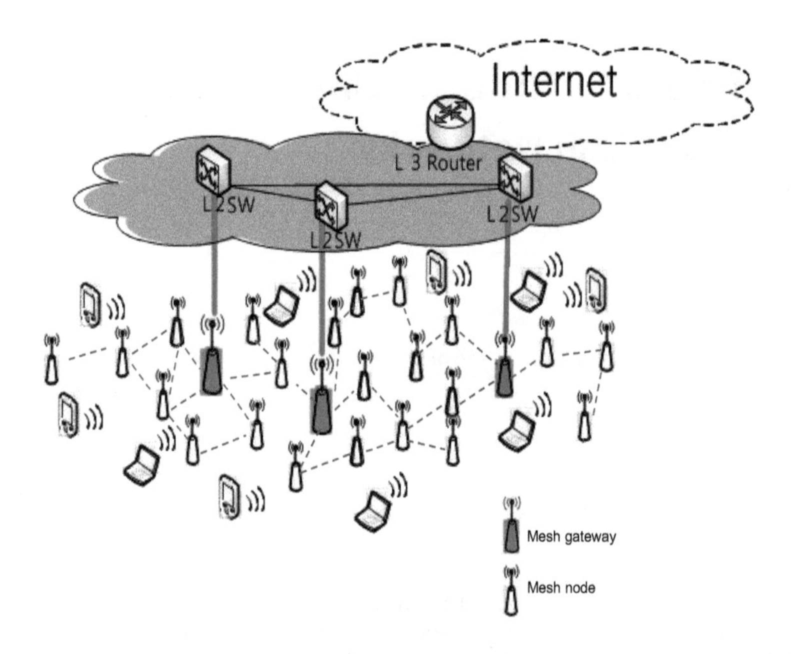

Figure 1. An Example of a Wireless Mesh Network.

To overcome the limitations of conventional approaches, we suggest a novel routing scheme inspired by electrostatic potential theory [12–15]. The main motivation comes from the fact that packet movements can be corresponding to electric charge behaviors governed by electrostatic potential. By constructing a virtual potential field for routing, our scheme forwards a packet following the steepest gradient direction towards any mesh gateway (a destination inside a WMN) as presented in Fig. 2. Interestingly, our scheme based on nature characteristics resembles a hybrid behavior of back-pressure and geographic routing schemes. With the help of numerical analysis techniques, our scheme operates in a distributed manner. Furthermore, our formula is equipped with a Gaussian function to adjust a routing reflecting ratio of back-pressure and geographic routing schemes for dynamic traffic environments.

Our work is relevant to recent approaches [17–19] motivated by physical systems of which system models have been studied for several centuries.

This chapter introduces a practical solution to develop large-scalable WMNs. The organization of this chapter is as follows. In section 2, we review relevant works and address distinguished features of our scheme. Section 3 provides a background of our work and designs a traffic-adaptive autonomous routing scheme for WMNs. In addition, we evaluate our scheme through simulations. (Section 4) Finally, we conclude this chapter in Section 5.

2. Related works

Our scheme is a family of gradient-based routing [16, 20–22]. In gradient-based routing, scalar values are assigned to each node to form a field gradient, so packets traverse followed by the

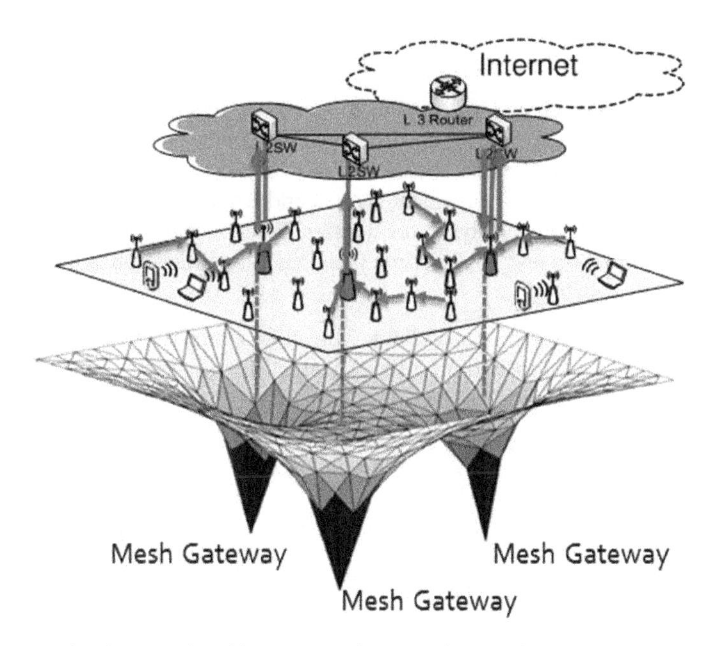

Figure 2. An Example of Potential Field Formation for a Wireless Mesh Network.

lowest gradient. Several proposed works differ by the target network, communication pattern, and parameters to determine scalar values for a field.

A. Basu et al. have introduced potential-based routing (PB routing) [20] for unicast Internet traffic. Their idea is to set potentials with queue lengths on the top of the standard shortest path (hop count) routing. Here, the steepest gradient routing enables routers to be less congested. Even though the aim of this work is quite similar to ours, it is for fixed wired networks and unicast with flooding overheads. Regarding an anycast, V. Lenders et al. propose a density-based strategy [21] for wireless ad hoc networks in the framework of gradient-based routing. Under the scheme, a field is constructed based on node density and routing is towards a dense group destination, which increases the success probability of packet transmission. However, the routing scheme cannot react to traffic congestion; hence, even a worse scenario could occur where traffic is densely populated. For an anycast in a WMN, R. Baumann et al. develop HEAT routing [22] where a temperature field is used for routing. In this work, two metrics are considered to influence a temperature value: one is the distance from a node to a gateway, and the other is the robustness of a path towards a gateway. Because this model is based on Laplace's equation, which is a special form of Poisson's equation, it cannot deal with traffic dynamics so that congested hot spots degrade the routing performance.

On the other hand, our scheme utilizes the numerical analysis techniques of a finite element method (FEM) [23, 24] and a local equilibrium method (LEM) [25], to achieve a distributed algorithm [12–15]. Similarly, a finite difference method routing (FDMR) is suggested in [26] reflecting link-diversity. It is also a gradient routing scheme based on Laplace's equation.

However, it is known that an FDM is restricted to a grid topology [23, 24]; hence, another algorithm is desired for arbitrarily-shape topologies.

Furthermore, we adopt a Gaussian function to tune our scheme for dynamic networking environments. Conventionally, Gaussian functions have been applied in the Gaussian filter [28] in signal processing, the Gaussian beam [29] in microwave systems, self-similar network traffic generation [30] in WAN (wide area network) and LAN (local area network), and other modeling researches [31]. Even though a Gaussian function has been variously applied in many areas such as statistics and engineering, there has been little approach to use a Gaussian function for routing modeling.

In conclusion, the novelties of our scheme are characterized by anycast capability in WMNs, load balancing, distributed algorithm, scalability with constant control overheads, self-adaptation, and random topology accommodation.

3. Traffic-adaptive autonomous routing

3.1. A hybrid routing inspired by electrostatics

Our aim is to combine geographic routing and back-pressure routing represented as:

$$\Pi = aD + (1-a)T, \tag{1}$$

or

$$\Pi' = D + \frac{1-a}{a}T, \tag{2}$$

where routing metric Π (or Π') is a linear combination of geometric distance D and traffic component T adjusted by ratio a. Interestingly, a distributed form of Poisson's equation [12] can be matching to (2), which describes the movement behaviors of electrostatic charges [12, 13]:

$$\phi(v) = \{\Sigma_{k=1}^{n} \frac{(\phi(p_{v,k+1})\vec{r}_{v,k} - \phi(p_{v,k})\vec{r}_{v,k+1}) \cdot (\vec{r}_{v,k} - \vec{r}_{v,k+1})}{A_k} + \alpha q(v)\} / \Sigma_{k=1}^{n} \frac{\|\vec{r}_{v,k} - \vec{r}_{v,k+1}\|^2}{A_k} \tag{3}$$

where routing metric potential of node v, $\phi(v)$, is obtained by consideration of neighbor nodes' potential $\phi(p_{v,k})$, distance component $\vec{r}_{v,k}$, and queue length $q(v)$ as shown in Fig. 3. In [12] and [13], we describe how to derive (3) by using an FEM. On the other hand, every node operates (3) in an iterative manner to get converged potential which is used as a routing metric under boundary conditions:

$$\pi(G) = Min, \tag{4}$$

$$\pi(B) = Max, \tag{5}$$

where G is the set of mesh gateways and B is the set of boundary nodes which are located at the outer boundary of a WMN. As a result of an LEM (refer to Fig. 4), a potential field is formed inclined from mesh nodes to mesh gateways in a range of [*Min*, *Max*]. According to a routing policy, a packet traverses following the steepest gradient field towards a mesh gateway which has the lowest potential. (See Algorithm 1) The steps for our routing scheme (ALFA-Advanced, ALFA-A) are as follows:

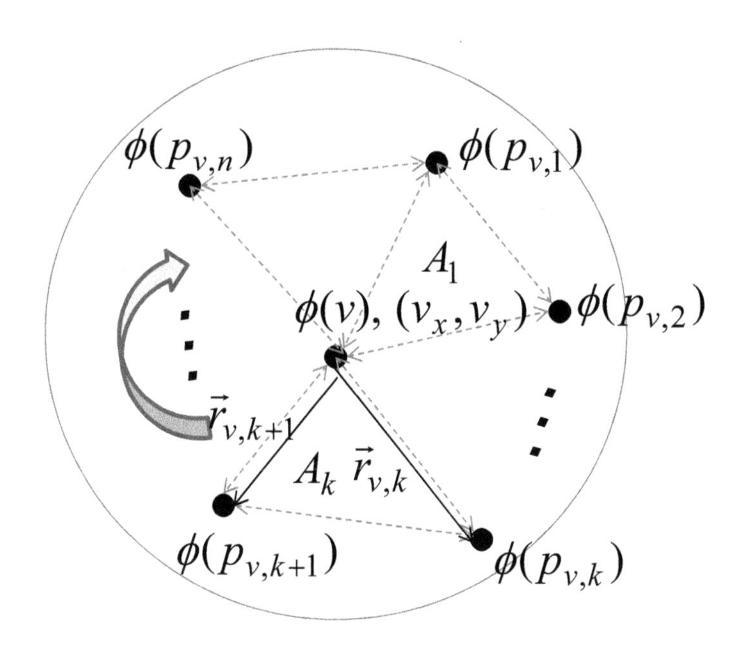

Figure 3. FEM Geometric Elements.

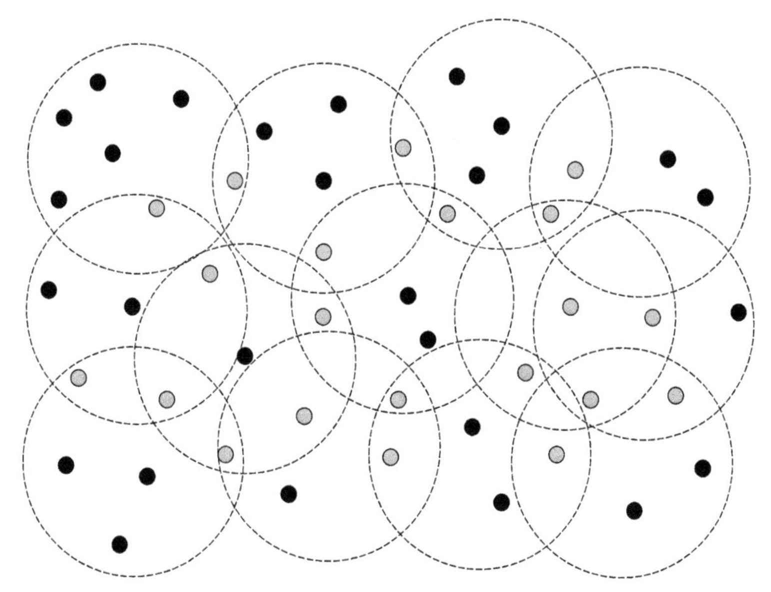

Figure 4. An example of an LEM. Local iterations of (3) reach a global solution of a WMN with the contributions of intermediate nodes represented by gray dots.

Algorithm 1 ALFA-A

```
1: //Set initial condition
2: if Mesh Gateway then
3:     Fix Potential -1
4: else
5:     if Boudnary Node then
6:         FixedPotential 0
7:     else
8:         Set Potential 0
9:     end if
10: end if
11: for i ∈ all do
12:     Advertise Potential;
13:     Result=ComuputePotential(Node);
14:     Set Potential Result;
15: end for
16: //Generate Hello Message
17: Advertise Potential;
18: Update NeighborNodePotentialList;
19: //ALFA-A Routing
20: if queue>0 then
21:     Result-ComputerPotential(Node);
22:     Set Potential Result;
23:     ForwardingNode=NeighborNode by (6);
24:     Send a packet to ForwardingNode;
25: end if
```

- Step 1. In an initial network deployment stage, boundary conditions are defined as (4) and (5).
- Step 2. Every mesh node v, which is non-mesh gateways and -boundary nodes assigns zero as its potential value.
- Step 3. Every mesh node v exchanges its potential with its neighbor nodes n via a hello message.
- Step 4. Every mesh node v calculates its potential with (3).
- Step 5. A forwarding node is determined by:

$$\arg \max_{n_{v,k} \in N_v} \frac{\phi(n_{v,k}) - \phi(v)}{||\vec{r}_{n,k} - \vec{r}_v||}, \tag{6}$$

On the other hand, the gradient field of ALFA-A is directed from mesh nodes to mesh gateways, a different routing scheme is required for downlink traffic. For downlink routing, ALFA-A adopts a kind of source-based forwarding. That is, paths used for recent uplink traffic are learned and used for downlink traffic routing. This behavior is achieved by recording the previous-hop and source addresses of every uplink packet in a downlink forwarding information database (FIB) at every mesh node and gateway. Since a usual IP header has no previous hop IP address information, a node retrieves it by using Reverse Address Resolution Protocol (RARP) with a previous hop MAC address. Then the node obtains a source IP address from an IP header and creates a new FIB entry. This downlink FIB is then used to determine the next-hop address of each downlink packet by looking up the downlink destination address among recorded uplink source addresses.

In cases of one-way traffic such as for IPTV, the downlink FIB may not be updated often enough because of the absence of uplink traffic, and, thus, dynamic load balancing cannot be achieved. This problem can be mitigated by each mesh node sending association refresh messages periodically to a gateway for a downlink path.

The principal ability of our scheme is autonomous load balancing among multiple mesh gateways as well as among mesh nodes. For example, when a congested hot spot forms due to an extensive local traffic area, it causes a potential increase of nodes around the hot spot due to the increase of the queue lengths. This potential increase prevents packets from ingressing the region; packets are forwarded away from the congested high-potential area as shown in Fig. 5 and 6.

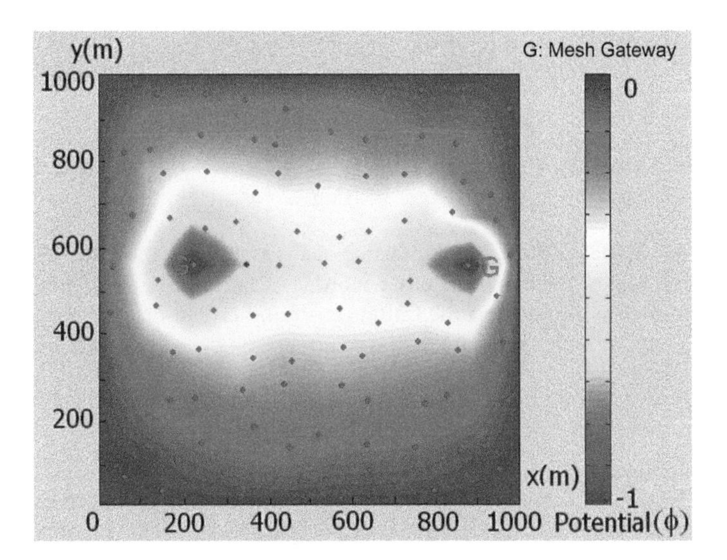

Figure 5. Potential Distribution: Uncongested Case.

3.2. Tuning for traffic dynamics

The characteristics of (3) are affected by tunable parameter α, which determines the reflecting ratios of geographic routing and back-pressure routing. That is:

- When α of (3) is large, the sensitivity to traffic congestion increases and the behavior of our scheme resembles that of back-pressure routing.

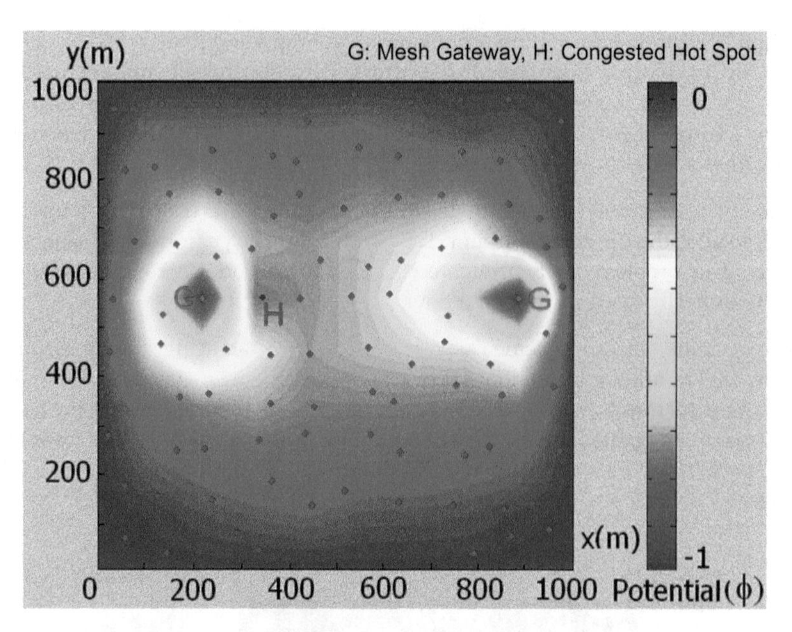

Figure 6. Potential Distribution: Congested Case.

- When α of (3) is small, the sensitivity to traffic congestion decreases and the behavior of our scheme is close to that of geographic routing.

In a previous work, ALFA [12], the constant value of α is assumed so that the reflection degrees of both routing schemes are fixed. In other words, ALFA is only optimized for a specific network environment with no dynamic traffic consideration. In our scheme, we develop a new form of Îś adopting a Gaussian function. The reasoning behind this setting is come from:

- If traffic load level is low under network capacity, a routing decision is not necessary to consider a traffic component (queue length) because the network resources are sufficient.
- If traffic load level is high more than network capacity, a routing decision is not necessary to consider a geographic factor because the short path marginally contributes to minimizing delay.

Interestingly, the shape of a Gaussian function can exhibit low reflection of the routing metric and high reflection of that via queue length as shown in Fig. 7. Therefore, we set α as:

$$\alpha = Ce^{-R(q(v)-Q)^2} \tag{7}$$

where C is a constant, R is a sensitivity for queue length, and Q is a bounded value from which queue length cannot affect the value of α. Because we can adjust C and R with respect to a network characteristic, our scheme flexibly adopts to a dynamic network scenario. Previously, we apply a Gaussian function to an FDM-based routing scheme [15]. As an extension, we introduce a Gaussian function applied to an FEM and evaluate our proposed scheme in the next section.

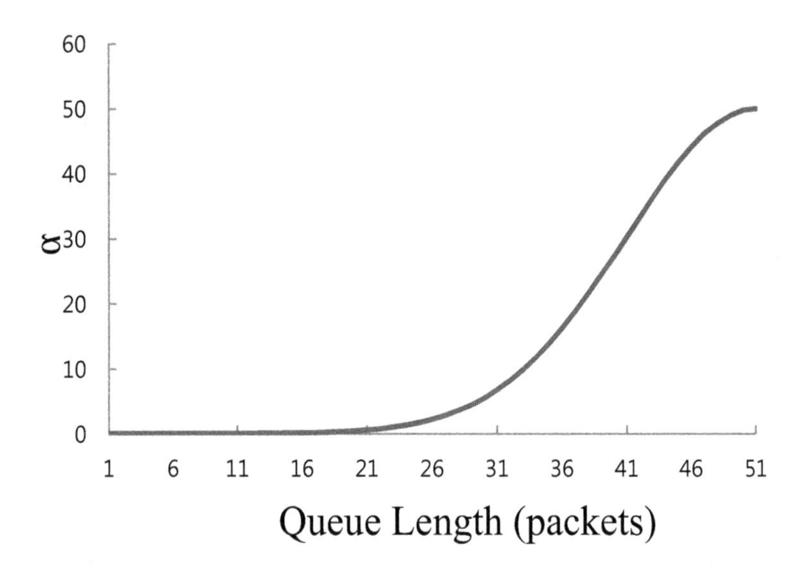

Figure 7. α with respect to queue length in a Gaussian function

4. Performance evaluations

In this section, we evaluate our scheme (ALFA-Advanced, AFLA-A) compared with conventional ALFA, geographic routing (GR), and back-pressure routing (BPR) using NS2 [33]. Incorporating the PHY/MAC models of IEEE 802.11, we conduct simulations with randomly deployed 100 mesh nodes and two mesh gateways in 1000m×1000m areas. The average distance between two mesh nodes is set to 200m. The transmission and interference ranges are set to 250m and 550m, respectively. We assume the two-ray ground model as a radio propagation model. We select 15 mesh nodes located almost equally far away from two mesh gateways as traffic source nodes (which generate traffic loads from 20-40Mbit/s) and 20 mesh nodes as back-ground traffic (10Mbit/s) generation nodes. The total mesh gateway capacity is 2Mbit/s channel bandwidth. The size of each data packet (UDP) is 2000 bytes long. The maximum queue size is set to 2000. Every mesh node utilizes RTS/CTS and hello-message-jittering. For boundary conditions, we assign -1 for the potential of mesh gateways and 0 for the boundary nodes. On the other hand, we modify GR and BPR to accommodate anycast communications, in such a way that a mesh node first checks which mesh gateway is the closest to itself. Removing the data of a transient period, we collect the data of the mid 1000s of out of 1600s simulations run period. In the simulations, we set α as 0.005 for conventional ALFA considering previous empirical results. For our scheme, we set 2 and 0.5 for C and R, respectively.

First, we observe the load balancing behaviors of the schemes with the aggregate throughput. Because our purpose is to increase the aggregate throughput in an entire network by efficiently sharing the loads, instead of just even distribution of the loads, which is unnecessary when there are sufficient network resources. Similar to BPR, ALFA utilizes multiple paths for mitigating traffic congestion but avoids unnecessary explorations of a large number of paths observed in BPR at some point. However, ALFA-A maintains stable performance regardless of traffic loads.

Second, we show the hop counts of utilized paths per source. Due to the genetic nature of ALFA and ALFA-A, the hop counts of those two tend to maintain small hop counts comparable with GR. Still, the strength of ALFA remains only for a specific environment.

Finally, we conduct simulations to investigate the robustness of our scheme to node failures. With the constant control overheads, ALFA and ALFA-A maintain the performance of a network under node failures.

4.1. Load balancing

In Fig. 8, we present aggregate throughput for each scheme. ALFA-A, ALFA, and BPR diversify paths so that they show relatively higher performance compared with GR. The difference between those three schemes comes from the number of unnecessary paths. ALFA is only optimized for a limited case and BPR is known to utilize tremendous number of paths which increases path-lengths. In case of ALFA-A, it dynamically adjust the level of traffic reflection in its routing metric considering the traffic load level, and thus, it shows the highest performance among four schemes.

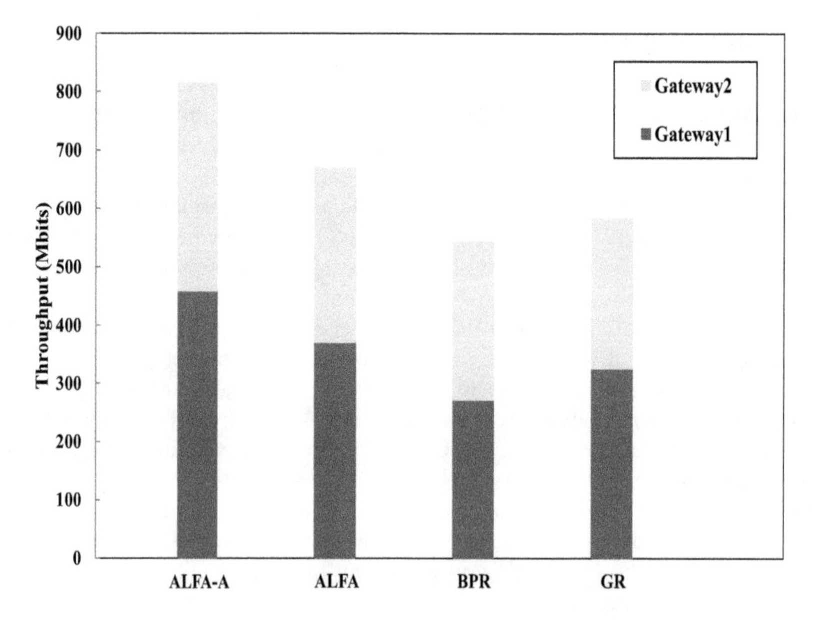

Figure 8. Aggregate Throughput under a Dynamic Load Scenario.

4.2. Hop counts

We illustrate the hop count distribution for each packet in Fig. 9. ALFA-A and ALFA behave as GR under lightly-loaded cases, whereas BPR traverses relatively long paths, as reported

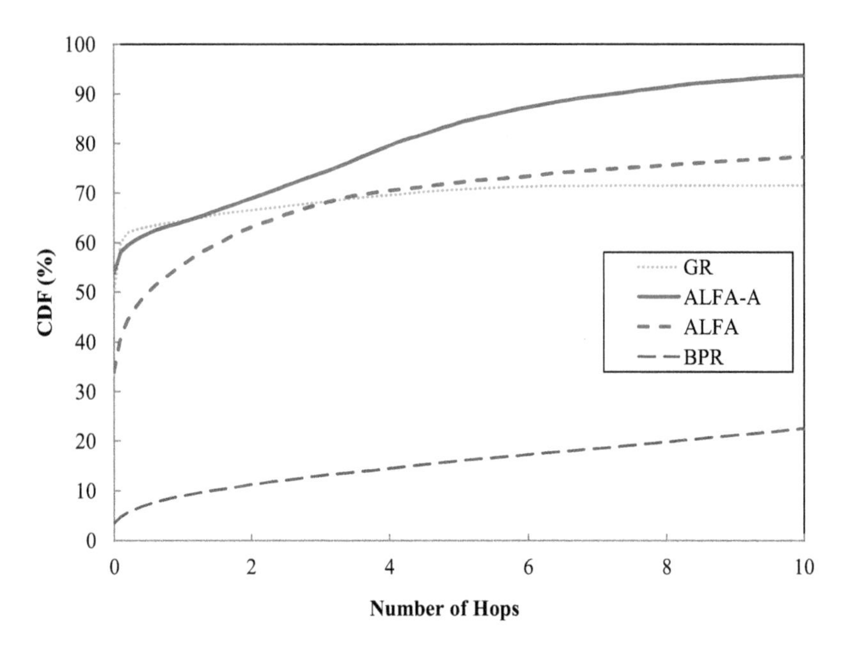

Figure 9. Hop Counts under a Dynamic Load Scenario.

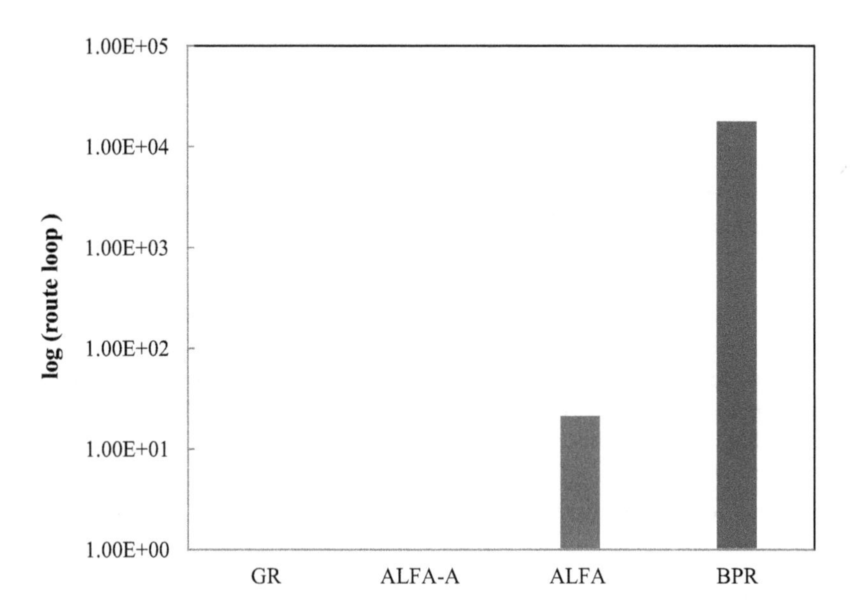

Figure 10. Number of Route Loop in Routing.

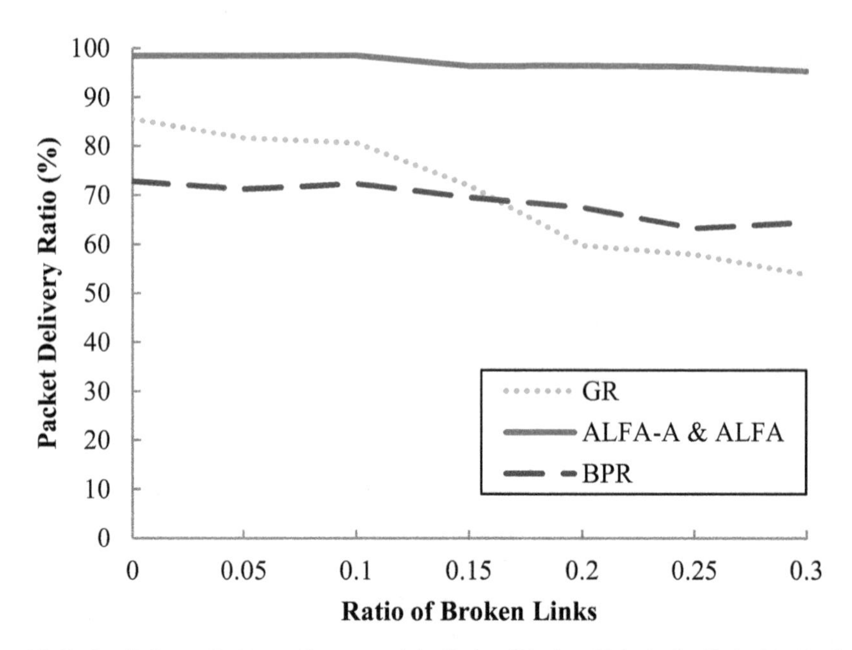

Figure 11. Packet Delivery Ratio as a Function of the Ratio of Broken Links in the Entire Network .

in [11]. Because BPR exploits unnecessarily long paths due to large hop counts, including frequent route loops, it is very harmful to delay-sensitive applications. BPR shows inferior performance compared with ALFA-A, ALFA, and GR. The result of high hop counts can be understood with long queueing delays at every hop. The long queueing delay of BPR is caused by the large queue lengths. As a queue increases, the queueing delay also increases. We observe that ALFA-A shows the shortest hop counts compared with those of others because ALFA-A tends to take short paths and to avoid nodes with large queue lengths. In GR, which maintains short path lengths, packet paths tend to be concentrated on a specific region and form a congested hot spot. Such a hot spot causes large queueing delays. In the case of BPR, it has a large number of hops and large queue sizes at the same time; hence, the delay performance is undesirable. This behavior is also originated from the number route looping in routing as shown in Fig. 10. Because GR does not consider detouring, it shows zero route-looping. Similarly, ALFA-A adapts its routing behavior appropriately to a network state, it also shows zero route-looping. Differently, ALFA and BPR generate frequent route-looping due to overestimate congestion degree in a network.

4.3. Robustness to node failures

Intrinsically, ALFA-A and ALFA adaptively utilize multiple paths in search of less congested areas toward a mesh gateway in a hop-by-hop fashion. This property also enables packet routing to be robust to node failures. Because ALFA-A and ALFA always maintain a field gradient toward mesh gateways, if one node disappears, they automatically select an alternative forwarding node with no additional route re-establishment. This feature corresponds to the stateless property similarly to geographic routing.

We present the robustness of ALFA-A as the same as ALFA against node failures in Fig. 11. As a performance metric, we consider a packet delivery ratio that is defined as the percentage of packets that have successfully arrived at destinations out of the total packets sent by sources. From the figure, we see that ALFA-A and ALFA can maintain approximately a 95% or higher packet delivery ratio when the ratio of broken links over the entire network varies from 0% to 30% of the total links. In contrast, the delivery ratio of GR drops from 86% to 52% when the ratio of broken links reaches 30% because GR experiences voids or dead-ends when nodes disappear. In the case of voids or dead-ends, GR conducts face routing to track an available alternative forwarding node, but it incurs longer delays due to a larger number of hops [34]. On the other hand, BPR experiences relatively low degradation (from 72% to 68%) due to its adaptive routing mechanism.

4.4. Control overheads

All routing schemes evaluated in this chapter, ALFA-A, ALFA, GR, and BPR, significantly reduce the control overheads as they require no state-vector flooding mechanism, which is used in traditional reactive routing protocols such as AODV [35]. Like the other two, ALFA requires just one-hop neighbor information for route decisions. That is, it only uses a 'hello' message with the interval of 1s, which is commonly used for all routing protocols to find one-hop neighbors. The 'hello' message delivers information for route decisions such as potential and location. This simple local behavior achieves scale-free routing, and simultaneously achieves network-wide load balancing routing in response to congestion. This merit is an advantage in the cases of congestion and node failures different from flooding-based schemes that consume large network resources with route re-establishment processes.

An underlying understanding of constant control overheads is that a local behavior is important and sufficient to make a routing decision, and we can avoid requiring an accurate solution for the global potential field. In fact, the computation cost of every update might be $O(|S|)$ within a small subset of nodes S, meaning that the impact of a local change is quickly seized by the neighbor nodes. A global impact is slowly gained through an LEM property and affects the routing as packets traverse a network with multiple hops. In other words, a dynamic queue length change at a far node affects little in choosing the next hop.

Compared to the routing cost of the traditional shortest-path routing, $O(|S^3|)$, such as in the Floyd Warshall's "all pairs shortest path" algorithm [32], ALFA-A is obviously a superior solution for scalable WMNs. In addition, the memory space requirement for routing is greatly reduced because every node stores only its location and potential and those of one-hop neighbors. The total storage required in an entire network is $O(|S|+|S|)$.

5. Conclusions

In this chapter, we deal with a potential-based anycast routing scheme for WMNs, which achieves autonomous traffic balancing and path-length reduction. By analogy with an electrostatic potential theory governed by Poisson's equation, we derive a routing metric that shows the hybrid behaviors of back-pressure routing and geographic routing. The beauty of our routing scheme is adopting the strengths of the two aforementioned routing schemes while overcoming the limitations of the two. We adopt an FEM and an LEM to design a distributed potential assignment as the routing metric; only one-hop neighbor information is required for scale-free global routing. The stateless property of our scheme contributes to maintaining robustness to node failures and eliminating requirement of flooding overheads for re-routing. Furthermore, our scheme utilizes a Gaussian function to dynamically adapt to rapidly changing environments. Using simulations, we investigate how our scheme behaves with respect to a tuning parameter, which characterizes a routing behavior similarly to back-pressure routing or geographic routing. In addition, we demonstrate the superior performance of our scheme compared with conventional schemes in the aspects of throughput, load balancing, and path lengths. Considering the implementation issues of a protocol in practical applications, our scheme is the appropriate solution combining the properties of back-pressure routing and geographic routing.

As a future work, applications of our scheme can be extended to other mesh networking areas based on sensor networks, machine-to-machine communications, and LTE-Advanced, with practical network service models.

Author details

Sangsu Jung
Future Internet Research Team, National Institute for Mathematical Sciences, Daejeon, S. Korea

6. References

[1] C. Metz (2002) IP anycast point-to-any point communication. IEEE Internet Computing. 6: 94-98.
[2] V. Lenders, M. May, B. Plattner (2008) Density-based anycast: a robust routing strategy for wireless ad hoc networks. IEEE/ACM Trans. on Networking. 16: 852-863.
[3] J. Wang, Y. Zheng, W. Jia (2003) An AODV-based anycast protocol in mobile ad hoc network. in: Proc. IEEE PIMRC.
[4] J. Wang, Y. Zheng, W. Jia (2003) A-DSR: a DSR-based anycast protocol for IPv6 flow in mobile ad hoc networks. in: Proc. IEEE VTC.
[5] U. C. Kozat, L. Tassiulas (2003) Network layer support for service discovery in mobile ad hoc networks. in: Proc. IEEE INFOCOM.
[6] J. Shin et al. (2005) Load balancing among Internet gateways in ad-hoc networks. in :Proc. IEEE VTC.
[7] A. Velmurugan, R. Rajaram (2006) Adaptive hybrid mobile agent protocol for wireless multihop Internet access. Journal of Computer Science. 2: 672-682.
[8] D. Nandiraju et al. (2006) Achieving load balancing in wireless mesh networks through multiple gateways. in: Proc. IEEE MASS.

[9] L. Tassiulas, A. Ephremides (1992) Stability properties of constrained queueing systems and scheduling polices of maximum throughput in multihop radio networks. IEEE Trans. on Automatic Control. 37: 1936-1948.
[10] B. Karp, H. T. Kung (2000) GPSR: Greedy Perimeter Stateless Routing for wireless networks. in: Proc. MOBICOM.
[11] L. Ying, S. Shakkottai (2009) A. Reddy, On combining shortest-path and back-pressure routing over multihop wireless networks. in: Proc. IEEE INFOCOM.
[12] S. Jung et al. (2009) Distributed potential field based routing and autonomous load balancing for wireless mesh networks. IEEE Comm. Lett.. 13: 429-431.
[13] S. Jung et al. (2009) Autonomous load balancing field based anycast routing protocol for wireless mesh networks. in: Proc. IEEE HotMESH.
[14] S. Jung et al. (2010) Greedy local routing strategy for autonomous global load balancing based on three-dimensional potential field. IEEE Comm. Lett.. 14:839-841.
[15] B. Jin, S. Jung (2012) On dynamics of field based anycast routing in wireless mesh networks. in: Proc. IEEE ICACT.
[16] Q. Liang et al. (2012) Potential field based routing to support QoS in WSN. Journal of Computational Information Systems. 8: 2375-2385.
[17] S. Toumpis (2008) Mother nature knows best: a survey of recent results on wireless networks based on analogies with physics. Computer Networks. 52: 360-383.
[18] E. Hyytia, J. Virtamo (2009) On the optimality of field-line routing in massively dense wireless multi-hop networks. Performance Evaluation. 66: 158-172.
[19] S. Toumpis, S. Gitzenis (2009) Load balancing in wireless sensor networks using kirchhoff's voltage law. in: Proc. IEEE INFOCOM.
[20] A. Basu, A. Lin, S. Ramanathan (2003) Routing using potentials: a dynamic traffic-aware routing algorithm. in: Proc. ACM SIGCOMM.
[21] V. Lenders, M. May, B. Plattner (2008) Density-based anycast: a robust routing strategy for wireless ad hoc networks. IEEE/ACM Trans. on Networking. 16: 852-863.
[22] R. Baumann, S. Heimlicher, B. Plattner (2008) Routing in large-scale wireless mesh networks using temperature fields. IEEE Network. 22: 25-31.
[23] L. J. Segerlind (1984) Applied finite element analysis. 2nd edition. John Wiley&Sons.
[24] J. L. Volakis, A. Chatterjee, L. C. Kempel (1998) Finite element method for electromagnetics. IEEE Press.
[25] Z. Li, M. B. Reed (1995) Convergence analysis for an element-by-element finite element method. Computer Methods in Applied Mechanics and Engineering. 123: 33-42.
[26] V. Lenders, R. Baumann (2008) Link-divesity routing: a robust routing paradigm for mobile ad hoc networks. in: Proc. IEEE WCNC.
[27] D. J. Griffiths (1998) Introduction to electrodynamics. 3rd edition. Prentice Hall.
[28] R. A. Haddad, A. N. Akansu (1991) A class of fast Gaussian binomial filters for speech and image processing. IEEE Trans. on Speech and Signal Processing. 39: 723-727.
[29] B. E. A. Saleh, M. C. Teich, B. E. Saleh (1991) Fundamentals of photonics. Wiley Online Library.
[30] T. M. Cover, J. A. Thomas (1991) Elements of information theory. Wiley Online Library.
[31] V. Paxson (1997) Fast, approximate synthesis of fractional Gaussian noise for generating self-similar network traffic. ACM SIGCOMM Com. Comm. Rev.. 27: 5-18.

[32] R. Ahuja, T. Magnanti, J. Orlin (1993) Network flows: theory, algorithms, and applications. Prentice hall.
[33] NS-2. http://www.isi.edu/nsnam/ns/.
[34] S. Jung et al. (2008) A geographic routing protocol utilizing link lifetime and power control for mobile ad hoc networks. in: Proc. ACM FOWANC.
[35] C. Perkins, E. Belding-Royer, S. Das. Ad hoc on-demand distance vector (AODV) routing. http://www.ietf.org/rfc/rfc3561.txt. RFC 3561.

Chapter 9

Achievable Capacity Limit of High Performance Nodes for Wireless Mesh Networks

Thomas Olwal, Moshe Masonta, Fisseha Mekuria and Kobus Roux

Additional information is available at the end of the chapter

1. Introduction

Research background: Next generation fixed wireless broadband networks have immensely been deployed as mesh networks in order to provide and extend access to the internet. These networks are characterised by the use of multiple orthogonal channels available within the industrial, scientific and medical (ISM) liscensed-free frequency bands. Nodes in the network have the ability to simultaneously communicate with many neighbours or stream different versions of the same data/information using multiple radio devices over orthogonal channels thereby improving effective "online" channel utilisation (Kodialam & Nandagopal, 2005). The ability to perform full duplex communication by individual multi-radio nodes without causing network interference has also been achieved through decentralized transmission power control schemes in (Olwal, 2010; Olwal et al., 2011). Allen et al. (2007) alluded that multiple radios that receive versions of the same transmission may together correctly recover a frame that would otherwise be lost based on multipath fading, even when any given individual radio cannot. Many such networks emerging from standards such as IEEE 802.11 a/b/g/n and 802.16 are already in use, ranging from prototype testbeds (Eriksson et al., 2006) to complete solutions (Mesh Dynamics, 2010).

The increasing question is how the theoretical capacity of such static multi-radio multichannel (MRMC) network scales with the node density, irregularity of the terrain and the presence of tree foliage (Intini, 2000). In their seminal work, Gupta and Kumar (2000) determined the capacity of single radio single channel networks. Their findings have been later extended to derive the capacity bounds of the MRMC configurations of a network scope by Kyasanur and Vaidya (2005). In addition, the link throughput performance parameters in IEEE 802.11 networks have also been discussed in Berthilson & Pascual (2007). However, the considered MRMC network architecture has so far been presented with a number of impractical assumptions. The first assumption asserts that the location of nodes and traffic patterns can be controlled in arbitrary networks. The second assumption claims that channel fading can be excluded in the capacity analysis such that each frequency channel can support a fixed data rate. Lastly, nodes are randomly located on the surface of a torus of unit area to avoid technicalities arising out of edge effects. However, in realistic networks, location of nodes is determined by the irregularity of the terrain, the presence of tree foliage (Tse & Viswanath, 2005), and users' needs and their locations (Makitla et al., 2010). Moreover, typical rural based wireless networks can be described by (i) long single hop links, (ii) limited and unreliable energy sources, and (iii) clustered distribution of Internet users (Ishamel et al., 2008).

In order to address some of these issues and obtain high network throughput performance, high performance nodes (HPNs)™ for community-owned wireless mesh networks, have been implemented in most parts of rural South Africa (Kobus, et al., 2009). The innovation as shown in Figure 1, has been developed by the CSIR Meraka Institute and it provides high network throughput (capacity). The HPN™ is an IEEE 802.11 based multi-interface node made up of three interfaces or radio devices and controlled by an embedded microcontroller technology (Makitla et al., 2010). To ensure high speed performance, the innovation has the first radio interface card attached to a 5 GHz directional antenna for backhaul mesh routing, the second interface card is connected to a 5 GHz omni-directional antenna for backhaul mesh connectivity and access. The third radio interface card is attached to a 2.4 GHz omni-directional antenna for mesh client access network. As shown in Figure 2, the HPN block diagram has a weather proof Unshielded Twisted Pair (UTP) connector at the bottom of the node that provides Power-Over-Ethernet (PoE) and Ethernet connectivity to the HPN. To attach the HPN to a pole or a suitable structure, a mounting bracket is fixed at the back of the router (See Makitla et al., 2010) for other operational details. The HPNs are often installed on roof tops, street poles and buildings of villages, local schools, clinics, museums and agricultural farmlands.

Figure 1. High performance node (HPN)™ (Makitla et al., 2010)

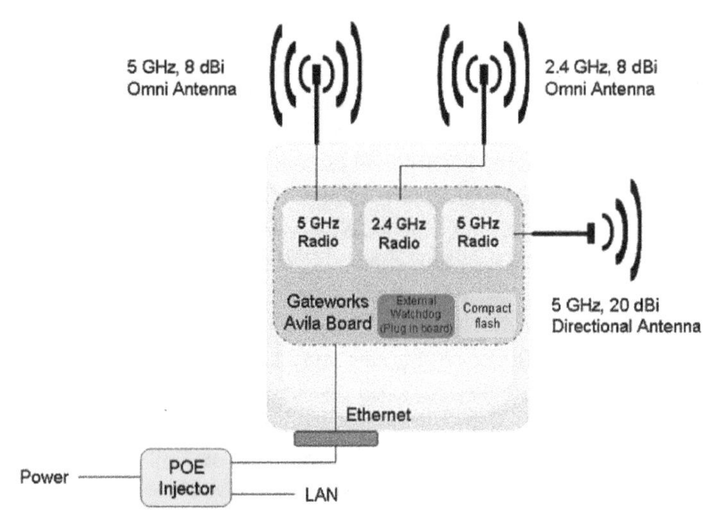

Figure 2. Block diagram of HPN™ (Makitla et al., 2010)

In this study, we shall concentrate on the backhaul terminal connectivity of the HPNs. The backhaul terminal connectivity offers aggregated traffic volumes of all flows within the network. The traffic flows traverse long links between any two HPNs and are faced with severe climatic conditions. Thus, evaluating the capacity limits of such links provides useful inputs toward optimal design of the cross-layer protocols. Figure 3 illustrates the broadband for all (BB4all™) architecture of a single wireless link based on two HPNs (that is, Node A and Node B) with end to end (E2E) Ethernet cable.

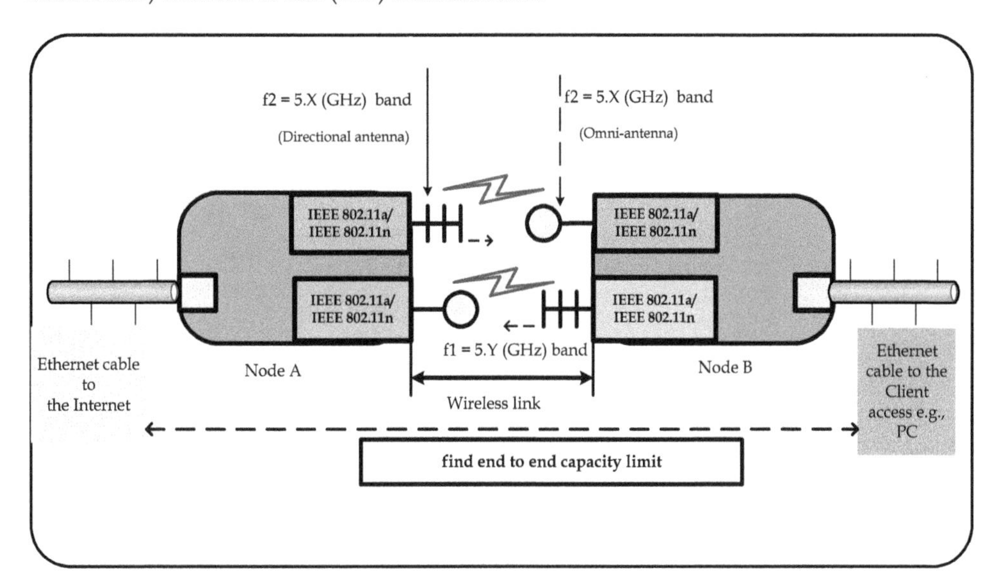

Figure 3. Single link architecture of HPNs

Research problem or questions: The main problem constitutes the need to increase capacity of community owned existing wireless broadband networks so that network users can scale without losing any connectivity and multimedia services can be provided in remote and rural areas (Mekuria et al., 2012). This problem is further subdivided into a number of research questions. Firstly, what is the achievable capacity limit of the HPNs based on IEEE 802.11a air interface under multipath fading channels (Tse & Viswanath, 2005)? Secondly, what is the achievable capacity limit of the HPNs based on IEEE 802.11n air interface under multi-input multi-output (MIMO) fading channels? Thirdly, what is the achievable end-to-end (E2E) capacity limit in HPNs in community mesh networks under: regular, irregular, and clustered node placements? The study assumes that there is no frequent channel switching even though the number of channels may be greater than the number of radios per node (Olwal, 2010). This implies that, non overlapping channels are assigned statically to available radio devices over a transmission period. Statically assigned channel over a given interval is a reasonable consideration since there is high probability that traffic volumes in rural areas are low compared to urban areas most of the times.

Research objectives: In order to investigate the capacity performance of the HPNs, the first aim of this chapter is to characterize the impact of multipath and MIMO fading channels on achievable theoretical capacity limits of single links IEEE 802.11a and IEEE 802.11n based standards. The second aim is to derive the impact of number of interfaces and channels per each HPN on the E2E capacity limits of BB4all™ mesh networks. This objective is achieved by considering a varying node density over a fixed deployment area, and the rate of a single wireless link that depends on the physical communication barriers.

Methodology: In order to achieve these objectives, firstly, the per link capacity limit under frequency selective channel is developed using conventional approaches in literature. The analytical capacity results of the BB4all™ architecture are numerically compared to IEEE 802.11a standard data sheet in order to understand the performance gain of HPNs. Secondly, the per link capacity limit for MIMO fading channels is developed and the results are numerically compared to IEEE 802.11n standard data sheet in order to show case the benefits of HPNs. Thirdly, given a typical rural community network with a pre-defined deployment area having varying node density, the impact of interfaces and channels per node on the capacity of BB4all™ mesh architecture is derived. The capacity limits of regular, irregular and cluster network topologies are obtained and compared with results from Kyasanur and Vaidya (2005) for arbitrary networks.

Research results: Analytical results indicate that the multipath fading channels and MIMO channels can be exploited to improve channel diversity in community mesh networks. Diversity improves capacity of wireless links over multiple paths and through multiple frequency channels. For regular, irregular and clustered node placements, the following analytical results were obtained for *the upper bound end-to-end capacity limit*, respectively,

$$\mathrm{O}\left(nR\sqrt{\frac{mc}{\delta}}\right),\ \mathrm{O}\left(Rn\sqrt{\frac{mc}{\delta p}}\right),\text{and } \mathrm{O}\left(R\sqrt{\frac{nmc}{1}\left(\frac{n_1}{\delta_1}+\frac{n_2}{\delta_2}\right)}\right).$$

Here, R is the single link rate in bits/s computed by taking into account multipath effects and innovative HPNs built-in structure, n is the number of HPNs, m is the number of radio interface cards per each HPN, c is the number of frequency channels that do not cause interference in duplex communication, $0 < p < 1$ is the irregularity rate (probability) of HPN placement, and δ is the HPN distribution density that is varied over a fixed deployment area.

The rest of the chapter is organised as follows. Section 2 provides a description of a typical rural community mesh network in which the BB4all™ architecture proposal can be applied. In Section 3, issues of theoretical capacity limits for single links are discussed. Section 4 analyses upper bound capacity limits for mesh networks in real deployments. Section 5 furnishes the numerical capacity limit of a selected real network in a given rural area size. The chapter is concluded with highlights of the main contribution of this study and future research and development (R&D) perspectives.

2. Rural community mesh network: A case of Peebles valley mesh in South Africa

Peebles valley mesh (PVM) is a typical rural community mesh network that is funded by the International Development Research Centre (IDRC) and is deployed in Mpumalanga province in South Africa (Johnson, 2007). The conventional PVM network, consists of nine (9) single radio nodes, and covers an area of about 15 square kilometres in Masoyi tribal land. The Masoyi tribal land is located at the North East of White River along the road to the Kruger National Park in South Africa. The land is hilly with some large granite outcrops and it has a valley that stretches from the AIDS care training and support (ACTS) clinic and divides the wealthy commercial farms from the poorer Masoyi tribal area. The Masoyi community is underserviced with lack of tarmac roads and most houses are lacking running water. However, there is unreliable electricity present in the Masoyi area. The power outages occur on average one outage in seven days and might even last up to a full day (i.e., 24 hours). Albeit the government subsidizes the cost of electricity, a large population cannot afford electricity fees due to the low economic levels of the area. ACTS clinic (a non-governmental organization sponsored clinic) provides medical services to AIDS patients, counseling, testing and Anti-retroviral (ARV) treatment (Johnson & Roux, 2008).

Figure 4 demonstrates architecture of the PVM network when HPNs are deployed. In this figure, the clinic connects to surrounding schools, homes, farms and other clinic infrastructure through a mesh network. The network is seen as community asset with some of the equipment at key nodes are actually belonging to the community. In this area mesh connectivity offers:

- Scalable connectivity to the hilly terrain, over multiple hops based long distances and through non line of sight (NLOS).
- Auto-configurable traffic routing mechanisms with minimal human interventions. This feature ensures network sustainability in an area with apparent low skilled technical-personnel who cannot regularly maintain the network.

- Auto-organizable connectivity against severe climatic conditions that commonly cause links, nodes, and network failures in the area.

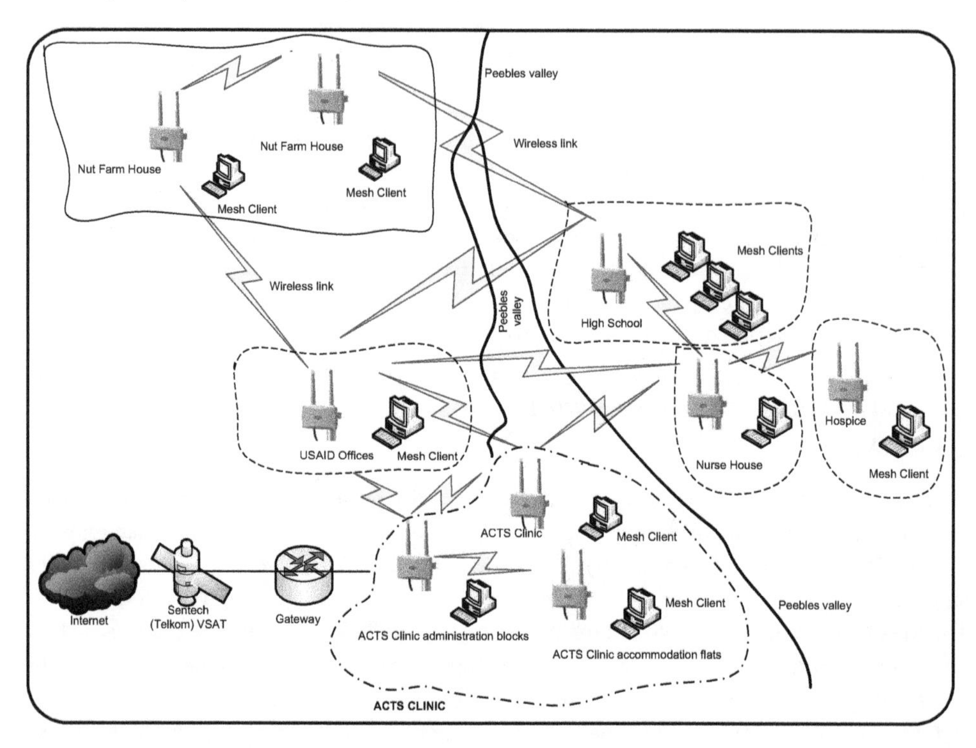

(Source: *http://wirelessafrica.meraka.org.za/*)

Figure 4. Mesh network architecture at Peebles valley in South Africa

Traditionally, the PVM is endowed with VSAT link that provides the network at the clinic with 2 Gbits per month at a download rate of 256 kbps and an upload rate of 64 kbps (Johnson & Roux 2008). The clinic provides 400 Mbps per month available to the single radio mesh network. The single radio mesh has nine users (mesh routers) so that each user (mesh router) receives about 44.4 Mbps per month on average. This traffic bandwidth drops downstream the network from the satellite gateway to the terminal users. This is due to lack of single radio network resiliency against effects of wireless multipath. However, in this document we believe that the design of the HPNs making the BB4all™ architecture can be a suitable candidate for improved capacity in multipath environment (BelAir Networks, 2006). As a result high data rates as the network scales away from the satellite gateway can be realized in the PVM deployment. The HPNs utilize the multiplicity of the low cost radio devices and non-overlapping channels to improve capacity delivered across the network.

Thus, the BB4all™ architecture constitutes a gateway connected to the internet via Sentech VSAT to the Peebles valley or ACTS clinic. Within the ACTS clinic there can be mesh servers, personal computers as the mesh clients and HPNs may be installed to serve as wireless routers that link ACTS clinic accommodation flats to USAID offices about 1 Km away. The HPN link can connect Legogote Hospice and USAID premises about 3.35 Km over the valley via the Nurse house. The link over the valley between the USAID and Sakhile high school is about 2.4 Km. The link from Sakhile high school to the Legogote Hospice is about 4.6 Km, and the distance from high school to the farmers' houses is about 5.55 Km over the Peebles valley. It is also anticipated that the mesh network will expand to public clinics and schools that are farther way even up to 25 Km from the ACTS clinic center in the near future.

In conclusion the rural PVM project has triggered further insights for newer research, development and innovation. The terrain irregularity, the long distances, the tree foliage coupled with the need for high capacity Internet provision in rural communities are the key drivers for the BB4all™ connectivity solution. It is also noted that clear understanding of classical physical channel models combined with innovative ICT products is expected to promote sustainable internet services to billions including previously disadvantaged subscribers (Mekuria et al., 2012).

3. Achievable capacity limit for a single link with multipath fading

In order to realize long distance coverages by single links with multipath effects in wireless mesh networks in rural areas, the IEEE 802.11a and IEEE 802.11n standards commodity devices can be used. This is because these devices are off-the-shelves, operate in multiple ISM channel bands and are affordable to the rural communities (Kyansanur & Vaidya, 2004). That is to say that only fewer radio interface cards at each node are needed than the number of non-overlapping frequency channels freely available. Kyasanur and Vaidya emphasized how expensive it could be to equip a node with one interface card for each frequency channel. The IEEE 802.11a standard, for example, offers 24+ non-overlapping channels and configuring a commensurate number of radio interface cards on each node might be unnecessary costly. As a result, many IEEE 802.11 interface cards can be switched from one channel to another, albeit at the cost of a switching delay. Moreover, the advantage of eliminating frame losses due to path-dependent (e.g., multipath fading effects), location-dependent (e.g., noise effects), and statistically independence between different receiving radios can be achieved by using multi-radio diversity principle (Miu, Balakrishnan & Koksal, 2007). The idea is that even when each individual reception of a data frame is erroneous, it might still be possible to combine the different versions to recover the correct version of the frame. In this study, the question to be addressed is that what is the capacity expression for single links with multipath effects in a rural based wireless mesh network. It is understood that most of previous studies solve capacity problem with simple channel models that may not reflect the true wireless channel conditions (Gupta & Kumar, 2000).

3.1. IEEE 802.11a air interface

The standard IEEE 802.11a specifies an over-the-air interface between two wireless routers or between a wireless client and a router. It provides up to 54 Mbps in the 5 GHz frequency band and uses an orthogonal frequency division multiplexing (OFDM) encoding scheme.

This implies that a frequency selective channel is the most suitable approach to model the IEEE 802.11a air interface (Tse & Viswanath, 2005). This is because a frequency selective channel perfectly captures effects of multipath on signal propagation (i.e., due to terrain irregularity and tree foliage). The OFDM scheme is basically the preferred method to the frequency hopping spread spectrum (FHSS) or direct sequence spread spectrum (DSSS) schemes due to its robust performance over multipath. In this context, the IEEE 802.11a radio interface cards (Intini, 2000) make use of OFDM to provide high capacity over parallel wireless channels. In their definition, Tse and Viswanath (2005) states that a parallel channel is a channel which consists of a set of non-interfering sub-channels, each of which is corrupted by independent additive white Gaussian noise (AWGN).

To obtain the capacity over single link wireless medium, we assume each mth sub-channel of a parrallel channel is allocated a waterfilling power p_m such that the average power constraint P is still met on each input OFDM symbol to the multipath channel. Also consider that the AWGN power level to a parallel channel is N_0 and the co-channel interference caused by neighbouring transmissions is denoted as I. These parameters may be held constant in practice considering that most rural network applications are characterized by constant and low interference levels (Ismael et al., 2008). Then, the maximum capacity per every OFDM symbol of a reliable communication over M_c parallel streams or subcarriers is given by:

$$R_{M_c} = \sum_{m=0}^{M_c-1} \log_2\left(1+\frac{p_m\left|\tilde{h}_m\right|^2}{(N_0+I)}\right), \text{ bits / OFDM symbol,} \tag{1}$$

whereby the achievable capacity per link in bits/s/Hz for each parallel stream is written as

$$R_{multipath} = R_{M_c} / M_c, \text{ bits / s / Hz.} \tag{2}$$

Resulting from (2), the link capacity of a propagating OFDM signal over a wireless multipath channel is expanded in terms of the exponential function of the channel gain:

$$\begin{aligned} R_{OFDM/multipath} &= \frac{1}{M_c}\sum_{m=1}^{M_c} \log_2\left(1+\frac{p_m}{(N_0+I)}\times\left|\tilde{h}_m\right|^2\right), \text{ bits / s / Hz} \\ R_{OFDM/multipath} &= \frac{1}{M_c}\sum_{m=1}^{M_c} \log_2\left(1+\frac{p_m}{(N_0+I)}\times\left|\sum_{l=1}^{L} h_l \exp\left(-\frac{j2\pi lm}{M_c}\right)\right|^2\right), \text{ bits / s / Hz} \end{aligned} \tag{3}$$

The achievable capacity of IEEE 802.11a air interface in terms of antenna gains (each antenna system for each radio interface), the range distances, the path loss exponent, the path multiplicity, and over the total bandwidth, W is defined as:

$$R_{OFDM/multipath} \approx W\log_2\left(1+\frac{P}{(N_0+I)}\times L^2\times\frac{\mathrm{K}_{antenna}d_0^{\alpha}}{d^{\alpha}}\right), \text{ bits / s.} \tag{4}$$

From (4), P is the maximum power allowed per sub-carrier, L is the number of paths associated to each sub-carrier and α is the path loss exponent. We denote $K_{antenna}$ as the combined antenna gain which is simply the product of the transmitter and the receiver antenna gains, d_0 is the reference distance (Abhayawardhana et al., 2007). The combined antenna gain is thus, expressed as:

$$K_{antenna} = K_{antennaTxT} \times K_{antennaRxV} \tag{5}$$

Inserting the result in (5) into the expression in (4) reveals that the higher the combined antenna gain, the higher the achievable link capacity. The improved antenna gain is the main attractive feature that the HPN based BB4all™ architecture offers to the conventional standards (Makitla, Makan & Roux, 2010). From (3), to view effects of frequency $f = nW / M_c$ on the time invariant wireless channel $\tilde{h}_m$, the known Fourier Transform (Bracewell, 1986) can be invoked:

$$\tilde{h}_m = \sum_{l=0}^{L-1} h_l \exp\left(-\frac{j2\pi lm}{M_c}\right) \Leftrightarrow H(f) = \sum_{l=0}^{L-1} h_l \exp\left(-\frac{j2\pi lf}{W}\right), \tag{6}$$

where $f \in [0, W]$. It should be deduced from the exponential relation that lowering f increases the gain in (6) that in turn increases the capacity limit in (4). Suppose we let that between any two mesh nodes directly connected there exists a clear line of sight (LOS) as it is the usual case in a mesh network. Then, the following multipath channel simplification can be made:

$$\left|\tilde{h}_m\right| = \left|\sum_{l=1}^{L} h_l \exp\left(-\frac{j2\pi lm}{M_c}\right)\right| = \sum_{l=1}^{L} |h_l|. \tag{7}$$

Based on these simplication, it is worthwhile noting that effects of multipath often produce inter-symbol interference (ISI), signal attentuation and multipath echoes. This leads to significant capacity drops. Fortunately, the OFDM communication exploits these channel diversity to improve capacity. Therefore, joint OFDM and HPN structural configurations can be utilised for capacity improvement in rural based networks.

3.2. IEEE 802.11n air interface

In the case of IEEE 802.11n air interface, the model of the wireless channel is characterized by antenna arrays with LOS and reflected paths as shown in Figure 5. The difference with IEEE 802.11a air interface is that multiple antennas are required at both transceivers when constructing the IEEE 802.11n HPNs. In this way, the LOS and reflected paths present wireless channel diversity that multi-input multi-output (MIMO) techniques need to exploit for channel capacity enhancement. In particular, if the direct path is denoted as *path 1* and the reflected path is denoted as *path 2*, then the channel $\mathbf{H}$ is given by the principle of superposition (Franceschetti et al, 2009):

$$\mathbf{H} = a_1^b \mathbf{e_r}(\Omega_{r1})\mathbf{e_t}(\Omega_{t1})^* + a_2^b \mathbf{e}_r(\Omega_{r2})\mathbf{e_r}(\Omega_{t2})^*, \text{for } i = 1, 2 \tag{8}$$

with, $a_i^b = a_i\sqrt{n_t n_r}\exp\left(-\frac{j2\pi d^{(i)}}{\lambda_c}\right)$, where n_t is the number of transmit antennas, n_r is the number of receive antennas and λ_c is the wavelength of the pass-band transmitted signal. The distance between the transmit antenna 1 and recieve antenna 1 along path i is denoted by $d^{(i)}$. Figure 5 illustrates transmit and receive antenna arrays that is seperated by a concatenation of two channel $\mathbf{H}'$ and $\mathbf{H}''$ with virtual relays A and B. According to expression (8) and Figure 5, the unit spatial signature in the directional cosine Ω (i.e., $\Omega = \cos\phi$) is defined as follows:

$$\mathbf{e}_r(.) = \frac{1}{\sqrt{n_r}}\begin{pmatrix} 1 \\ \exp(-j2\pi\Delta_r\Omega) \\ \exp(-j2\pi 2\Delta_r\Omega) \\ . \\ . \\ . \\ \exp(-j2\pi(n_r-1)\Delta_r\Omega) \end{pmatrix}, \quad \mathbf{e}_t(.) = \frac{1}{\sqrt{n_t}}\begin{pmatrix} 1 \\ \exp(-j2\pi\Delta_t\Omega) \\ \exp(-j2\pi 2\Delta_t\Omega) \\ . \\ . \\ . \\ \exp(-j2\pi(n_t-1)\Delta_t\Omega) \end{pmatrix}, \tag{9}$$

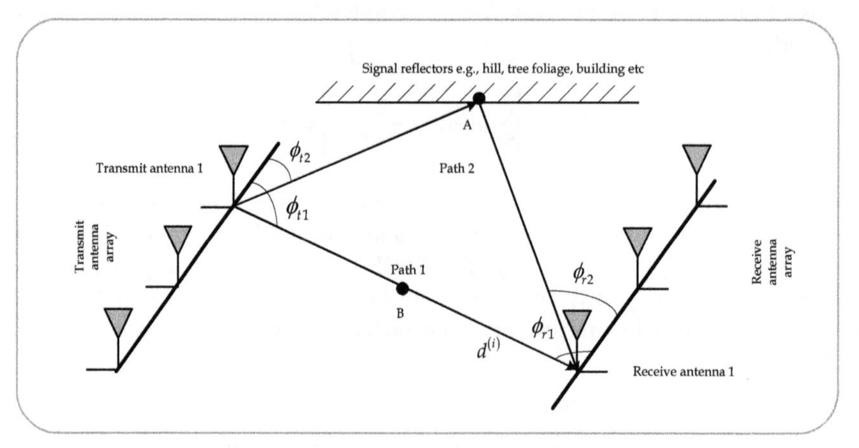

Figure 5. A MIMO channel with a direct path 1 and a reflected path 2. The channel is a concatenation of two channels H' and H'' with virtual relays A and B

From (9), the notation ϕ is the angle of incidence of the LOS onto the receive antenna array and a is the signal attenuation. In a reasonable sense, the condition that as long as

$$\Omega_{t1} \neq \Omega_{t2} \mod \frac{1}{\Delta_t} \text{ and } \Omega_{r1} \neq \Omega_{r2} \mod \frac{1}{\Delta_r}$$

then the matrix $\mathbf{H}$ is of rank 2 holds. That is, the maximum number of independent rows or columns of the matrix is 2. Based on the defined condition, we let $\Delta_t = L_t \div n_t$ and $\Delta_r = L_r \div n_r$ whereby L_t and L_r are normalized lengths of transmit and receive arrays, respectively. As a subsequence, the implication is explained as follows. When the number of antennas in each HPN is increased for any fixed normalized length of arrays, the factor denoted by Δ will decrease and from the modulo operation, the directional cosine denoted by Ω will increase proportionately. This might cause ill-conditioned $\mathbf{H}$ with impossible inverse. To make $\mathbf{H}$ well-conditioned so that its inverse can be computed, the angular separations Ω_t and Ω_r should satisfy the following:

$$\Omega_t = \cos\phi_{t2} - \cos\phi_{t1},\ L_t = n_t\Delta_t \text{ and } \Omega_r = \cos\phi_{r2} - \cos\phi_{r1},\ L_r = n_r\Delta_r. \tag{10}$$

Moreover, in order to view the influence of multipath, $\mathbf{H}$ is re-written as $\mathbf{H} = \mathbf{H}''\mathbf{H}'$, where $\mathbf{H}'$ is a 2 by n_t matrix while $\mathbf{H}''$ is an n_r by 2 matrix. Consequently, the capacity limit of the channel with LOS and reflected paths in an HPN based mesh link is given as:

$$\begin{aligned} R_{LoS+\mathrm{Re}\,flect} &= \log_2\left(1 + SINR_j \times \left\|\mathbf{H}\right\|^2\right), \text{ bits / s / Hz} \\ R_{LoS+\mathrm{Re}\,flect} &= \log_2\left(1 + SINR_j \times \left\|\mathbf{H}''\mathbf{H}'\right\|^2\right), \text{ bits / s / Hz} \\ &= \log_2\left(1 + SINR_j \times 2 \times \left(\left(a_1^b\right)^2 + 2a_1^b a_2^b + \left(a_2^b\right)^2\right)\right), \text{bits / s / Hz,} \end{aligned} \tag{11}$$

where

$$\left(a_1^b\right)^2 = a_1^2 n_t n_r,\ \left(a_2^b\right)^2 = a_2^2 n_t n_r,$$

Suppose fading channels are assumed, then the single link of IEEE 802.11n based HPNs can be characterized by stochastic channel behaviours. Statistical MIMO channel models are adopted to capture the key properties that enable spatial multiplexing (Tse & Viswanath, 2005). For instance, given an arbitrary number of physical paths between the transmitter and the receiver, the channel matrix $\mathbf{H}$ may be written as:

$$\mathbf{H} = \sum_i^{no\ of\ multipaths} a_i^b \mathbf{e}_r\left(\Omega_{ri}\right) \mathbf{e}_t\left(\Omega_{ti}\right)^*, \tag{12}$$

where a_i^b, $\mathbf{e}_r(\Omega)$ and $\mathbf{e}_t(\Omega)$ take the definition provided in (9). From (11) and without loss of generality, the capacity limit of the MIMO multipath fading channel can similarly be written as:

$$\begin{aligned} R_{multipath} &= \log_2\left(1 + SINR_j \times \left\|\mathbf{H}\right\|^2\right), \text{ bits / s / Hz} \\ &= \log_2\left(1 + SINR_j \times \left\|\sum_i^{multipaths} a_i^b \mathbf{e}_r\left(\Omega_{ri}\right) \mathbf{e}_t\left(\Omega_{ti}\right)^*\right\|^2\right), \text{ bits / s / Hz} \end{aligned} \tag{13}$$

In conclusion, if the number of physical paths is two then the expression of capacity over multipath phenomenon (13) simply reduces to the expression (11) of the direct and reflected paths. Clearly from expression (13), one notes that increasing the multiplicity of paths of a single wireless link and the number of antennas at each HPN in (11) or (13) will increase the capacity limit of the wireless mesh links, depicted in (4) due to MIMO technology benefits.

4. Achievable capacity limit of HPNs in a mesh network

4.1. Practical considerations

In order to obtain the achievable capacity bound for the HPN (the dual channel dual radio) based mesh network we consider a typical static wireless mesh network. Suppose the network is assumed to consist of varying n number of HPNs upto 50 nodes with a fixed area of deployment region (i.e., 5 Km by 5 Km). Also to generalize our derivations and only apply specific cases later with numerical examples, we employ the approach presented by Kyasanur and Vaidya (2005) in order to investigate the impact of number of channels and interfaces on the capacity of multi-channel wireless networks. In our derivations, the term "channel" will refer to a part of frequency spectrum with some specified bandwidth and the term "radio" will mean the network interface card. Let us assume that the HPNs based mesh network has c channels and every node is equipped with m interfaces so that the relation between the number of interface cards and channels is $2 \leq m \leq c$. Each interface card can only transmit and receive data on any one channel at a given time, that is half-duplex communication. Thus, the mesh network of m interfaces per node, and c channels will be noted as (m,c)-network. Suppose each channel can support a fixed data rate of $R = R_{multipath}$, independent of number of non-overlapping channels of the network. Then, the total data rate possible by using all c non-overlapping channels is Rc. The number of non-overlapping channels can be increased by utilizing extra frequency spectrum of the standard technologies. For example, IEEE 802.11a standard technology uses 5 GHz band and has a capability of 24+ non-overlapping channels (c = 24+) each of 20 MHz bandwidth size (W = 20 MHz). Moreover, the IEEE 802.11n standard technology implements MIMO channels with bandwidth size of 40 MHz (Cisco systems, 2011).

4.2. Capacity limit for regular placement in real network

Consider Figure 6 that shows the topology of HPNs up to a maximum of 50 nodes placed regularly in a 5km by 5km of an area. This network scenario reflects a typical wireless mesh network set-ups in rural and remote areas where inter node distance is large and the land-scape affects network performance. It should be seen that the separation distance between the source and the destination HPNs are assumed to take the longest route with a mean line joining the two nodes computed to be 6505 m. The regular placement of nodes ensures that there are no any two HPNs that are placed within a radius less than 700 m. The main reason for this decision is to avoid interference between close neighbours. It will be discussed in detail how this placement criteria is ensured using the carrier sense multiple access with collision avoidance (CSMA/CA) in IEEE 802.11 standards (IEEE 802.11 standard working group, 1999). In this topology setting, the regular placement of nodes on a fixed area will be termed as an arbitrary network. That is, the location of nodes and traffic patterns can be controlled as introduced by Gupta and Kumar (2000). Controlling nodes' placement locations and the traffic patterns makes the derived capacity bounds to be viewed as *the best case* capacity bounds with results remaining applicable to any network. As introduced by Gupta and Kumar, the aggregate *end to end* network throughput over a given flow or a set of flows is measured in terms of "bit-meters/sec". That is the network is said to transport one "bit-meter/sec" when one bit has been transported across a distance of one meter in one second.

Theorem 1: The E2E upper bound on capacity of a statically assigned channel network of type (m,c)-arbitrary regular placement of nodes when, $\frac{c}{m} = O(n)$, is given as $O\left(nR\sqrt{\frac{mc}{\delta}}\right)$, bit-meters/sec.

Proof: For the best case capacity limit, let's assume that multiple interfaces of HPNs receive and transmit on interference free channels. This assumption is reasonable with the HPNs that transmit directionally but receive and ensure connectivity omnidirectionally. As the number of channels is much larger than the number of interfaces. Thus, given that each HPN has a constant radio range, the spatial reuse is considered to be proportional to the physical area of the network. Let the node density δ be uniform with distribution regularity equals to one (i.e., probability equals to one) throught the deployment area.

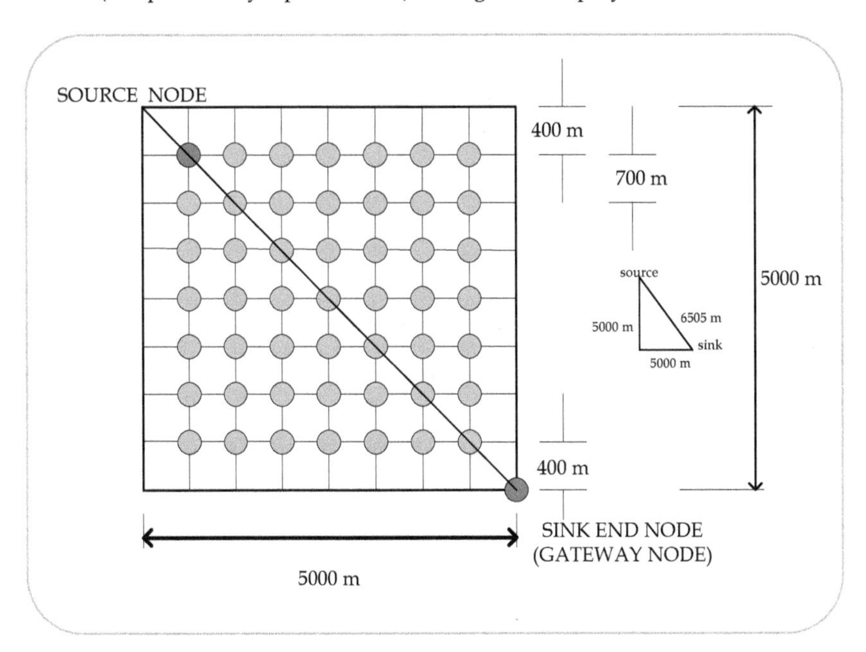

Figure 6. Regular placement of HPNs in a 5 km x 5 km

The physical area of deployment, A, can be related to the total number of HPNs by $A=\frac{n}{\delta}$. Consider also that, the capacity of each channel, R, is proportional to the physical area in accordance to the relation, $R=kA=k\frac{n}{\delta}$ for some constant k (in bits/s/square meters). Suppose each source HPN can generate packets from higher layers protocol at a rate of λ bits/sec and the mean seperation distance between the source and destination HPN pairs is $\overline{L}$ meters (via multiple hops), then the E2E network capacity of the network is (Gupta & Kumar, 2000):

$$\lambda n\overline{L},\ \text{bit}-\text{meters}/\text{sec} \tag{14}$$

The expression in (14) is evaluated without taking into account the lower layer number of frequency channels, interference, path loss effects and number of interface cards. In addition, in order to relate this high level network capacity with actual number of hops in a multi-hop wireless network, the overall bits transported in the network can be evaluated as follows. Suppose bit b, $1 \le b \le \lambda n$ (bits/sec), traverses $h(b)$ hops on the path from its source to its destination, where the hth hop traverses a distance of r_b^h, then the overall bits transported in the network in every second is summed and is related to (14) as:

$$\lambda n\overline{L} \le \sum_{b=1}^{\lambda n}\sum_{h=1}^{h(b)} r_b^h,\ \text{bit}-\text{meters}/\text{sec}, \tag{15}$$

The inequality in (15) holds since the mean length of the line joining the source and destination, is at most equal to the distance traversed by a bit from its sources to its destination (Kyasanur & Vaidya, 2005).

Let us define X to be the total number of hops traversed by all bits in a second, i.e., $\mathrm{X}=\sum_{b=1}^{\lambda n}\mathrm{X}(b)$. Therefore, the number of bits transmitted by all nodes in a second (including bits forwarded) is equal to X (bits/sec). Since each HPN node has m interfaces, and each interface transmits over a frequency channel of bandwidth W, with a data rate R possible per channel, the total bits per second that can be transmitted by all interfaces is at most $\frac{Rmn}{2}$ (transporting a bit across one hop requires two interfaces, one each at the transmitting and the receiving nodes). Consequently, the relation between a single channel single link rate, the number of interface cards creating single links, the number of nodes in the network and the total number of hops traversed by all bits in every second is given by,

$$\mathrm{X} \le \frac{Rmn}{2},\text{bits}/\text{sec} \tag{16}$$

It should be noted that under the interference protocol model (Gupta & Kumar, 2000), a transmission over a hop of length r in a path loss link is successful only if there can be no active transmitter within a distance of $(1+\Delta)r$. In IEEE 802.11a/b/g/n standards the medium access control (MAC) layer protocols execute carrier sense multiple access with collision avoidance (CSMA/CA) mechanism that ensures that this condition is satisfied. Figure 7 illustrates this type of collision avoidance mechanism. To illustrate this concept further, suppose node A is transmitting a bit to node B, while node C is simultaneously transmitting a bit to node D and both sessions are over a common frequency channel, W. Then, using the interference protocol model and the geometry sufficient for successful reception, node E cannot transmit at the same time. Mathematically, one has

$$d(C,B) \geq (1+\Delta)d(A,B) \text{ and } d(A,D) \geq (1+\Delta)d(C,D) \tag{17}$$

Adding the two inequalities together, and applying the triangle inequality to (17), we can obtain the inequality in (18),

$$d(B,D) \geq \frac{\Delta}{2}\left(d(A,B) + d(C,D)\right) \tag{18}$$

Therefore, in collision avoidance (CSMA/CA) principle of IEEE standards , expression (18) can be viewed as each hop covering a disk of radius $\frac{\Delta}{2}$ times the length of the hop around each receiver. As shown by Figure 7, the total area covered by all hops must be bounded above by the total area of the deployment (domain, A). The seperation distance between receiver B and transmitter C is at least $(AB+\Delta AB)$ and that of transmitter A and receiver D is at least $(CD+\Delta CD)$.

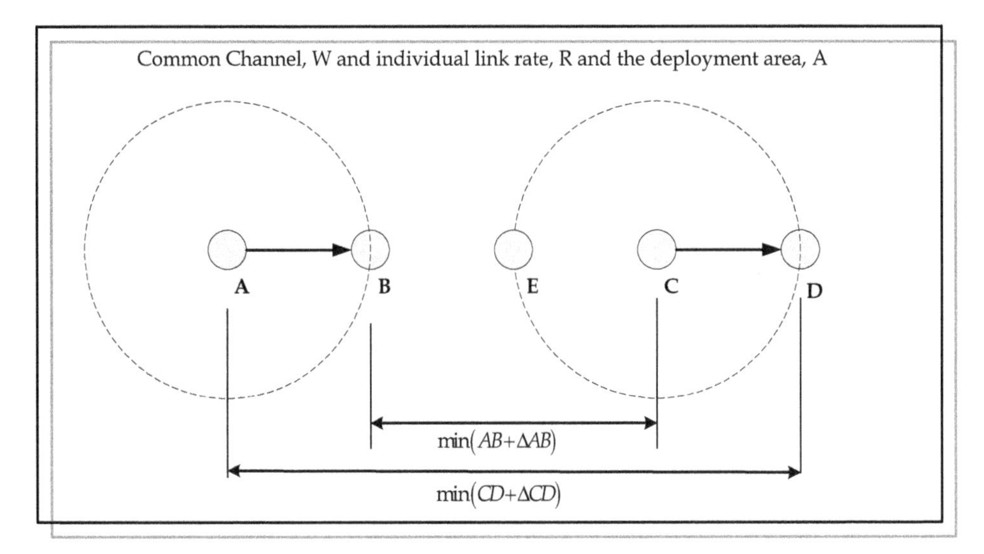

Figure 7. Topology of HPNs and Geometry

From the geometry of Figure 7, we sum over all channels (which can potentially transport Rc bits per second) and obtain the constraint formulated as,

$$\sum_{b=1}^{\lambda n}\sum_{h=1}^{h(b)}\frac{\pi\Delta^2}{4}\left(r_b^h\right)^2 \leq ARc, \tag{19}$$

this can be rewritten as,

$$\sum_{b=1}^{\lambda n}\sum_{h=1}^{h(b)}\frac{1}{\mathrm{X}}\left(r_b^h\right)^2 \leq \frac{4ARc}{\pi\Delta^2\mathrm{X}} \tag{20}$$

Since the expression on the left hand side in (20) is convex, one obtains,

$$\left(\sum_{b=1}^{\lambda n}\sum_{h=1}^{h(b)}\frac{1}{\mathrm{X}}r_b^h\right)^2 \leq \sum_{b=1}^{\lambda n}\sum_{h=1}^{h(b)}\frac{1}{\mathrm{X}}\left(r_b^h\right)^2 \tag{21}$$

Therefore, from (20) and (21) one gets,

$$\sum_{b=1}^{\lambda n}\sum_{h=1}^{h(b)} r_b^h \leq \sqrt{\frac{4ARc\mathrm{X}}{\pi\Delta^2}} \tag{22}$$

Substituting for X from (16) in (22), and using expression (15) we have,

$$C_{mesh} \leq nR\sqrt{\frac{2mc}{\delta\pi\Delta^2}},\ \text{bit} - \text{meters / sec} \tag{23}$$

Therefore, the E2E asymptotically upper bound capacity limit for a scaling number nodes with node density δ, and static channel assignment without channel switching mechanisms in HPN network is given by

$$\begin{aligned} \lambda n\overline{L} &= \mathrm{O}\left(nR\sqrt{\frac{mc}{\delta}}\right),\ \text{bit} - \text{meters / sec}\ \textit{for varying node density} \\ \lambda n\overline{L} &= \mathrm{O}\left(nR\sqrt{mc}\right),\ \text{bit} - \text{meters / sec}\ \textit{for constant node density} \end{aligned} \tag{24}$$

4.3. Capacity limit for irregular placement in real network

Consider Figure 8 that shows the topology of HPNs up to a maximum of 50 nodes placed irregularly in a 5km by 5km of an area. This network scenario reflects typical wireless mesh network set-ups in rural and remote areas where inter node distance is large and the land-scape affects network performance. To avoid interference, it is assumed that no any two HPNs are placed within a radius less than 400 m at the edge and less than 700m toward the centre of the deployment area. However, between any two HPNs the largest separation distance is allowed as much possible as the size of the area can accommodate. The diagram indicates one of the possible settlement distribution patterns of the Internet users in community based networks such as the case of Peebles valley mesh (PVM) networks.

Theorem 2: The E2E upper bound on capacity of a statically assigned channel network of type (m,c)-arbitrary irregular placement of nodes when, $\frac{c}{m} = O(n)$, is given as, $\lambda n\overline{L} = O\left(Rn\sqrt{\frac{mc}{\delta p}}\right)$ bit-meters/sec.

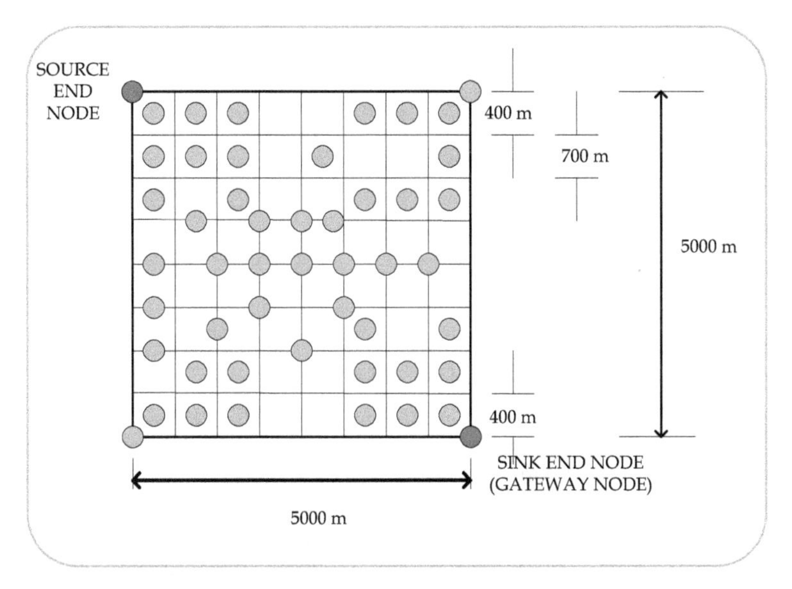

Figure 8. Irregular placement of HPNs in a 5 km x 5 km

Proof: Let us consider that in irregular static networks, the node density δ varies over space (i.e., an area) but stays constant at any given time since nodes are taken to be static. Suppose we let the node density δ to vary over space with irregularity rate (probability), $0 < p < 1$ then the area A is defined as $A = \frac{n}{\delta p}$. Therefore, capacity of the network will depend on the expected average node density, δp of an irregular placement as well as the number of nodes, n. Additionally, HPN nodes have m interfaces per node and with a data rate of R possible per channel. Then, the total bits per second that can be transmitted by all interfaces in the network and all channels is at most $\frac{Rnmc}{2}$.

If we let $X = \sum_{b=1}^{\lambda n} X(b)$ as the number of bits transmitted by all nodes in a second (including bits forwarded). From (22), we found out that

$$\sum_{b=1}^{hn}\sum_{h=1}^{h(b)} r_b^h \leq \sqrt{\frac{4ARRnmc}{2\pi\Delta^2}} \tag{25}$$

Alternatively, it has been established that

$$\sum_{b=1}^{hn}\sum_{h=1}^{h(b)} r_b^h = \lambda n\overline{L} \tag{26}$$

We have

$$\begin{aligned} \lambda n\overline{L} &\leq R\sqrt{\frac{2Anmc}{\pi\Delta^2}} = Rn\sqrt{\frac{2mc}{\delta p\pi\Delta^2}} \\ \lambda n\overline{L} &= O\left(Rn\sqrt{\frac{mc}{\delta p}}\right), \text{bit}-\text{meters / seconds} \end{aligned} \tag{27}$$

4.4. Capacity limit for clustered placement in real network

Suppose that n nodes are arbitrarily located (a cluster fashion) on a square of a fixed area with LOS ensured between any two neighbouring nodes shown in Figure 9. Note that the deployment area is fixed to 5 km by 5 km. To avoid interference no two HPNs can be placed at a radius less than 400 m near the edge and less than 700 m toward the centre of the deployment area. Thus, within a cluster a minimum separation distance of 700 m is considered, while any largest separation distance possible is considered between clusters. Figure 9 shows the regularly clustered topology indicating how far as possible the separation distance between the source and destination HPNs. The diagram depicts typical rural community networks such as Peebles valley mesh (PVM) networks. The community mesh network is considered to adopt such a distribution pattern and the goal would be to find the achievable capacity over wireless mesh networks.

Theorem 3: The E2E upper bound on capacity of statically a signed channel network of type (m,c)-arbitrary clustered placement of nodes when $\frac{c}{m} = O(n)$ is given as $\lambda n\overline{L} = O\left(R\sqrt{\frac{nmc}{1}\left(\frac{n_1}{\delta_1}+\frac{n_2}{\delta_2}\right)}\right)$ in bit-meters/sec, where R is the min (R_1, R_2), n_1 are number of nodes in a regular cluster and n_2 are number of clusters in the network.

Proof: We assume *a clustered placement* of the mesh network as *a special case of the regular HPNs placement*. However, in this case the node densities are respectively, $\delta_1 = \frac{n_1}{A_1}$ as the density of nodes within a cluster consisting of n_1 nodes occupying A_1 geographical area and $\delta_2 = \frac{n_2}{A_2}$ as the density of clusters consisting of n_2 clusters occupying A_2 of an area. This assumption is reasonable since HPNs within a cluster use a shorter transmission range compared to that range that is being used by nodes while communicating between clusters. The application layer generates the E2E capacity according to Gupta and Kumar model. This capacity depends on the number of nodes and can be simplified as $\lambda n\bar{L}$, bit-meters/sec.

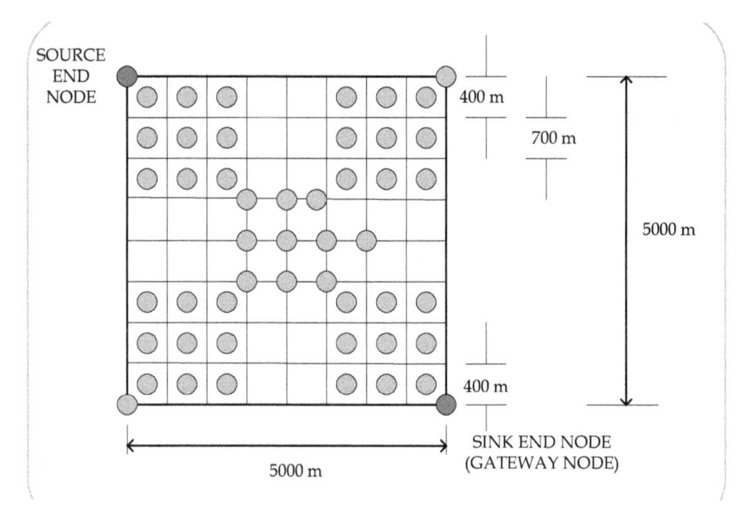

Figure 9. Regularly clustered placement of HPNs in a 5 km x 5 km

Suppose bit b, $1 \le b \le \lambda n$ (bits/sec), traverses $h(b)$ hops on the path from its source to its destination, where the h-th hop traverses a distance of $r_b^h = r_b^h(\delta_1) + r_b^h(\delta_2)$ (having intra-cluster and inter-cluster hop distances), then one obtains by summing over all bits in the network:

$$\lambda n\bar{L} \le \sum_{b=1}^{\lambda n}\sum_{h=1}^{h(b)} r_b^h, \text{ bit} - \text{meters / sec} \tag{28}$$

Let us define X to be the total number of hops traversed by all bits in a second, i.e., $X = \sum_{b=1}^{\lambda n} X(b)$. Therefore, the number of bits transmitted by all nodes in a second (including bits forwarded) is equal to X (bits/sec). It is known that each HPN node has m interfaces, and each interface transmits over a frequency channel of bandwidth W, with a data rate R per channel, the total bits per second that can be transmitted by all interfaces is at most $\frac{Rmn}{2}$ (transporting a bit across one hop requires two interfaces, one each at the transmitting and the receiving nodes). But in clustered networks where bits traverse the intra cluster hops and inter cluster hops with R_1 and R_2 rates respectively. R takes the minimum rate since R drops with distance. Consequently, we have,

$$X \le \frac{Rmn}{2}, \text{bits / sec} \tag{29}$$

Therefore, using similar arguments and steps provided in the *Proof of Theorem 1*, the interference constraint protocol of the clustered mesh network will still hold. The derived E2E capacity limit is then upper bound as

$$\lambda n\bar{L} \leq R\sqrt{\frac{2nmc}{\pi\Delta^2}\left(\frac{n_1}{\delta_1}+\frac{n_2}{\delta_2}\right)}, \text{ bit}-\text{meters / sec} \tag{30}$$

Hence, the asymptotic end to end upper bound capacity limit for a scaling number of nodes with node density δ_1 and cluster density δ_2, and statically assigned channels in HPN network is given by

$$\lambda n\bar{L} = \mathrm{O}\left(R\sqrt{\frac{nmc}{1}\left(\frac{n_1}{\delta_1}+\frac{n_2}{\delta_2}\right)}\right), \text{ bit}-\text{meters / sec} \tag{31}$$

5. Numerical examples using the Peebles valley mesh

Tables 1 and 2 shows useful data that can be used to determine the achievable capacity limit over long links with direct LOS of about 6.505 km from one end of the network to the other. The data mimics the physical scenario of PVM as compared to the data sheet values of IEEE 802.11a and IEEE 802.11n air interfaces, respectively. The computed capacity values assume CSMA/CA and the protocol model whereby a transmission over one link is successful only if there is no active transmitter within a distance of $(1+\Delta)d$. That is, the distance d is the range between a transmitter and receiver, and Δ signifies a fraction of one hop distance needed to ensure collision-free transmission. The assumption of protocol model is reasonable in sparsely placed nodes in rural set-up whereby interference effects can be neglected without loss of generality. Furthermore, let the size of data carriers in OFDM scheme be, $M_c = 48$ with each HPN having an 8dBi omni-directional antenna and 20 dBi being the directional antenna (i.e., the combined antenna gain is 630.96 (6.3096 times 100) and α in a hilly and foliage area is taken approximately to be three between the frequency channels 5.15 GHz and 5.85 GHz (Durgin et al., 1998). Then, from the capacity limit expression in Section 3, practical data rates can be obtained.

5.1. Single link achievable capacity

Table 1, lists parameters needed to evaluate the achievable data rate for all wireless streams of a single link IEEE 802.11a radios. A comparison is made between the achievable data rate computed using data from the IEEE 802.11a air interface data-sheet and the data rate of the IEEE 802.11a air interface constructed from BB4all™ architecture. It is observed that while specifications of other parameters are kept the same in both cases, the combined antenna gain is taken to be 9 dBi from the data-sheet and that of BB4all™ architecture to be 28 dBi. With these antenna gains, the achievable data rate in standard architecture is 60.792 Mbps compared to 183.30 Mbps for the HPNs. This numerical result is explained as follows. When transmission power is kept the same for both cases, an increase in antenna gain due to more focussed beams increases capacity substantially. Thus, the HPNs have capacity gains over the standard IEEE 802.11a devices.

Parameter	*Data sheet*	*BB4all ™ architecture*
Maximum single link range (metres), d	~6505	~6505
Modulation scheme	OFDM	OFDM
RF Industrial, Science & Medical (ISM) band (GHz), f	5	5
Number of spatial streams, L	2	2
Combined antenna gain (dBi), $K_{antenna}$	9	28
Channel width (MHz) of IEEE 802.11 a, W	20	20
Maximum output power (mWatts) of IEEE 802.11a radio, P	50	50
Reference distance (metres), d_0	5	5
AWGN (mWatts), N_0	1e-10	1e-10
Achievable data rate (Mbps) for all streams, R	60.792	183.30

Table 1. IEEE 802.11a air interface single link capacity

Table 2 lists parameters needed to evaluate the achievable data rate for all wireless streams of a single link IEEE 802.11n radios. A comparison is made between the achievable data rate computed using data from the IEEE 802.11n air interface data-sheet and the data from the BB4all™ architecture. It is observed that while specifications of other parameters are kept the same in both cases, the datasheet combined antenna gain is taken to be 7 dBi and that of BB4all™ architecture to be 28 dBi. With these antenna gains, the achievable data rate is 291.33 Mbps in standard architecture compared to 570 Mbps for the HPNs. This numerical result is explained as follows. When transmission power and the size of the MIMO are constant, an increase in antenna gain increases capacity substantially. Thus, the HPNs have capacity gains over the standard IEEE 802.11n devices.

Parameter	*Data sheet*	*BB4all™ architecture*
Maximum single link range (metres), d	~6505	~6505
Modulation scheme	OFDM	OFDM
RF Industrial, Science & Medical (ISM) band (GHz), f	5	5
Number of spatial streams (2xMIMO), L	2	2
Combined antenna gain (dBi), $K_{antenna}$	7	28
Channel width (MHz) of IEEE 802.11 a, W	40	40
Maximum output power (mWatts) of IEEE 802.11a radio, P	100	100
Reference distance (metres), d_0	5	5
AWGN (mWatts), N_0	1e-10	1e-10
Achievable data rate (Mbps) for all streams, R	291.33	570

Table 2. IEEE 802.11n air interface single link capacity

5.2. End to end achievable capacity under different HPN placements

Tables 3 and 4 show the E2E numerical values of capacity, right from the ethernet at one end of the network to ethernet at the other end of the network. Consider a wireless mesh network made up of IEEE 802.11a and IEEE 802.11n (Cisco systems, 2011) HPNs. Suppose typical information available are: the radio interfaces $m=2$, the orthogonal channel $c=2$, the deployment area $A = 5000m \times 5000m$ and the bandwidth $W=20\ Mhz$ and carrier frequency of 5.85 GHz. Assume that Carrier sense multiple access with collision avoidance (CSMA/CA) protocol is employed in order to identify pairs nodes that can simultaneously transmit (Kodialam and Nandagopal, 2005). In this protocol, neighbours of both an intended transmitter and receiver have to refrain from both transmission and reception in order to avoid collisions. Practically, we can let Δ =10% of one hop distance to be sufficient enough to prevent neighbouring nodes from transmitting on the same subchannel at the same time. One hop distance is approximately 2100 m. This study also assummed an optimized link state routing (OLSR) protocol that proactively maintains fresh lists of destinations and their routes (Johnson, 2007). These routing tables are periodically distributed in the network. The protocol ensures that a route to a particular destination is immediately available. Couto et al (2005) proposed an expected transmission count (ETX) metric to calculate the expected number of retransmissions that are required for a packet to travel to and from a destination. ETX metric is adapted in this study as a default routing metric to determine the amount of successful packets at any receiver node from a transmitting neighbour within a window period. ETX metric is also viewed as a high-throughput path metric for multi-hop wireless mesh network (Couti et al., 2005). Using such information, we can illustrate the end to end

(E2E) capacity limit with practical examples of network deployments. In particular, consider the following cases:

a. Regular pattern when $n = 10$ and when $n = 50$, the node density is distributed with uniform probability of one.
b. Irregular pattern when $n = 10$ and when $n = 50$. Assume that the average distance of source-destination pair is 6505 m. The value enables the computation of achievable capacity over direct LOS path (i.e., without multi-hops) between the source and destination nodes. Nodes are assummed to be placed irregularly with a rate (probability) p, taken arbitrally as 0.9. Note that $0 < p < 1$. The choice of p depicts the severeness of the irregular placements of HPNs, with smaller values of p depicts more irregular environment and larger value shows that the placement of nodes in an area is carefully planned.
c. Regularly clustered pattern above when $n = 10$ and 5 clusters each of 2 nodes, as well as when $n = 50$ with 5 clusters each of 10 nodes.

HPNs placement in a 5 km x 5 km area	*No. of HPNs*	*Achievable link capacity (Mbps)*	*E2E achievable capacity (Mbps)*
Regular at p = 100%	10	**R(2100 m) = 281.12**	**0.5192**
	50	**R(700 m) = 376.22**	**0.9322**
Irregular at p = 90%	10	**R(2100 m) = 281.12**	**0.5473**
	50	**R(700 m) = 376.22**	**0.9827**
Clustered	10	**R_1(700 m) = 376.22** **R_2(4200 m)= 221.13** **R = min (R_1, R_2)**	**0.4202**
	50	**R_1(700 m) = 376.22** **R_2(1400 m)= 316.22** **R = min (R_1, R_2)**	**0.5374**

Table 3. IEEE 802.11a of HPNs of BB4all ™ architecture

HPNs placement in a 5 km x 5 km area	*No. of HPNs*	*Achievable link capacity (Mbps)*	*E2E achievable capacity (Mbps)*
Regular at p = 100%	**10**	**R(2100 m) = 722.24**	**1.3339**
	50	**R(700 m) = 912.44**	**2.2609**
Irregular at p = 90%	**10**	**R(2100 m) = 722.24**	**1.4061**
	50	**R(700 m) = 912.44**	**2.3832**
Clustered	**10**	**R_1(700 m) = 912.44** **R_2(4200 m)= 602.24** **R = min (R_1, R_2)**	**1.1443**
	50	**R_1(700 m) = 912.44** **R_2(1400 m)= 792.44** **R = min (R_1, R_2)**	**2.0201**

Table 4. IEEE 802.11n of HPNs of BB4all™ architecture

Table 5 illustrates the achievable E2E capacity results of the BB4allTM architecture compared to closely related work on dual radio dual channel analytical results by Kyasanur and Vaidya (2005).

Dual-radio dual-channel mesh network	*Consists of IEEE 802.11a HPNs: regularly placed*	*Consists of IEEE 802.11n HPNs: regularly placed*	*Arbitrary network of dual radio dual channel (Kyasanur and Vaidya, 2005)*
Upper bound capacity value (of 50 nodes) in Mbps in a 5 km x 5 km	0.9322	2.2609	0.01

Table 5. E2E achievable capacity gain of BB4all™ architecture

5.3. Discussions on E2E achievable capacity

It should be noted from both Tables 3 and 4 that in a fixed area of 5 km by 5 km, the E2E achievable capacity evaluated shows that there is lower capacity when number of HPNs is ten than when the number is 50 in all node placement scenarios. The main reason is that a series of long links created between any two immediate nodes degrades the achievable E2E capacity. This was proven by single link capacity models in Section 3. For instance, at ten HPNs in the fixed sized network, the hop distances are much larger than the case for 50 HPNs. In each hop, the propagating signal faces path loss effects due to terrain irregularity, foliage and wireless medium conductivity. The implication is that signal traversing longer hop distances are faced with higher attenuation and lower E2E capacity than signal propagating over shorter hops. The numerical results plotted in the tables 3 and 4 also showed that HPNs distributed with irregularity rate (probability) of 90% provides the highest E2E achievable capacity limit compared to the three node placement scenarios. This means that when HPNs are distributed with irregularity rate of 90%, the probability of finding some nodes in some areas will likely reduce by 10%. However, with the same number of nodes and fixed area of deployment, the inter hop distances where nodes occur will be much smaller by 10% than in regular placements. But shorter hops imply higher capacity if and only if there is no interference as we have noted with single links. Moreover, according to Li et al. (2001), increasing or keeping constant the number of nodes placed in a fixed area automatically increases or keeps constant the average node density. The average node density is inversely proportional to the E2E capacity according to *Theorem 2*. Thus, a lower average density in an irregular node placement for the same number of nodes will yield a higher E2E capacity if and only if the area of deployment is fixed or decreased. Using similar argument, when values of p is decreased (i.e., 0.8, 0.7, 0.6, etc), the average δ decreases proportionately and if the area of deployment is fixed or reduced then for the same number of nodes, the capacity will increase. Interestingly, Tables 3 and 4 showed that in regularly clustered placements, the E2E capacity limit values are least compared to other placement scenarios. The explanation is motivated by viewing that there is long distances between clusters and shorter distances between HPNs within a cluster. While, the former situation exacerbates achievable capacity, the latter improves capacity. The contributing factor within a cluster is then the inter-cluster distances that degrades the overall capacity that can be achieved.

At n = 50 the achievable E2E capacity for clustered placement is almost two times less than one related to regular or irregular patterns in the case of IEEE 802.11a air interface network, but in the case of IEEE 802.11n air interface network it becomes more or less comparable. In Section 3, characterization of influence of multipath, multiple antennas, and hop distance on the link capacity revealed that multipath and distance predominantly affect capacity in IEEE 802.11a air interface, while multipath and the number of antennas predominantly influence the achievable capacity in IEEE 802.11n air interface. Because clustered placements irrespective of the number of antennas per HPN provide longer hop distances between one cluster and other, one expects much worse E2E capacity value in a clustered IEEE 802.11a air interface compared to regular and irregular placements.

It was also noted that network throughput dropped significantly from source HPN to the destination HPN or gateway. In particular, the drop was by about 99% across 3 long distance hops and by about 99% across 3 long distance hops considering regularly deployed HPNs from Tables 3 and 4, respectively. The general explanation is that, the channel gain drops with increase in propagation distance, and there are also overhead losses associated with medium access control (MAC) and the multi-hop routing such that the number of packets sent is not equal to the number of packets received successfully. Despite this observation, HPNs derived from IEEE 802.11n radios have a better E2E capacity achievable mainly due to the MIMO technologies that are capable of combating multi-path fading (Franceschetti et al., 2009).

In arbitral network, with a combined antenna gain of 9dBi, hop distance of 700 m, bandwidth of 20 MHz, transmitted power output of 100 mWatts and 1e-10 Watts, the conventional analytical results of Kyasanur and Vaidya (2005) was compared with the HPNs of the BB4all™ architecture. Data from Table 5 shows that HPNs of the latter with special radios and antenna arrangements is more superior to the HPNs with standard antenna gains. While all cases considered dual radio dual channel specifications, the HPNs of the BB4all™ architecture have higher throughput antenna configurations than the work proposed by Kyasanur and Vaidya (2005).

6. Conclusion and future work

The BB4all™ architecture makes use of omni-directional antennas to maintain mesh connectivity, while directional antennas support information relay over long distances with high power gains. It was found that the impact of multipath and MIMO of IEEE 802.11a/n air interfaces on achievable capacity can be characterized by OFDM modulation scheme, antenna configurations, and multiple streaming of frames or packets. Both the analytical and numerical results showed that the higher the dimensions of these parameters, the higher the achievable capacity due to benefits derived from channel diversity. It was also confirmed based on related previous works that increasing the number of interfaces per HPN and channels in the network does increase the achievable E2E capacity in any arbitral network placement. One of the contribution of this study was the innovation constructed to improve performance of the commercially available WLAN devices. The pillar of innovation was that increasing the antenna gains could improve capacity of real networks even without increasing the power settings of the transmitter.

The CSIR Meraka Institute, South Africa, through living lab initiatives, are currently gathering field data regarding end-to-end capacity that is experienced by rural community Internet users. The findings will be assessed with a view of considering possible improvements of future network architectures that can provide high data rates. Other possible exploration of increasing capacity of community networks (i.e., Peebles valley mesh in South Africa) include utilization of unused frequency (TV white space) spectrum. The TV white spaces spectrum fosters high capacity signal transmissions over long distances in rural terrains. Thus, cognitive and foraging radio techniques are promising tools toward spectrum and energy efficient network management for the next billion internet users. It should also be noted that, although the theoretical derivations were applied to the PVM network, they could also be applied to other rural deployments as well.

Author details

Thomas Olwal, Moshe Masonta, Fisseha Mekuria and Kobus Roux
CSIR Meraka Institute, South Africa

Acknowledgement

Authors would like to acknowledge the CSIR Meraka for financial support, Ishmael Makitla for the HPN diagram and the Book Publishing Editor for considering the chapter proposal for submission.

7. References

Abhayawardhana, V.S.; Wassell, I. J.; Crosby, D.; Sellars, M. P. & Brown, M. G., (2004). Comparison of empirical propagation path loss models for fixed wireless access systems, *In Vehicular Technology Conference,* 2005. ISSN: 1550-2252.

BelAir Networks. (2006). Capacity of wireless mesh networks: understanding single radio, dual radio and multi-radio wireless mesh networks, *White paper*, pp. 1-16, BDMC00040-C02.

Berthilson, L. & Pascual, A. E. (2007). Link performance parameters in IEEE 802.11: How to increase the throughput of a wireless long distance link, *White paper*, April 2007.

Bracewell, R. N. (1986). *The Fourier transform and its applications*, McGraw Hill, 1986.

Cisco Systems Inc. (2011). *Channels and maximum power settings for Cisco Aironet lightweight access points*. Pp. 1-130, No: OL-11321-08.

Couto, D. S. J. D.; Aguayo, D; Bicket, J. & Morris, R. (2005). A high-throughput path metric for multi-hop wireless routing. *Wireless Networks Journal*, Vol. 11, No. 4, pp. 419-434.

Durgin, G.; Rappaport, T. S. & Xu, H. (1998). Measurements and models for radio path loss and penetration loss in and around homes and trees at 5.85 GHz. *IEEE Transactions on Communications*, Vol. 46, No. 11, pp. 1484-1496.

Eriksson, J.; Agarwal, S.; Bahl, P. & Padhye, J. (2006). Feasibility study of mesh networks for all-wireless offices, In Proceedings of MobiSys'06, June 19-22, 2006, Sweden, pp. 69-82: ISBN: 1-59593-195-3.

Franceschetti, M,; Migliore, M. D. & Minero, P. (2009). The capacity of wireless networks: information-theoretic and physical limits. *IEEE Trans. on Information Theory*. Vol. 55. No. 8, pp. 3413-3424.

Gupta, P., and Kumar, P. R. (2000). The capacity of wireless networks. *IEEE Transactions on 4 Information Theory*, Vol. 46, no. 2, pp. 388-404, March 2000: ISSN: 0018-9448.

Intini, A. L. (2000). OFDM wireless networks: Standard IEEE 802.11a. *Technical report at the University of California Santa Barbara.*

Ishmael, J., Bury, S., Pezaros, D., and Race, N. (2008). Deployment rural community wireless mesh networks. IEEE Internet Computing. Vol. (2): 22-29: DOI: 1089-7801/08/2008 IEEE.

IEEE 802.11 Standard Working Group, (1999). Wireless LAN medium access control (MAC) and physical layer specifications: high speed physical layer in the 5 GHz band, IEEE 802.11a standard.

Johnson, D. (2007). Evaluation of a single radio rural mesh network in South Africa, *Proceedings of International Conference on Information and Communication Technologies*, Bangalore, India, December 2007, ISBN:

Johnson, D. & Roux, K. (2008). Building rural wireless networks: lesions learnt and future directions, *Proceedings of 2008 ACM Workshop on Wireless Networks and Systems for Developing Regions*, San Francisco, California, USA, 19 September 2008, pp. 17- 22, ISBN: 978-1-60558-190-3

Kobus, R., et al. (2009). Broadband for all (BB4all)™: CSIR Science Scope, Magazine, Magazine, Vol. 3, no. 3, January 2009, pp. 16-17.

Kodialam, M., and Nandagopal. T. (2005). Characterizing the capacity region in multi-radio multi-channel wireless mesh networks. *MobiCom'05,* August 28-September 2, 2005, Cologne, Germany: ISBN: 1-59593-020-5.

Kyasanur, P. & Vaidya, N. H. (2004). Routing and interface assignment in multi-channel multi-interface wireless networks, *Technical Report of Department of computer science at the University of Illinois at Urban-Champaign*, pp. 1-7, ANI-0125859.

Kyasanur, P. & Vaidya, N. H. (2005). Capacity of multi-channel wireless networks: impact of number of channels and interfaces, *Technical Report of Department of computer science at the University of Illinois at Urban-Champaign*, March 2, 2005.

Li, J.; Blake, C.; De Couto, D. S. J.; Lee, H. I. & Morris, R. (2001). Capacity of Ad hoc Wireless Networks, *Proceedings of MobiCom Conference,* July 2001, Rome, Italy, pp. 61-69, ISBN: 1-58113-486-X

Mesh Dynamics Inc., (2010). Wireless mesh networks that scale like switch stacks, Available at *http://www.meshdynamics.com*.

Makitla, I. ; Makan, A. & Roux, K. (2010). Broadband provision to underprivileged rural communities. *In Proceedings of CSIR 3rd Biennial Conference 2010.* Also available at www.csir.co.za : Reference no: HE04-PO-F.

Mekuria, F.; Masonta, M. T., and Olwal, T.O. (2012). Future networks to enable wireless broadband technologies and services for the next billion users, *In Proceeding Future Network & Mobile Summit 2012,* Berlin, Germany, 4-6 July 2012.

Miu, A., Balakrishnan, H., and Koksal, C. E. (2007). Multi-radio diversity in wireless networks, *Wireless Networks*, pp.13:779-798. DOI:10.1007/s11276-006-9854-2.

Olwal, T. O. (2010). *Decentralised dynamic power control for wireless backbone mesh networks,* PhD Thesis, University of Paris-EST and Tshwane University of Technology.

Olwal, T. O et al. (2011). Optimal control of transmission power management in wireless backbone mesh networks, In: *Wireless Mesh Networks,* Funabiki, N (Eds). PP. 3-28, InTech, ISBN: 978-953-307-519-8, Croatia.

Tse, D., and Viswanath, P. (2005). *Fundamentals of wireless communication,* Chapter 5 and Chapter 7, University of California Berkley.

Chapter 10

High Throughput Path Establishment for Common Traffic in Wireless Mesh Networks

Hassen A. Mogaibel, Mohamed Othman, Shamala Subramaniam and Nor Asilah Wati Abdul Hamid

Additional information is available at the end of the chapter

1. Introduction

Recently, Wireless Mesh Networks (WMNs) technology has gained a lot of attention and become popular in the wireless technology and the industry fields. This rising popularity is due to its low cost, rapid development and ability to offer broadband wireless access to the internet in places where wired infrastructure is not available or worthy to be deployed [2].

Wireless Mesh Networks (WMNs) consist of mesh routers that collect and forward the traffic generated by mesh clients. Mesh routers are typically fixed and equipped with multiple radio interfaces. Mesh clients are mobile, and data are forwarded by mesh routers to the intended destination. One or more mesh routers may have gateway functionality and provide connectivity to other networks such as internet access, as shown in Fig. 1. In the WMNs, most of the flows are between the mesh client and the gateway; this kind of traffic is called internet traffic which is the common WMNs traffic as users need to access wired resources.

Gateway discovery approaches in multihop wireless mesh network can be categorized into three categories as follows:

1. Proactive approach: The proactive gateway discovery is initiated by the gateway itself. The gateway periodically broadcasts a gateway advertisement (GWADV). The mesh nodes that receive the advertisement create or update the route entry for the gateway and then rebroadcast the message. Therefore, each node in the WMN are registered with a gateway [11, 15]. The proactive approach provides a good network connectivity and good handoff before losing the connectivity to their original gateway and the gateway routes are always available at all times, which reduces the routing discovery latency. However, this approach imposes a high overhead due to flooding the GWADV message throughout the network.
2. Reactive approach: The reactive gateway discovery is initiated by the mesh router that creates or updates a route to a gateway. The mesh router node broadcasts a Route REQuest (RREQ) message with an "I" flag (RREQ_I) to the gateways. Thus, only the gateways are addressed by this message, and only they process it. When a gateway receives a

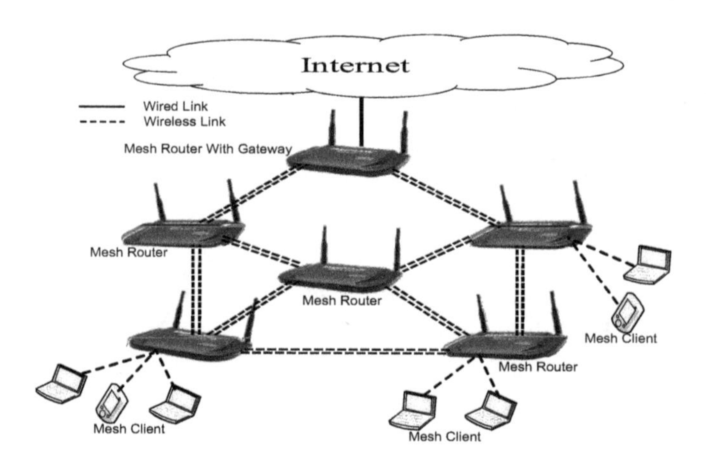

Figure 1. Multi-radio wireless mesh networks.

RREQ_I, it unicasts back a Route Reply (RREP) message with an "I" flag (RREP_I), which, among other things, contains the IP address of the gateway [5, 13, 36]. The advantage of this approach is that the control messages are only generated when a mesh node needs information about a reachable gateway. However, this approach may increase the packet end-to-end delay since the external path is not always available.

3. Hybrid approach: To consider the advantages of the proactive and reactive approaches, they can be combined into a hybrid proactive/reactive method for gateway discovery. For mesh router nodes in a certain range around a gateway, the proactive gateway discovery is used while the mesh router nodes residing outside this range use the reactive gateway discovery to obtain information about the gateway [18, 27, 39]. The approach provides good network connectivity while reducing the overhead. However, the main issue is the optimal value of the advertisement zone.

WMN capacity is reduced by interference from concurrent transmissions. There are two types of interference that affect the throughput of WMN, intra-flow and inter-flow interferences. The intra-flow interference refers to the interference between intermediate nodes sharing the same flow path, whereas, inter-flow interference refers to the interference between neighboring nodes competing on the same busy channel. These come from the half duplex of the radio and the broadcast nature of the wireless medium [21, 35].

Several approaches have been proposed to improve the WMN capacity. One approach is that each mesh router uses a single radio interface that dynamically switches to a wireless channel with a different frequency band to communicate with different nodes [4, 31]. However, this approach increases the routing overhead due to a switching delay. A more practical approach uses multiple radio interfaces that are dedicated to non-overlapping channels [1, 19, 32].

The IEEE 802.11 b/g and IEEE 802.11a standards define three and twelve non-overlapping channels (frequencies) [1, 8, 17]. One of the most important issues for the design of multi-radio multi-channel networks is how to bind the radio interface to a channel in a way that maintains network connectivity. Three approaches have been proposed t solve the channel assignment problem in multi-radio multi-channel WMNs which can be described as flowing:

1. Static channel assignment: In a static channel assignment approach, each interface is assigned to a channel for long time durations. Static assignment can be further classified into two types:
   (a) Common channel approach: In this approach, the radio interfaces of all nodes in the network are assigned to common channels [10]. For example, if two interfaces are used at each node, then the two interfaces are assigned to the same two channels at every node.
   (b) Varying channel approach: In this approach, the radio interfaces in different nodes may be assigned to different channels [19, 26]. With this approach, it is possible that the length of the routes between nodes may increase, also, the network partitions may arise due to the inability of different neighbors to communicate with each other unless they assign a common channel.
2. Dynamic channel assignment: The dynamic channel assignment approach allows any interface to be assigned to any channel, and interfaces can frequently switch from one channel to another [31]. Therefore, a network using such a strategy needs some kind of synchronization mechanism to enable communication between nodes in the network. The benefit of dynamic assignment is the ability to switch an interface to any channel, thereby, offering the potential to use many channels with few interfaces. However, the key challenges are channel switching delays, and the necessity for coordination mechanisms to switch between node channels.
3. Hybrid channel assignment: In the hybrid approach, all the nodes are equipped with multi-radio interfaces in which the multiple radios are divided into two groups, fixed group and switchable group. In the fixed group, each radio interface is assigned a fixed channel for receiving packets, thereby, ensuring the network connectivity, while the switchable group can dynamically switch among the other data channels [17].

However, most of the previous research focuses on how to answer this question without considering the unique properties of WMNs, which include the following:

- Most of the traffic in WMNs is designated at the gateways as the users need to access the internet or wired resources. This kind of traffic can be considered as a multi-source single destination traffic.
- The local traffic and internet traffic must pass through the backbone nodes to reach their destination. Thus, improving the backbone performance will increase the WMN's performance.
- Availability of multi-links between adjacent mesh routers makes the mesh routers support simultaneous multi-flow transmission for both kinds of the traffic.

The unique characteristics of WMNs motivated us to developed On-demand Channel Reservation Scheme (AODV-MRCR) with aims to establish high throughput path for the gateway traffic, reduces the interference caused by local traffic, supports full duplex node and only assigns channel to the active node. We achieve these objectives by integrated the reactive routing protocol with channel distribution.

The reactive approach is choosing in order to establish high throughput paths for the gateway traffic and assigns channel to active node. This meaning that all nodes will statically assign common channels to their interfaces, and only the node that has gateway traffic allowed to switch some of its interfaces to the selected channels. Our contributions are as follows:

(i) Enhance the capability of the node to receive and transmit concurrently by ensure that distinct channel should be reserved for the reverse and forward routing entry for each node involve in path establish process during the gateway discovery process.

(ii) Integrated the channel assignment and distribution with the reactive gateway discovery process in order to efficiently utilize the limited number of non-overlapping channel and establish high throughput paths for the gateway traffic.

(iii) Developed a hybrid interface assignment that reduces the packet collision for the gateway traffic due to the broadcast nature of the wireless medium and existing of local traffic. This done by proposed static and dynamic channel assignment. Static interface assign to static channel and used to support the local traffic while the dynamic interface only assign to active node during the gateway discover process. This interface used to supported gateway traffic.

(iv) Developed channel assignment that simple (reduce the channel assignment complexity) and independent of any particular profile such as traffic, interference, and topology profile.

The remainder of the chapter is organized as follows. Section two discusses relevant work. The AODV-MRCR protocol is explained in section three. In section four, we provide the details of our simulation environment. Simulation results and their analysis are presented in section five, with concluding remarks in section six.

2. Related works

A major problem facing multi-hop wireless networks is the interference between adjacent links. The throughput of a single-radio single-channel wireless network has been studied in [37]. The authors formalized it as a multi-commodity flow problem with constraints from conflict graph, which is NP hard, and gave an upper bound and a lower bound of the problem.

There have been many studies on how to assign limited channels to network interfaces in a multi-radio multi-channel wireless mesh network as to minimize interference and maximize throughput. They differ in several assumptions made in WMNs, and therefore in the models and related solutions.

One approach assumes a known traffic profile in the network, because the aggregate traffic load of each mesh router changes infrequently. The authors of [26] proposed an iterative approach to solve the joint routing and channel assignment problem. Heuristic techniques are used to estimate the traffic load in each link. The algorithm starts with an initial estimation of the expected traffic load and iterates over both channel assignment and routing until the bandwidth allocated to each virtual link matches its expected load. While this scheme presents a method for channel allocation that incorporates connectivity and traffic patterns, the assignment of channels on links may cause a ripple effect whereby already assigned links have to be revisited, thus, increasing the time complexity of the scheme. Moreover, this approach is performed during the network plan and assumes that the traffic profile is known. The centralized flow-based and rate channel assignment algorithm is proposed in a paper by [3]. The agreed heuristic algorithm is used for channel assignment rate. [6] enhances the [26] centralized algorithm to support automatic and fast failure recovery. The failure recovery mechanism is located at the gateway and all nodes send periodic messages to the gateway. In the case where the gateway does not receive a message during a period of time from node x, it deletes the corresponding information, node id, position, rate, and then runs the algorithm to update the gateway tables. Based on the new tables, it recalculates the link ranking and channel assignment. However, in general, the centralized approach causes a high computation overhead at the centric node and it is unwieldy in use due to the need for gathering network information. Moreover, most of them are static assignment which is not optimally utilizing the limited number of available non-overlapping channels. In contrast, our approach is a more dynamic approach, which is performed during the real-time networking; in addition, no prior knowledge of the traffic profile is needed.

Other studies assume that the traffic profile of each mesh router is not known, and usually consider channel assignment and routing separately. The authors of [25] assumed that the traffic from the Internet gateway to clients is dominant, and thus proposed distributed channel assignment based on spanning tree topology, where the gateway is the root of the spanning tree. The protocol dedicates one interface channel for communication with its parent node on the tree, and the other interfaces are configured as children for communication with their child nodes. Hence, the protocol divides the node interfaces into two subsets - downlink and uplink interfaces. The uplink interfaces are used to connect the node with its parent node while the downlink is used to connect the node with its child nodes. The node can only switch its child. For channel assignment, the channel assignment strategy starts from the root of the tree. Each node switches its parent interfaces to the parent node child interface and selects a new channel for its child interfaces. One drawback of this protocol is that it only considers the common traffic where data are transmitted from the source to gateway and vice versa.

Multi-channel routing protocol (MCR) [17] the peer-to-peer traffic was assumed to be dominant in the network. The authors first constructed a k-connected backbone from the original network topology, and then assigned channels on the constructed topology. The MCR classified the node interfaces into fixed or switchable interfaces. The protocol assigns a fixed channel to the fixed interface for communication between neighbors, and the remaining interfaces are considered as switchable interfaces. When a node wants to communicate with others, it looks in its table to find the destination's fixed channel and switches one of the switchable interfaces to that channel. To exchange fixed channels between neighbors, MCR uses a "hello" message to carry the fixed channel information. However, this protocol may not work well in a multi-flow transmission because of high switching interfaces and because it does not utilize all the non-overlapping channels as the static channel assignment uses.

Although there are many distributed solutions proposed in literature [7, 9, 16, 23, 25, 38]. In [16], the authors proposed the Local Channel Assignment (LCA) algorithm, which adopts a tree-based routing protocol for common traffic similar to Hyacinth. The LCA algorithm solved the Hyacinth interface-channel assignment conflict problem which is caused when a parent switches to the least load channel that may be in use by one of its children. The interface-channel assignment problem may cause recursive channel switching and delays. LCA solved this problem by dividing the non-overlapping channel into groups and making each parent interface belong to one group different from its child interface group. The paper of [7] proposed a distributed joint channel assignment and routing protocol for multi-radio multi-channel ad hoc network. The scheme dedicates one interface for the control message and another interface for data transmission. The control interface is assigned to a common channel while the data interfaces could work as a fixed or switchable interface based on the receiving call direction. However, in this approach, the control interface becomes the bottleneck, especially in high-density networks.

In the paper of [9], the authors proposed a hybrid multi-channel multi-radio wireless mesh network architecture, which combines the advantages of both static and dynamic channel allocation strategies. The architecture is similar to Hyacinth architecture [25]; it classifies the interface to work as a fixed interface or a switchable interface. The protocol only considers one interface to work as a switchable interface. This interface has the ability to switch channels frequently, while the remaining interfaces are considered as fixed interfaces that work on fixed channels. The channel allocation of static interfaces aims at maximizing the network throughput from end-users to the gateway, while the dynamic interface is used to communicate with the neighbor node that has a different fixed channel on-demand fashion. Two dynamic interfaces that are within radio transmission range of each other are able to communicate by switching to the same channel when they have data to transmit.

In [23], the authors proposed a learning based approach for distributing channel assignments. It uses a learning based algorithm to determine the best channels to assign its own interfaces based on collecting information from the neighbor nodes. Hence, each mesh node periodically sends a "hello" message in order to discover its neighbors and the channel usage in its neighborhood. The algorithm achieves effective channel usage, and also adapts well to the change of network topology. [38] proposed a distributed channel assignment for uncoordinated WMNs to minimize the interference with adjacent access points. The algorithm assigns the least interference channel to the access point interference according to the gathered channel information from neighboring access points and associated clients. Both the protocols discussed earlier assign channels from node to node, and each node in the WMNs assigns a fixed channel, which makes it different from our approach. In our approach, channel assignment is based on data flow such that a channel is only assigned to a node if it has data to send or forward to the gateway.

The authors of [34] and [7] proposed algorithms to minimize network interference. The first one is interference-aware because it visits the links in decreasing order of the number of links falling in the interference range and it selects the least used channel in that range. Assuming the set of connection requests to be routed, both an optimal algorithm based on solving a Linear Programming (LP) and a simple heuristic are proposed to route such requests, given the link bandwidth availability as determined by the computed channel assignment. The algorithm considers minimum-interference channel assignments that preserve k-connectivity. The algorithm proposed in [7] uses a genetic approach to find the largest number that makes the whole network connected while minimizing network interference. However, such approaches only focus on minimizing the network interference that may decrease the network connectivity. In contrast to the above mentioned approaches, our approach is based on eliminating the interference for the common traffic on WMNs while maintaining network connectivity. Besides static channel assignment algorithms, which assign channels to interfaces without change for a long time, there have been several dynamic channel allocation algorithms proposed, which allow interfaces to switch channels frequently.

The authors of [29] proposed an on-demand channel allocation protocol in a wireless mesh network, where each node has two interfaces. In their framework, one interface of each node is devoted to controlling channel negotiation only while the other interface is used for data transmission. On the other hand, the frameworks proposed in [30] and [4] do not require a separate control interface, and the channel negotiation happens on the same interface for data transmission.

The Channel Assignment Ad hoc On-demand Distance Vector routing (CA-AODV) [12], has been proposed to assign channels within K hops in an ad hoc network, allowing for concurrent transmission on the neighboring links along the path and effectively reducing the intra-flow interference. Similar to CA-AODV, [33] proposed to join the channel assignment with the

AODV. The source node needs to ensure the channel selection by sending two messages. The first message is to inform neighbors about the selected channel and the second message is sent by the neighbors to confirm the channel. In case the channel is in use, the node should be waiting until the channel is free or selects a new channel. However, such an approach may not work well in WMNs where most of the traffic is directed toward the gateways and must pass through mesh routers.

3. Multi-radio ad hoc on-demand distance vector routing with channel reservation scheme

The Multi-radio ad hoc on-demand distance vector routing with channel reservation scheme (AODV-MRCR) protocol is a multi-radio on demand distance vector routing protocol, which is proposed to establish high throughput paths for the gateway traffic in WMNs.

Our scheme uses on-demand reactive routing protocol to distribute the reserved channels list among all the nodes along the path from the source to the gateway. The source node that does not has fresh route to the gateway, it sends RREQ with flag set to one which means only the gateway can reply this message. Once the gateway receives a new RREQ_I, it selects a reserved channel list and attaches to the RREP_I message. The message is sent back to the source node. During the RREP_I stage, each intermediate node selects its recommend channel based in the hop count index[20]. The intermediate node reserves at least two interfaces for the gateway path, one link for the forward path and other for reverse path.

We assume that m channels are available that can be used in a wireless area without interfering. In addition, k channels of available channel are statically assigned to i interfaces, half of k channels are used as "used channels", and the $m - k/2$ channels will be considered as "unused channels". The i available interfaces at each node can be classified as:

- Fixed interfaces: Some n of the i interfaces at each node are assigned for long intervals of time to k channels, we designate these interfaces as "fixed interfaces", and the corresponding channels as "used channels". These interfaces are used to keep the network connectivity as well as support the local traffic. Therefore, they are not allowed to switch.
- Switchable interface: The reaming $i - n$ interfaces are switched to the selected channels for long intervals of time. These interfaces are assigned to a channel that is selected from the range $m - k/2$ channels during the RREP_I message.

For example, if the mesh router has four interfaces, two channels of twelve (1, 2) will be considered as used channels. The reaming channels will be considered as unused channels. Moreover, the interfaces one and two will be considered as fixed interfaces while the interfaces three and four are switch- able interfaces. The scheme consists of two parts. The first part is carried out at the gateway, which is used to reserve a unique list of channels for each RREQ_I received at the gateway. The second part is carried out when the intermediate nodes along the path back to the source receive the RREP_I message. Following is a clarification of the procedures.

3.1. Channel reservation scheme

Channel reservation scheme is carried out into two stages. First stage is carried out by the gateway, which is used to reserve a unique list of channel for each received RREQ_I message, see algorithm 1. The second stage is carried out when the intermediate nodes along the path back to the RREQ_I source node receive RREP_I message, which is aims to distributed the channel along the active nodes, see algorithm 2.

Once the gateway receives a new RREQ_I message, it checks the channel reservation table for the RREQ_I source address entry. Each table entry contains the source node address and reserved channel list for each received RREQ_I.

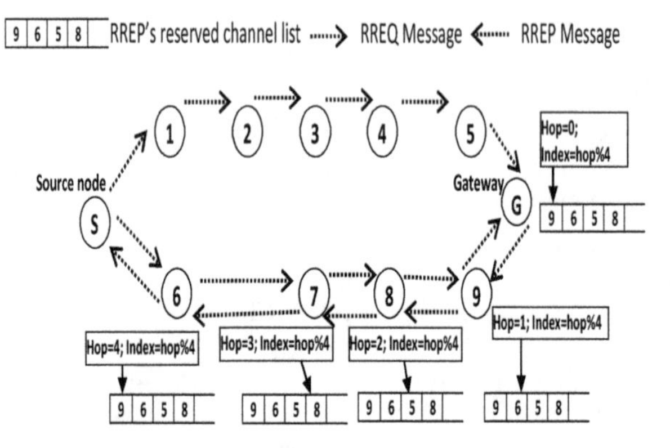

Figure 2. Example of channel reservation scheme.

When the intermediate node receives the RREP_I message, it creates/updates the forward route entry with the recommended channel as indicated in RREP_I's recommended channel. Then, it selects a channel from the RREP_I's list based on the hop count. The intermediate node selects the channel only if it satisfied the following constrains.

1. The RREP's reserve channel list is not empty
2. At least there is one interface work on fixed channel to support the local traffic.

If the intermediate nodes along the forward path are satisfied the above constraints, a recommended channel will be selected based on the hop count modular the number of channel in the reserve channel list as shown in Fig. 2.

In case it matches, the corresponding entry is attached to the RREP_I and send back along the reverse path to the RREQ_I source node. However, if no entry found, then the gateway randomly select a new reserved channel list from the unused channel list, update the channel reservation table and attach the list to the unicast RREP_I.

2 explains the channel selection at the gateway. Two RREQ_I messages from source node (S) received at the gateway through different paths. The path with minimum hop count will be selected as the best path toward the source node (S). The gateway (G) checks the channel reservation table for the source node address (S). If a match is found, the corresponding entry is attached to the RREP_I message. If not found, a new reserve channel list will be selected and attached to the RREP_I message. The maximum number of the channel in the reserve channel list is four channels [20]. The reserve channel list along with recommend channel is attached to the RREP_I message and is sent back to the RREQ_I's source node.

Algorithm 1 The selection of the reserved channel list at the gateway

Require: $RREQ\_I$: Routing Request message with an "I" flag
$RREP\_I$: Routing Reply message with an "I" flag
RC: Recommended Channel
$hops$: Hop count field in RREQ_I message
$\mathcal{A}$: Unused channels
RCL: Reserved Channel List
$max\_channel$: Maximum number of channels in each list
no_iface: Number of interfaces per mesh router
{ initialize variable }
1: $RCL(i) \leftarrow 0, \forall i \in \{0,..,4\}$ {4 means the maximum number of channel in RCL}
2: $RC \leftarrow 0$
3: BEGIN
4: **if** $(RREQ\_I.hops > 3)$ **then**
5: $max\_channel \leftarrow 4$
6: **else**
7: $max\_channel \leftarrow RREQ\_I.hops$
8: **end if**
9: **if** (*gateway has been assign channel to theRREQ_I source node*) **then**
10: $RCL[] \leftarrow Lookup_{c}hannelTable(RREQ\_I.source\_address)$
11: **else**
12: $i \leftarrow 0$
13: **while** $(i < max\_channel)$ **do**
14: $RC \leftarrow rand(\mathcal{A})$
15: check RCL for RC duplicate value.
16: **if** $(duplicate\ not\ found)$ **then**
17: $RCL(i) \leftarrow RC$
18: $i \leftarrow i+1$
19: **end if**
20: **end while**
21: **end if**
22: $RC \leftarrow RCL(0)$
23: update RC of the reversed routing entry that belong to the $RREQ\_I$ source node with RC value.
24: $RREP\_I.RC \leftarrow RC$
25: $RREP\_I.RCL \leftarrow RCL$
26: send $RREP\_I$
27: END

In this figure, the gateway selects the channel at location zero. This is because the hop count at the gateway is zero. Once the message receives at the next hop, the hop count will be one and the result of modular operation point to the location one in the reserve channel list, and so. When the hop count exceeds the maximum number of channel in the reserve channel list, such as node six, the recommended channel will be selected based on the modular operation which will point to location zero. However, In case the above constraints are not satisfied, the node sets the RREP's recommended channel to zero and forward the message to next hop, see algorithm 2.

Algorithm 2 Channel selection at an intermediate node

Require: $RREQ\_I$: Routing Request message with an "I" flag
$RREP\_I$: Routing Reply message with an "I" flag
RC: Recommended Channel
$hops$: Hop count field in RREQ_I message
RCL: Reserved Channel List
$max\_channel$: Maximum number of channels in each list
no_iface: Number of interfaces per mesh router
$sw\_interface$: Number of interfaces that has been switching
1: BEGIN
2: **if** $((RREP\_I.RC \neq 0)$ **and** $(sw\_interface \leq no\_iface/2))$ **then**
3: update RC of the forward path that belong to the $RREP\_I$ source node with the $RREP\_I.RC$
4: $sw\_interface \leftarrow sw\_interface + 1$
5: **end if**
6: **if** $((RREP\_I.RCL \neq 0)$ **or** $(sw\_interface > no\_iface/2))$ **then**
7: $RC \leftarrow 0$
8: $RREP\_I.RC \leftarrow 0$
9: **else**
10: $index \leftarrow (RREP\_I.hops \bmod max\_channel)$
11: $RC \leftarrow RREP\_I.RCL(index)$
12: $RREP\_I.RC \leftarrow RC$
13: update RC of the reversed routing entry that belong to the $RREQ\_I$ source node with the new RC value.
14: **end if**
15: send $RREP\_I$
16: END

3.2. Channel switching and negotiation

The proposed protocol is an across layer protocol that integrated the channel assignment and routing protocol with the MAC layer. The routing protocol is used to select the reserved channel list for each received RREQ_I at the gateway and distribute them among the nodes along the path from the source to the destination. Moreover, the four handshake, rts-cts-data-ack, MAC messages are used by the proposed protocol in order to coordinate the channel switching between pair of nodes and ensure that both nodes have been switched to the selected channel (recommended channel) same as proposed in [8, 29].

Fig. 3 describes the channel switching and negotiation, when the intermediate node receives the RREP_I message, it sends the message to the MAC layer using the reverse route entry interface field. Then the MAC layer sends a RTS message including the channel information to the neighboring nodes. The neighboring nodes, which work on the same channel as the sender, will receive the RTS message; upon receiving the RTS message, the receiver sends back a CTS message. Once the source node receives the CTS message, it starts sending the RREP_I packet. If the node receives ACK or the transmission time has expired, it switches the interface to the new channel. At the receiver side, the receiver replies with an ACK and waits for the transmission time to expire when it receives data. After that, it switches the interface to the new channel.

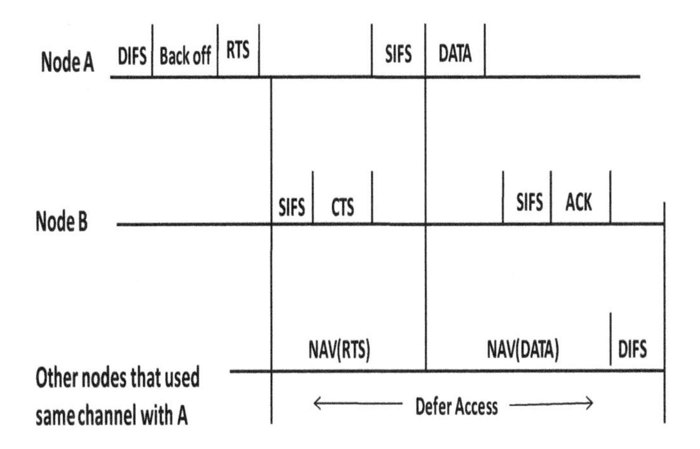

Figure 3. Channel switching and negotiation.

4. Simulation environment

Since our protocol is multi-radio mutli-channel routing protocol that take into consideration the availability of multi-radio per mesh router as well as the multi-channel, We evaluate the performance of AODV-MRCR compared to AODV-MR [24] and MCR [17] using ns-2 [22]. The former AODV-MR is developed to enhance the reactive routing protocol to support the multi-radio and take into account the advantage of multi-links between mesh routers. The protocol uses ICA channel assignment where each mesh router has multi-radio interfaces, which statically dedicated to non-overlapping channel. The latter MCR routing protocol is developed to utilize the multi-channel where each node randomly selects a channel from the none-overlapping channels. The selected channel is assigned to an interface. This interface is considered as fixed interface for communication between neighbors and the remaining interfaces are considered as switchable interfaces. The switching mechanism is used to exchange the message between neighbors. When a mesh router has a data to send, it switches one of its switchable interfaces to the distentionŠs fixed channel. AODV-MR and MCR protocols distribute the channels based on unknown knowledge profile similar to our protocol. Moreover, they are based on the reactive routing discovery process in order to discover the path. When the node has data to send, it sends RREQ message and waits for receiving RREP message. Every intermediate node receives RREQ, it creates a reverse path and forwards the message to all interfaces. Once the destination receives the RREQ message, it replies with RREP message back to the RREQŠs source node.

A mesh network on an area of 1000×1000 meter^2 a 2-dimensional open area, without any building or mountain, was established using a random distribution mesh routers. Hence, a simple two-ray-ground propagation model [28] is used. all nodes are randomly distributed and each node is equipped with multiple wireless interfaces that statically turned to non-overlapping channels using the same channel allocation scheme. At the MAC layer, IEEE 802.11 DCF with RTS/CTS collision avoidance is used since we used the four hand check to distribute the channel between pair nodes along the path from source to destination. The channel switching latency is set to $80\mu s$ [12]. The common parameters for all the simulations are listed in Table 1.

Simulation time	250 second
Simulation area	1000×1000 meter^2
Transmission range	250 meter
Traffic type	Constant Bit Rate (CBR)
Packet size	512 bytes
Packet rates	20 packet per second
Number of nodes	100
Number of connection	50

Table 1. Simulation parameters

4.1. Performance metrics

The simulation provides the following five performance metrics:

1. Packet Delivery Ratio(PDR): The ratio between the number of data packets successfully received by the destination nodes and the total number of data packets sent by the source nodes.
2. Aggregate goodput: The total number of application layer data bits successfully transmitted in the network per second.
3. End-to-end delay of data packets: The delay between the time at which the data packet originated at the source and the time when it reaches the destination. It includes all of the possible delays caused by queuing for transmission at the node, buffering the packet and retransmission delays. This metric represents the quality of the routing protocol.
4. Routing overhead: The ratio of the total number of packets generated to the total number of data packets that are successfully received.
5. Packet loss: The number of packets that were lost due to unavailable or incorrect routes, MAC layer collisions or through the saturation of interface queues.

5. Simulation results and discussion

To evaluate our protocol, we carried out two simulations with different scenarios, as follows.

5.1. Simulation 1:Compare our protocol with multi-radio routing protocol

In this simulation, the efficiency of AODV-MRCR compared to AODV-MR [24] was evaluated using ns-2 [22]. Each mesh router is equipped with four wireless interfaces that statically turn to non-overlapping channels using the same channel allocation scheme. Concurrent UDP flows are established between the randomly selected source and the gateway while keeping the other parameters, as in Table 1. We run AODV-MRCR with twelve non-overlapping channels, and AODV-MR with four channels statically dedicated to the node interfaces. We carry out different scenarios to show the various factors affecting the performance of our protocol. The scenarios are: varying the number of flows, effect of number of radio per node, and evaluating the performance of the protocol using TCP traffic.

5.1.1. Scenario 1: Varying number of the flow

In this scenario, we investigate the impact of traffic load in the network. The number of generated flows varied from 10 to 50 flows with increments of 10. The packet size for each connection was fixed to 512 bytes with a 20 packet per second data rate. Fig. 4 (a-d) shows the packet loss, aggregated goodput, PDR and end-to-end delay performance. As observed from Fig 4, performance of the AODV-MRCR is comparable to that of the AODV-MR at a lower traffic load, such as 10 or 20. This is because AODV-MR can utilize the four channels to minimize the number of links on the same channel within the path as well as create channel diversity along the path.

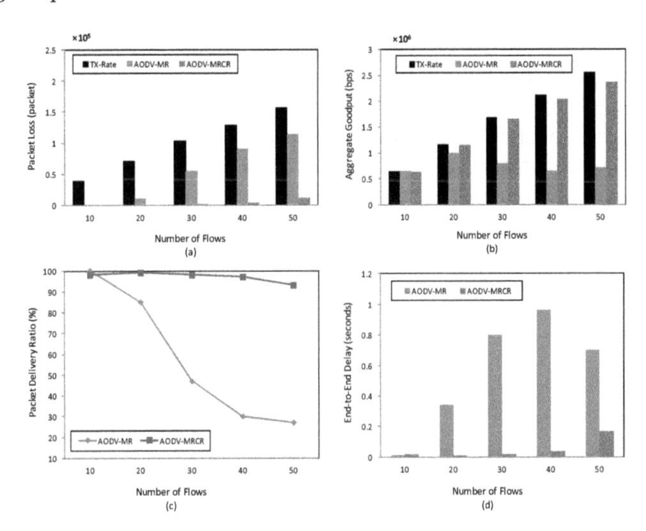

Figure 4. Simulation 1: results for scenario 1 (Varying the number of flows).

However, when the flow increases, the network becomes saturated. Thus, the AODV-MR end-to-end delay and packet loss increase as shown in Fig. 4(a, d) while the PDR and aggregated goodput decrease (Fig. 4(b, c). This is because AODV-MR is unable to avoid the congestion area as it does not have an intelligent way to route the packet through a less congested area. Moreover, minimum channel diversity cannot achieve due to using the hop count as a metric. In contrast, 4 shows an improvement in the performance of the AODV-MRCR under increasing traffic load relative to the AODV-MR. In addition, it shows that the lower packet losses incurred by our protocol enables it to achieve a significantly higher packet delivery ratio.

This is because the protocol assigns a unique list of channels to every RREQ_I received at the gateway during the route-establishing stage. Hence, the single collision domain divided to many collision domains as the number of flows increase.

Fig. 4(d) As the number of the flows increase, our protocol achieves better average end-to-end delay than the AODV-MR. This is because our reservation scheme allows the AODV-MRCR protocol to assign different non-overlapping channels for each node for the gateway traffic. Resulting in, reduces the contention time at the MAC layer as well as allowing the node to receive and send simultaneously. Moreover, It is interesting to note that at high traffic load such as 50 flows, the AODV-MR's end-to-end-delay (Fig. 4(d)) reduces compare to other flows such as 30 and 40 flows. The reason for that is at high traffic load the network is saturated and the aggregated throughput exceeds the actual bandwidth, hence the collision probability of the multi-hop packets becomes high as the number of flows increases. Accordingly, a few multi-hop packets will be received at the destination while most of the received packets are single hop packets.

5.1.2. Scenario 2: Effect of number of radio per node

5 depicts the performance of AODV-MR and proposed algorithm versus the number of radio. The number of radios in each mesh router is varied from three to eight in increments of one along with all other simulation parameters, as per Table 1. This scenario carried out to determine the optimal number of radios to be placed in each mesh router. The results shown in Fig. 5 show that AODV-MR has an improvement in performance up to eight radios in each mesh router. The reason for this is that AODV-MR cannot utilize all the non-overlapping channels, unless they are assigned to the interface. Hence, adding a new interface to the mesh router means adding a new channel to the AODV-MR spectrum utilization.

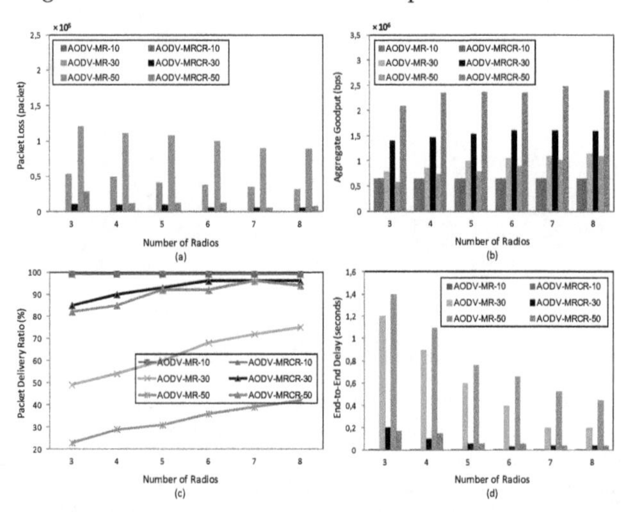

Figure 5. Simulation 1: results for scenario 2 (Varying the number of radios in each mesh router).

Fig. 5 also shows that a limited number of radios per node, about 7, are sufficient for AODV-MRCR to achieve its maximum performance improvements. However, increasing the number of radios beyond seven seems to only achieve a marginal improvement. This result obviously depends on the number of available channels, which are used to choose the reserved channel list for each RREQ_I received at the destination. The AODV-MRCR statically assigns a unique channel for each mesh router interface and half of these channels consider as used channel. Consequently, increasing the number of interfaces per mesh router, minimizes the non-distributed channels in the unused channel list, which increase the packet loss ratio, see Fig. 5(a). However, increase the number of interfaces per node lead to minimize the number of channel in the unused channel list which leads to increase the inter-flow and intra-flow interferences for the gateway traffic. Moreover, increasing the intra-flow and inter-flow interference leads to increase in the packet end-to-end delay due to an increase in the MAC contention time and packet retransmission at the MAC layer, see Fig. 5(d). In all of the cases considered, AODV-MRCR performs significantly better than the AODV-MR routing protocol.

5.1.3. Scenario 3: Varying TCP traffic

We study the performance of the AODV-MRCR and AODV-MR when there are 100 nodes distribute on area of 100 1000 $\times$ 1000 meter^2 and twelve of non-overlapping channels. Table 1 shows the simulation parameters for this scenario.

We analyze the performance of the proposed protocol in two main scenarios. In the first scenario, we study the effective of TCP traffic load by varied the number of the flow from 10 to 50, and in the second scenario, we varied the packet size from 128 to 1440 bytes.

Fig. 6 shows the performance of the proposed protocol and the AODV-MR. The results of Fig. 6 show that the AODV-MR achieves poor performance as the number of flows or the packet size increases. This result occurs because AODV-MR suffers from many problems. First, it used hop count metric as path selection metric, hence, AODV-MR not being able to avoid the hot spot area, it may also select paths with small channel diversity and route the packet through high-congestion areas. The result is that the links frequently get saturated and suffer from multi-flow interference.

Second, the AODV-MR using the same route and same channel to forward the TCP data and the acknowledgement, which leads to an increased Round Trip Time (RTT). In contrast, The AODV-MRCR minimizes the packet round trip time by assignees different channels for reverse and forward routes per node, which allows the node to become a full-duplex node. This procedure reduces the contention and transmission time at the MAC layer. Moreover, the AODV-MRCR assigns a unique list of channels for each flow received at the destination, which leads to minimizing the interference and reducing the packet drops due to packet collision,

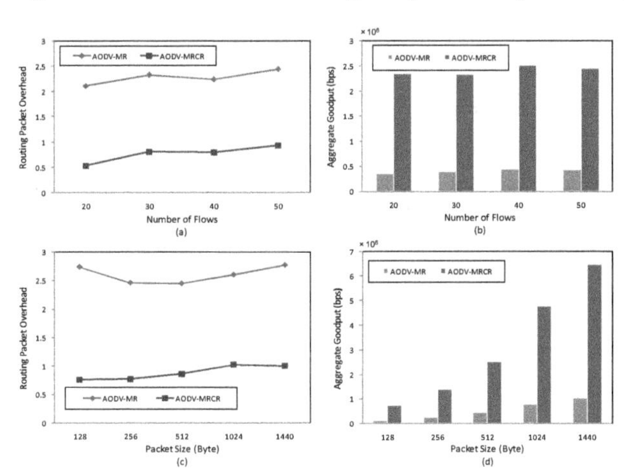

Figure 6. Simulation 1: results for scenario 3 test TCP traffic (Varying the number of flows and Varying packet size).

enabling the route to be effective for long durations while minimizing the number of route discovery messages.

Fig. 6(c, d) shows the performance of both protocols when varying the packet sizes. Our protocol outperforms AODV-MR in terms of aggregate throughput and routing overhead as the packet size increases. This is because our protocol reduces the round trip time at the intermediate node and reduces the transmission time. Reducing the RTT time leads to an increase in the network throughput due to the TCP packet generation depends on successfully receiving ACKs. Furthermore, similar to [14], Fig. 6 shows that a larger packet size can lead to an increase in link failure at the MAC layer. Similarly, a small packet size can reduce the duration of capture, resulting in frequent opportunities for channel access. However, a small packet size also increases the control overhead and can increase the number of collisions at the link layer.

5.2. Simulation 2: Compare our protocol with multi-radio multi-channel routing protocol

In this simulation, we compare the performance of our protocol with the performance of the MCR routing protocol [17]. Many centralized approaches have been proposed to assign a channel to node interfaces in multi-radio multi-channel networks. Most of these approaches need a global view of the network such as traffic profile or node position. Even though, our approach is a centralized approach that establishes high throughput paths for the gateway traffic, it does not require prior knowledge about the network. The MCR routing protocol is multi-radio multi-channel routing protocol that is similar to our protocol in some aspects, such as, it randomly selects a channel with no prior knowledge about the network and uses the routing management messages to inform the neighbors about the selected channel. For all scenarios, varying number of flows, varying packet size, studying the impact of node density, studying the impact of local traffic, and varying the number of non-overlapping channels available in this simulation, we keep the number of interfaces per mesh router to three interfaces and each interface is statically dedicated to a channel using the common channel assignment approach. We carried out different scenarios as described below:

5.2.1. Scenario 1: Varying number of the flows

In this scenario, we evaluate the impact of varying the number of the flows in the network. The number of CBR flows is varied from 10 to 50 with an increment of 10 flows. We keep the other parameters as in Table 1. Under heavy traffic load beyond twenty flows, the proposed approach performs better than MCR as shown in Fig. 7. Moreover, with high traffic load, the MCR end-to-end delay increases as the number of concurrent flows increase see Fig. 7(c). This is because the MCR routing protocol adds extra delay overheads for every received packet. In contrast, our protocol reduces the packet end-to-end delay. This is due to the fact that in proposed scheme, a node becomes full duplex transmission. Furthermore, the scheme only assigns channel to the active nodes, which considerably reduces the interference and contention time at the MAC layer and hence the delay reduces. The delay increment has direct impact in the aggregated goodput and number of packet loss see figure. Hence, increase the number of packet loss lead to increase in the of control message. Since the MCR routing protocol sends a copy of the broadcasting message on every channel, the MCR overhead (RREQ, RREP, HELLO) increases as the number of the flows increases as show in Fig. 7(d).

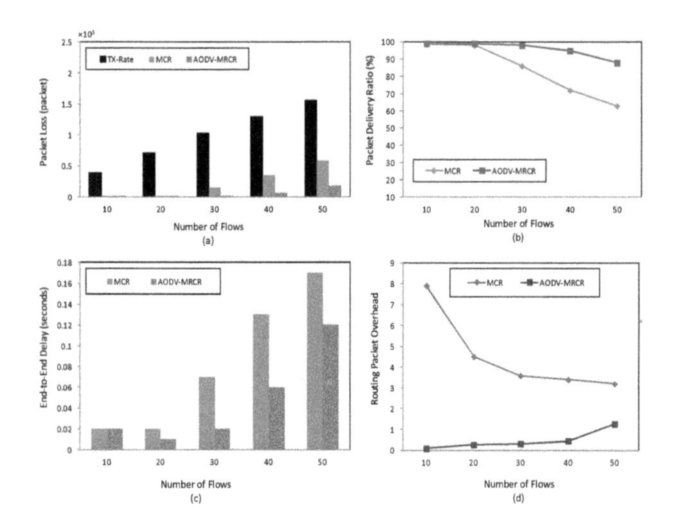

Figure 7. Simulation 2: results for scenario 1 (Varying the number of flows).

However, our protocol sends a copy of broadcasting message on all node's interfaces; in a multi-channel network the number of channels are more than the number of interfaces. Moreover, our protocol does not use the hello message to inform the neighbors about the node channel, as in the MCR. This is because our scheme distributes the channel to the nodes involved in the path during the routing discovery process.

5.2.2. Scenario 2: Varying the number of available channels

To investigate the impact of the number of channel in AODV-MRCR, we varied the number of available channels from 5 to 12. The other simulation parameters are fixed as in Table 1.

Fig. 8 shows the results of the MCR with respect to the AODV-MRCR for a different number of available channels. As observed from Fig. 8(b), AODV-MRCR shows higher PDR than MCR regardless of the number of channels. The performance difference becomes large with the increase in the number of channel. The reason of this performance improvement of AODV-MRCR is can be explained by the fact that adding more channel will increase the number of concurrent transmission. Moreover, using more channels allows our protocol to maximize the channel diversity along the path as well as minimize the channel use between multiple concurrent transmission flows. However, as the number of channels increases, the node interface becomes congested which can limit the MCR performance. This is because that the MCR routing protocol spent more time by the switchable interface in sending broadcast packets.

It is interesting to note that at 5 available channels, our protocol has a higher routing overhead than the MCR routing protocol. This is because our protocol selects the reserved channel list from the unused channel list. However, small available channels while keeping the number of interfaces per mesh router at three will minimize the number of channels in the unused channel list. For example, at 5 available channels the unused channel list will be four channels, as at least one channel will be used to keep the network connectivity and to support the local traffic.

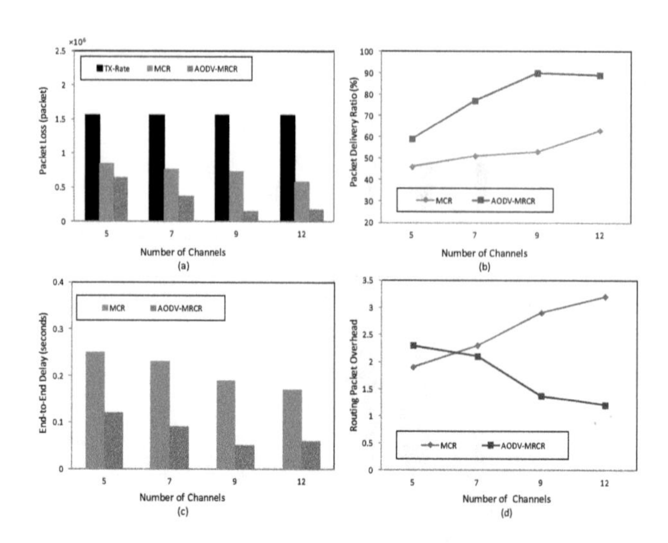

Figure 8. Simulation 2: results for scenario 2 (Varying the number of channels).

5.2.3. Scenario 3: Impact of local traffic

To investigate the impact of the local traffic on the gateway traffic, we varied the number of the peer-to-peer traffic from 5 to 20 flows. The number of the gateway traffic is the subtraction of the total number flows (50 flows) in the network and number of the local traffic per each scenario. The simulation parameters for this scenario are as shown in Table 1. Fig. 9 shows the end-to-end delay and PDR for both peer-to-peer and gateway traffic. This figure Shows

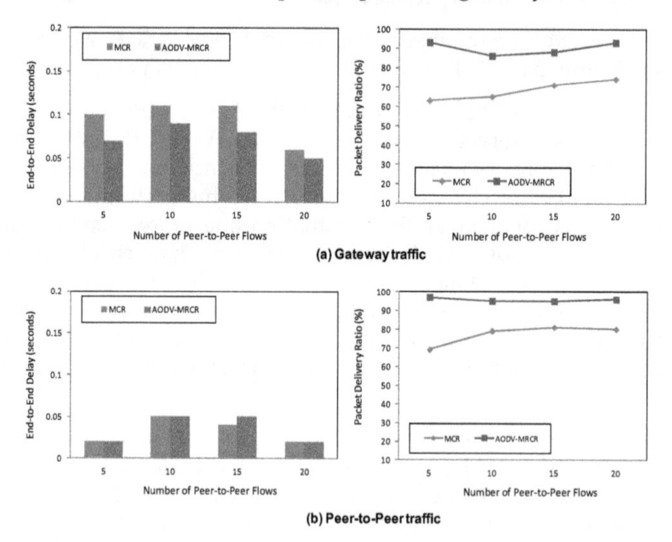

Figure 9. Simulation 2: results for scenario 3 (Impact of local traffic).

that as the local traffic increase, the AODV-MRCR improves both type of the traffic and still can get higher results than MCR. The figure also shows that the AODV-MRCR's performance metric for gateway traffic does not decrease when the number of the local traffic increase. The improved performance of AODV-MRCR can be explained as following. Our protocol differentiates between the local traffic and internet traffic by reserve a list of channels for the gateway traffic and these channels cannot be used to transmit the peer-to-peer traffic. Moreover, the AODV-MRCR protocol assigns different channel for the reverse and forward path which lead to better channel diversity than MCR routing protocol. Fig. 9 also shows that for the local traffic(peer-to-peer) of ten flows, the performance of both protocols in term of PDR and end-to-end delay are decreased. The reason for that is the randomly distributed traffic between mesh router nodes as well as the node random distribution may cause the traffic to be located in the same area.

6. Conclusion

In this chapter, we proposed a channel reservation scheme, which establishes a high throughput path for the gateway traffic by utilizing the WMN characteristics, such as multi-radio mesh router and most of the traffic toward the gateway. The channel reservation and assignment are integrated with the gateway routing discovery process. This scheme reduced the influence of local traffic on the performance of the gateway traffic. Moreover, the scheme minimized the number of nodes using the channel by only assigning channels to the node that is involved in the gateway path route discovery process. The performance of the proposed scheme is evaluated with respect to the metrics, such as packet delivery ratio, end-to-end delay, aggregate throughput, packet loss and routing overhead. The results obtained show that the proposed scheme is better than the existing schemes with respect to these metrics. Currently, the protocol designed in this chapter is mainly for infrastructure wireless mesh networks, which assumes that most of the traffic is towards the gateway. This proposed protocol could be further enhanced to support a more general wireless network, such as hybrid wireless mesh network. Moreover, it can be enhanced by assigning the channel to traffic based on the traffic load. Another possible extension is to consider multiple transmission rates. Different transmission rates can be achieved by using different modulation schemes, for example, IEEE 802.11b transmissions support four different data rates 1Mbps, 2Mbps, 5.5Mbps, and 11 Mbps. Finally, the proposed protocol can be further investigated with different MAC protocols for different radio interfaces.

Acknowledgement

The research was partially supported by the Research University, FRGS /1/11/SG/UPM/01/1

Author details

Hassen A. Mogaibel, Mohamed Othman, Shamala Subramaniam and Nor Asilah Wati Abdul Hamid

Department of Communication Technology and Network, Universiti Putra Malaysia, 43400 UPM, Serdang, Selangor D.E., Malaysia

7. References

[1] Adya, A., Bahl, P., Padhye, J., Wolman, A. & Zhou, L. [2004]. A multi-radio unification protocol for ieee 802.11 wireless networks, *Broadband Networks, 2004. BroadNets 2004. Proceedings. First International Conference on*, pp. 344–354.

[2] Akyildiz, I. F., Wang, X. & Wang, W. [2005]. Wireless mesh networks: a survey, *Comput. Netw. ISDN Syst.* 47: 445–487.

[3] Avallone, S., Akyildiz, I. & Ventre, G. [2009]. A channel and rate assignment algorithm and a layer-2.5 forwarding paradigm for multi-radio wireless mesh networks, *Networking, IEEE/ACM Transactions on* 17(1): 267–280.

[4] Bahl, P., Chandra, R. & Dunagan, J. [2004]. Ssch: slotted seeded channel hopping for capacity improvement in ieee 802.11 ad-hoc wireless networks, *Proceedings of the 10th annual international conference on Mobile computing and networking*, MobiCom '04, ACM, New York, NY, USA, pp. 216–230.

[5] Broch, J., Maltz, D. A. & Johnson, D. B. [1999]. Supporting hierarchy and heterogeneous interfaces in multi-hop wireless ad hoc networks, *Proceedings of the 1999 International Symposium on Parallel Architectures, Algorithms and Networks*, ISPAN '99, IEEE Computer Society, Washington, DC, USA, pp. .370 – 375.

[6] Chaudhry, A., Hafez, R., Aboul-Magd, O. & Mahmoud, S. [2010]. Fault-tolerant and scalable channel assignment for multi-radio multi-channel ieee 802.11a-based wireless mesh networks, *Technical report*, GLOBECOM Workshops (GC Wkshps), 2010 IEEE.

[7] Chen, J., Jia, J., Wen, Y., zhao, D. & Liu, J. [2009]. A genetic approach to channel assignment for multi-radio multi-channel wireless mesh networks, *GEC '09: Proceedings of the first ACM/SIGEVO Summit on Genetic and Evolutionary Computation*, ACM, New York, NY, USA, pp. 39–46.

[8] Chiu, H. S., Yeung, K. & Lui, K.-S. [2009]. J-car: An efficient joint channel assignment and routing protocol for ieee 802.11-based multi-channel multi-interface mobile ad hoc networks, *Wireless Communications, IEEE Transactions on* 8(4): 1706–1715.

[9] Ding, Y., Pongaliur, K. & Xiao, L. [2009]. Hybrid multi-channel multi-radio wireless mesh networks, *Quality of Service, 2009. IWQoS. 17th International Workshop on*, pp. 1–5.

[10] Draves, R., Padhye, J. & Zill, B. [2004]. Routing in multi-radio, multi-hop wireless mesh networks, *MobiCom '04: Proceedings of the 10th annual international conference on Mobile computing and networking*, ACM, New York, NY, USA, pp. 114–128.

[11] Elizabeth, Y. S., Sun, Y., Belding-Royer, E. M. & Perkins, C. E. [2002]. Internet Connectivity for Ad hoc Mobile Networks, *International journal of wireless information networks* 9(2): 75–88.

[12] Gong, M. X., Midkiff, S. F. & Mao, S. [2009]. On-demand routing and channel assignment in multi-channel mobile ad hoc networks, *Ad Hoc Netw.* 7(1): 63–78.

[13] Hamidian, A. A. [2003]. *A study of internet connectivity for mobile ad hoc networks in ns2*, Master's thesis, Department of Communication Systems, Lund Institute of Technology, Lund University,.

[14] Jiang, R., Gupta, V. & Ravishankar, C. V. [2003]. Interactions between tcp and the ieee 802.11 mac protocol, *DARPA Information Survivability Conference and Exposition*, 1: 273–282.

[15] Jönsson, U., Alriksson, F., Larsson, T., Johansson, P. & Maguire, Jr., G. Q. [2000]. Mipmanet: mobile ip for mobile ad hoc networks, *Proceedings of the 1st ACM international symposium on Mobile ad hoc networking & computing*, MobiHoc '00, IEEE Press, Piscataway, NJ, USA, pp. 75–85.

[16] Kim, S.-H. & Suh, Y. J. [2008]. Local channel information assisted channel assignment for multi-channel wireless mesh networks, *Vehicular Technology Conference (VTC) Spring 2008, IEEE, 2008*, pp. 2611–2615.
[17] Kyasanur, P. & Vaidya, N. H. [2006]. Routing and link-layer protocols for multi-channel multi-interface ad hoc wireless networks, *SIGMOBILE Mob. Comput. Commun. Rev.* 10: 31–43.
[18] Lee, J., Kim, D., Garcia-Luna-Aceves, J., Choi, Y., Choi, J. & Nam, S. [2003]. Hybrid gateway advertisement scheme for connecting mobile ad hoc networks to the internet, *Vehicular Technology Conference, 2003. VTC 2003-Spring. The 57th IEEE Semiannual*, Vol. 1, pp. 191–195.
[19] Marina, M. K., Das, S. R. & Subramanian, A. P. [2010]. A topology control approach for utilizing multiple channels in multi-radio wireless mesh networks, *Comput. Netw.* 54: 241–256.
[20] Mogaibel, H. A., Othman, M., Subramaniam, S. & Hamid, N. A. W. A. [2012]. On-demand channel reservation scheme for common traffic in wireless mesh networks, *Journal of Network and Computer Applications* (0): –.
URL: *http://www.sciencedirect.com/science/article/pii/S1084804512000318*
[21] Nandiraju, D. S., Nandiraju, N. S. & Agrawal, D. P. [2009]. Adaptive state-based multi-radio multi-channel multi-path routing in wireless mesh networks, *Pervasive Mob. Comput.* 5: 93–109.
[22] *NS. The Network Simulator, http://www.isi.edu/nsnam/ns/* [1989].
[23] Pediaditaki, S., Arrieta, P. & Marina, M. [2009]. A learning-based approach for distributed multi-radio channel allocation in wireless mesh networks, *Network Protocols, 2009. ICNP 2009. 17th IEEE International Conference on*, pp. 31–41.
[24] Pirzada, A. A., Portmann, M. & Indulska, J. [2006]. Evaluation of multi-radio extensions to aodv for wireless mesh networks, *MobiWac '06: Proceedings of the 4th ACM international workshop on Mobility management and wireless access*, ACM, New York, NY, USA, pp. 45–51.
[25] Raniwala, A. & Chiueh, T. [2005]. Architecture and algorithms for an ieee 802.11-based multi-channel wireless mesh network, *INFOCOM 2005. 24th Annual Joint Conference of the IEEE Computer and Communications Societies. Proceedings IEEE*, Vol. 3, pp. 2223–2234.
[26] Raniwala, A., Gopalan, K. & Chiueh, T. [2004]. Centralized channel assignment and routing algorithms for multi-channel wireless mesh networks, *ACM Mobile Computing and Communications Review* 8: 50–65.
[27] Ratanchandani, P. & Kravets, R. [2003]. A hybrid approach to internet connectivity for mobile ad hoc networks, *Wireless Communications and Networking, 2003. WCNC 2003. 2003 IEEE*, Vol. 3, pp. 1522–1527.
[28] Sarkar, T., Ji, Z., Kim, K., Medouri, A. & Salazar-Palma, M. [2003]. A survey of various propagation models for mobile communication, *Antennas and Propagation Magazine, IEEE* 45(3): 51–82.
[29] Shui, G. & Shen, S. [2008]. A new multi-channel mac protocol combined with on-demand routing for wireless mesh networks, *Computer Science and Software Engineering, 2008 International Conference, vol. 4, 2008*, pp. 1036–1039.
[30] So, J. & Vaidya, N. [2004a]. Multi-channel mac for ad hoc networks: Handling multi-channel hidden terminals using a single transceiver, *In ACM MobiHoc*, pp. 222–233.
[31] So, J. & Vaidya, N. [2004b]. A routing protocol for utilizing multiple channels in multi-hop wireless networks with a single transceiver, *Technical report*.
[32] Subramanian, A., Buddhikot, M. & Miller, S. [2006]. Interference aware routing in multi-radio wireless mesh networks, *Wireless Mesh Networks, 2006. WiMesh 2006. 2nd IEEE Workshop on*, pp. 55–63.

[33] Sun, W., Cong, R., Xia, F., Chen, X. & Qin, Z. [2010]. R-ca: A routing-based dynamic channel assignment algorithm in wireless mesh networks, *Ubiquitous, Autonomic and Trusted Computing, Symposia and Workshops on* 0: 228–232.

[34] Tang, J., Xue, G. & Zhang, W. [2005]. Interference-aware topology control and qos routing in multi-channel wireless mesh networks, *MobiHoc '05: Proceedings of the 6th ACM international symposium on Mobile ad hoc networking and computing*, ACM, New York, NY, USA, pp. 68–77.

[35] Tarn, W.-H. & Tseng, Y.-C. [2007]. Joint multi-channel link layer and multi-path routing design for wireless mesh networks, *INFOCOM 2007. 26th IEEE International Conference on Computer Communications. IEEE*, pp. 2081–2089.

[36] Wakikawa, Perkins, C., Nilsson, A. & Tuominen, A. J. [2001]. Global Connectivity for IPv6 Mobile Ad Hoc Networks, IETF Internet Draft, Work in process.

[37] Wan, P.-J. [2009]. Multiflows in multihop wireless networks, *Proceedings of the tenth ACM international symposium on Mobile ad hoc networking and computing*, MobiHoc '09, ACM, New York, NY, USA, pp. 85–94.
URL: *http://doi.acm.org/10.1145/1530748.1530761*

[38] Yue, X., Wong, C.-F. & Chan, S.-H. G. [2011]. Cacao: Distributed client-assisted channel assignment optimization for uncoordinated wlans, *Parallel and Distributed Systems, IEEE Transactions on* 22(9): 1433–1440.

[39] Zhuang, L., Liu, Y. & Liu, K. [2009]. Hybrid gateway discovery mechanism in mobile ad hoc for internet connectivity, *Proceedings of the 2009 International Conference on Wireless Communications and Mobile Computing: Connecting the World Wirelessly*, IWCMC '09, ACM, New York, NY, USA, pp. 143–147.

Permissions

The contributors of this book come from diverse backgrounds, making this book a truly international effort. This book will bring forth new frontiers with its revolutionizing research information and detailed analysis of the nascent developments around the world.

We would like to thank Andrey V. Krendzel, for lending his expertise to make the book truly unique. He has played a crucial role in the development of this book. Without his invaluable contribution this book wouldn't have been possible. He has made vital efforts to compile up to date information on the varied aspects of this subject to make this book a valuable addition to the collection of many professionals and students.

This book was conceptualized with the vision of imparting up-to-date information and advanced data in this field. To ensure the same, a matchless editorial board was set up. Every individual on the board went through rigorous rounds of assessment to prove their worth. After which they invested a large part of their time researching and compiling the most relevant data for our readers. Conferences and sessions were held from time to time between the editorial board and the contributing authors to present the data in the most comprehensible form. The editorial team has worked tirelessly to provide valuable and valid information to help people across the globe.

Every chapter published in this book has been scrutinized by our experts. Their significance has been extensively debated. The topics covered herein carry significant findings which will fuel the growth of the discipline. They may even be implemented as practical applications or may be referred to as a beginning point for another development. Chapters in this book were first published by InTech; hereby published with permission under the Creative Commons Attribution License or equivalent.

The editorial board has been involved in producing this book since its inception. They have spent rigorous hours researching and exploring the diverse topics which have resulted in the successful publishing of this book. They have passed on their knowledge of decades through this book. To expedite this challenging task, the publisher supported the team at every step. A small team of assistant editors was also appointed to further simplify the editing procedure and attain best results for the readers.

Our editorial team has been hand-picked from every corner of the world. Their multi-ethnicity adds dynamic inputs to the discussions which result in innovative outcomes. These outcomes are then further discussed with the researchers and contributors who give their valuable feedback and opinion regarding the same. The feedback is then collaborated with the researches and they are edited in a comprehensive manner to aid the understanding of the subject.

Apart from the editorial board, the designing team has also invested a significant amount of their time in understanding the subject and creating the most relevant covers. They scrutinized every image to scout for the most suitable representation of the subject and create an appropriate cover for the book.

The publishing team has been involved in this book since its early stages. They were actively engaged in every process, be it collecting the data, connecting with the contributors or procuring relevant information. The team has been an ardent support to the editorial, designing and production team. Their endless efforts to recruit the best for this project, has resulted in the accomplishment of this book. They are a veteran in the field of academics and their pool of knowledge is as vast as their experience in printing. Their expertise and guidance has proved useful at every step. Their uncompromising quality standards have made this book an exceptional effort. Their encouragement from time to time has been an inspiration for everyone.

The publisher and the editorial board hope that this book will prove to be a valuable piece of knowledge for researchers, students, practitioners and scholars across the globe.

List of Contributors

Gustavo Vejarano
Department of Electrical Engineering and Computer Science, Loyola Marymount University, Los Angeles, CA, USA

Fawaz Bokhari and Gergely Záruba
Department of Computer Science and Engineering, The University of Texas at Arlington, Texas, USA

Stefan Pollak and Vladimir Wieser
Department of Telecommunications and Multimedia, University of Zilina, Slovakia

Svilen Ivanov
rt-solutions.de GmbH, Oberländer Ufer 190a, D-50968 Cologne, Germany

Edgar Nett
Institut of Distributed Systems, Otto von Guericke University of Magdeburg, Universitätsplatz 2, 39106 Magdeburg, Germany

Doug Kuhlman
Motorola Mobility 600 US Hwy 45 E1-40Y Libertyville, IL 60048

Ryan Moriarty
Computer Science Department, University of California at Los Angeles

Tony Braskich, Steve Emeott and Mahesh Tripunitara
Motorola, Schaumburg, IL 60196

Sangsu Jung
Future Internet Research Team, National Institute for Mathematical Sciences, Daejeon, S. Korea

Thomas Olwal, Moshe Masonta, Fisseha Mekuria and Kobus Roux
CSIR Meraka Institute, South Africa

Hassen A. Mogaibel, Mohamed Othman, Shamala Subramaniam and Nor Asilah Wati Abdul Hamid
Department of Communication Technology and Network, Universiti Putra Malaysia, 43400 UPM,
Serdang, Selangor D.E., Malaysia